Essential Concepts in Metamaterials

Online at: https://doi.org/10.1088/978-0-7503-1131-1

Essential Concepts in Metamaterials

Jensen Li
Department of Physics and Astronomy, University of Exeter, Exeter, United Kingdom

IOP Publishing, Bristol, UK

ISBN 978-0-7503-1131-1 (ebook)
ISBN 978-0-7503-1132-8 (print)
ISBN 978-0-7503-1861-7 (myPrint)
ISBN 978-0-7503-1133-5 (mobi)

DOI 10.1088/978-0-7503-1131-1

Version: 20251201

IOP ebooks

British Library Cataloguing-in-Publication Data: A catalogue record for this book is available from the British Library.

Published by IOP Publishing, wholly owned by The Institute of Physics, London

IOP Publishing, No.2 The Distillery, Glassfields, Avon Street, Bristol, BS2 0GR, UK

US Office: IOP Publishing, Inc., 190 North Independence Mall West, Suite 601, Philadelphia, PA 19106, USA

To Shan, for her patience, understanding, and unwavering support.

Contents

Foreword

This is the latest of several books devoted to metamaterials. Rapid growth of the field, especially in recent years, demands fresh texts up to date on new developments. This book, written by one of the leading experts, supplies that need.

Metamaterials first came to the fore around the turn of the century addressing a need in electromagnetism for material properties not found in Nature. A good example would be negative refractive index which had long been sought, to be realised in practice only after advent of the metamaterial concept. Metamaterials can also reproduce more commonplace properties, but do so more efficiently than conventional solutions. Recent advances include construction of graded index lenses that are extremely thin and very light, enabling deployment where weight is a crucial factor.

To those of us who were present at the start of the field, it is with pleasure and not a little astonishment to see how the field has grown, expanding beyond electromagnetism into acoustics and other areas where there is a need to control waves with materials. Indeed, this book provides an extensive chapter on acoustics, a subject to which the author has contributed many research papers. Particularly gratifying is to see the technology embedded in many commercial developments ranging from 5G and coming 6G networks, to control of sound through metamaterial screens. Such applications demonstrate the need for metamaterial-skilled graduates and a corresponding need for books such as this.

J B Pendry
Imperial College London
2025

Preface

I entered the field of metamaterials soon after the first demonstrations of negative-index media that bend light in unexpected ways. These materials derive their properties from subwavelength structures rather than chemical composition, enabling responses beyond those of conventional composites. Over the past two decades the field has broadened rapidly, spanning effective-medium theories, transformation optics, bulk and metasurface implementations, acoustics and elasticity, and, more recently, time-varying and programmable platforms that actively shape wave behaviour. Connections to non-Hermitian and topological physics have further enriched the landscape and opened new directions for research.

Looking back, I have worked on this manuscript intermittently for more than ten years, during which I sought to broaden my experience in the field as much as possible while hoping to make the textbook comprehensive—an aim that ultimately proved difficult to fully achieve. A change of pace—relocating to Exeter—finally gave me the continuity needed to complete it. With the field now extending from fundamental studies to active and time-modulated systems and an expanding range of applications, it feels timely to provide an introductory text for first-year postgraduate students, helping them grasp the basic ideas and move more quickly toward research level.

This book is not meant to be comprehensive. It reflects my own experience in introducing techniques I consider foundational and questions I believe to be important. Each chapter presents key principles followed by applications, centred on two themes: modelling metamaterials through effective-medium approaches and designing devices based on those models. The chapters begin with basic ideas and end with problems that encourage reflection and practice, offering a concise foundation for approaching the research literature with confidence.

I would like to express my gratitude to Che Ting Chan, Sir John Pendry, Xiang Zhang, and Shuang Zhang for their help throughout my career. I am indebted to my former student, Fu Liu, whose early assistance in collecting materials for problem sets made this project possible. I am also grateful to Randy Stefan Tanuwijaya and to the many current and former members of my research group for their thoughtful suggestions and careful reading in the later stages. My deepest thanks go to Shan for her unwavering support. Finally, I would like to remember the late Geoff Stephenson, whose gift of his book really sparked my desire to write; his memory has stayed with me as I finish this work.

Author biography

Jensen Li

Jensen Li is Professor of Computational Engineering and Metamaterials at the University of Exeter in the United Kingdom. He received his MPhil and PhD in Physics from the Hong Kong University of Science and Technology, followed by postdoctoral work with Sir John Pendry at Imperial College London and with Xiang Zhang at the University of California, Berkeley. He later returned to Hong Kong as a faculty member before joining Exeter in 2024.

His research interests span metamaterials, transformation optics, acoustic and elastic wave control, and non-Hermitian wave systems. He enjoys exploring the connections between theoretical wave physics and experimental realisations, and has worked on topics including the carpet cloak, space-coiling acoustic structures, elastic Willis media, programmable meta-atoms, and holographic wave manipulation.

Professor Li is a Croucher Senior Research Fellow and an Optica Fellow. He continues to investigate new ideas for wave control, with current interests extending into time-varying media and quantum-inspired approaches to photonics and acoustics.

IOP Publishing

Essential Concepts in Metamaterials

Jensen Li

Chapter 1

A new era of tailor-made optical media

In the last two decades, we have witnessed a significant shift in material science, moving from traditional methods to obtaining tailor-made optical properties by structuring materials rather than simply mixing them in a conventional composite style. This has led to the emergence of new types of materials, including photonic crystals, metamaterials, and the latest nano-photonic structures. In this chapter, we will provide a brief introduction to motivate this subject area, an overview of current developments, some interesting examples, and their state-of-the-art experimental realizations before delving into a systematic introduction in the following chapters.

1.1 Structural evolution: from composites to artificial atoms

Major leaps in technology often follow advances in materials. Civilizations themselves are sometimes named after the dominant materials of their time: the Stone Age, the Bronze Age, the Iron Age, and, more recently, the Silicon Age, which underpins today's electronic revolution. Each era was shaped by new ways of discovering, processing, and exploiting materials.

In the modern era, it has become routine to create new materials with tunable mechanical, thermal, or electronic properties by combining existing ones to form composites. The same idea also applies to optics. For example, doping germanium into a glass fiber produces a radial profile of refractive index. This engineered profile supports guided optical modes, forming the backbone of optical communication systems. Such optical fibers underlie our global internet infrastructure, allowing high-speed data transmission across continents.

Let us illustrate the composite route with a simple model. Suppose we have two materials with positive permittivities ϵ_1 and ϵ_2. By mixing them with a controllable filling fraction, we obtain an effective medium whose permittivity lies somewhere between these two values. A number of effective medium theories have been developed for such problems. The Maxwell–Garnett formula (Garnett, 1905), for example, describes dilute inclusions embedded in a host material, while the

doi:10.1088/978-0-7503-1131-1ch1

Bruggeman formula (Bruggeman, 1935) applies to more symmetric mixtures. Beyond these heuristic models, mathematicians have also established rigorous bounds on the effective permittivity of mixtures. The Wiener bounds provide the simplest limits, while the tighter Hashin–Shtrikman bounds account for microstructural effects. In general, the effective permittivity of a mixture of positive dielectrics always remains between the minimum and maximum values of the constituents, regardless of the filling fraction. This picture represents the traditional composite materials route, which works more as an averaging or interpolation scheme.

A qualitatively new situation arises when metallic constituents are introduced. Metals can be described by a Drude-type permittivity that becomes negative below the plasma frequency. Embedding metallic inclusions in a dielectric host invalidates the picture underlying the classical bounds, particularly the idea that the electric field varies slowly across the units of the composite. At finite frequencies, strong plasmonic resonances develop at the metal–dielectric interfaces, concentrating the fields and producing sharp dispersive behavior. As a result, the effective permittivity is no longer restricted to lie between the constituent values. It can greatly exceed that of the dielectric host, or even cross zero and become negative. In practice, these resonant enhancements are always accompanied by absorption, since metals have finite conductivity. Even so, the dispersive response of metal–dielectric composites provides the essential ingredient for plasmonics and, more generally, for metamaterials, where the crucial physics is the ability to move beyond the traditional interpolation picture.

Figure 1.1(a) illustrates a whispering-gallery mode supported by a glass microsphere with a diameter on the order of 100 μm. Such dielectric resonators are well known for their extremely high quality factors, producing very sharp resonances at telecom wavelengths, here shown at 1.55 μm. In contrast, figure 1.1(b) shows a plasmonic mode confined on a gold nanorod. Although the quality factor of this metallic resonance cannot approach that of the dielectric microsphere, the confinement of the mode is much stronger. The nanorod length is only about 50 nm, more

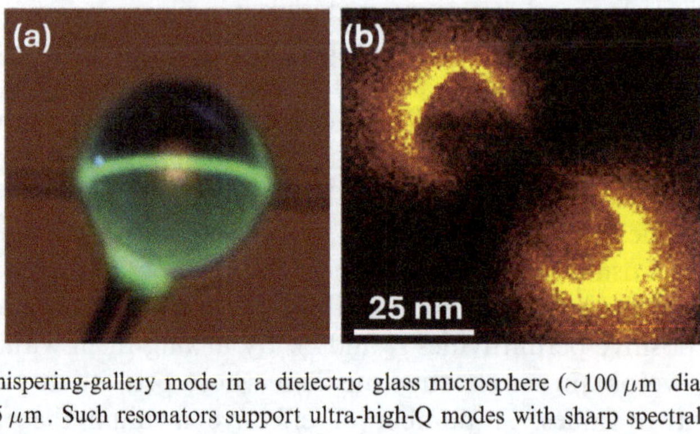

Figure 1.1. (a) Whispering-gallery mode in a dielectric glass microsphere (~100 μm diameter) excited at a wavelength of 1.55 μm. Such resonators support ultra-high-Q modes with sharp spectral features (Cai *et al* 2000). (b) Plasmonic resonance on a gold nanorod, imaged at $\lambda = 730$ nm. Scale bar: 25nm. Reproduced from Bosman *et al* (2007).

than an order of magnitude smaller than the free-space wavelength of $\lambda = 730$ nm. This comparison highlights two complementary regimes: large dielectric resonators supporting ultra-high-Q modes, and nanoscale plasmonic resonators where modes can be engineered deep inside the subwavelength domain with the introduction of metallic elements.

Rather than relying on random mixtures of metallic particles, one can take a more deliberate approach and design the resonant structures directly. By patterning metals into specific geometries, such as thin wires, split-ring resonators, or periodic lattices, it becomes possible to tune the resonant frequencies and tailor the effective medium response. In this way, the local resonance that first appears in simple composites is transformed into a powerful design principle. The field of metamaterials builds precisely on this idea: carefully engineered subwavelength structures that use resonance as a tool to achieve responses far beyond what any natural material can offer.

It is not a coincidence that the rise of metamaterials followed advances in nanofabrication. What began as a phenomenon in random nanocomposites can now be intentionally designed, thanks to fabrication tools with nanometer precision. Focused ion beam (FIB) milling, electron-beam lithography, and related methods can sculpt features only a few tens of nanometers wide, with precise control over size, shape, and arrangement. This new level of freedom allows us to design structures that act as *artificial atoms*: each unit cell of the metamaterial supports a tailored resonance in much the same way that a natural atom supports electronic transitions. Together, these artificial atoms can be assembled into artificial media with optical responses that have no counterpart in natural materials. This marks the beginning of the structural route to material design, known as *metamaterials*.

What becomes most prominent in this new approach is the structure itself, rather than the intrinsic chemistry of the ingredients. The starting materials need not be exotic: silicon, gold, silver, and quartz—already common in industry—are sufficient as building blocks. The key lies in how they are arranged. By carefully controlling geometry, height, orientation, and periodicity, nanostructures can be endowed with optical properties that far exceed those of the bulk materials. In metamaterials, function is dictated less by composition and more by architecture at the subwavelength scale.

In the next section, we will quantify this shift more precisely using the ratio λ/a, which compares the operating wavelength with the structural scale. This simple parameter will reveal when a composite behaves like an ordinary effective medium and when new optical physics emerges.

1.2 The λ/a rule: when structure matters

Placing structures into a periodic array creates a new way to make materials. In fact, structure plays an essential role in all materials, natural or engineered. Consider diamond: it is a vital optical material because of its high refractive index and transparency. However, the atomic structural details of diamond are far too fine to be resolved by the human eye. Our eyes perceive diamond as a homogeneous medium because the wavelength of visible light is much larger than the atomic spacing between carbon atoms. The wavelength λ is in the order of half a micron, while the atomic feature size a is about 0.1 nm. Thus, the collective response of all atoms averages into a smooth, effective medium described by a single parameter—the refractive index n. This is the regime where $\lambda \gg a$, often called the effective medium regime.

On the other hand, if we decrease the wavelength to the ultraviolet or x-ray regime, such that $\lambda \approx a$, the wave can resolve the atomic structure. In this case, the incident beam undergoes Bragg diffraction, producing discrete diffraction spots in the far-field pattern (Bragg 1929). These two limits, effective medium behavior and Bragg diffraction, are both dictated by the same simple parameter: the ratio λ/a. Figure 1.2 summarizes this classification, with natural and artificial examples placed along the λ/a-axis.

The first class of man-made materials that deliberately exploit this principle are photonic crystals (Yablonovitch 1987; John 1987; Joannopoulos *et al* 1995). They are essentially artificial analogs of Bragg diffraction in natural crystals, created using periodic structures fabricated at micron or submicron scales. A famous example is the three-dimensional (3D) woodpile structure, shown in figure 1.2. By engineering diffraction from different lattice planes to interfere destructively, one can block

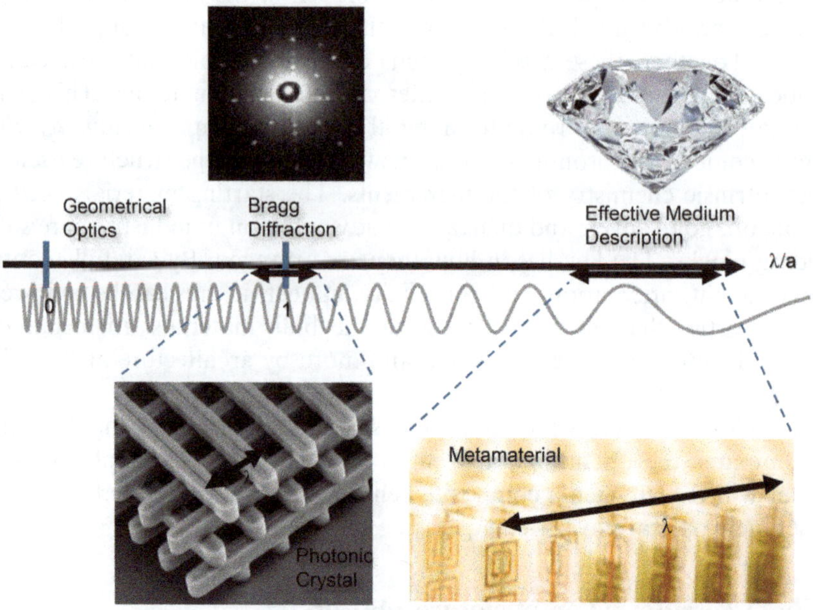

Figure 1.2. Structures and materials classified by the ratio of wavelength λ to structural scale a. When $\lambda \gg a$, waves cannot resolve microscopic details, and materials can be described by effective parameters such as a refractive index, as in diamond or in metamaterials. When $\lambda \approx a$, Bragg diffraction dominates, leading to photonic band gaps in artificial photonic crystals or diffraction in natural crystals. When $\lambda \ll a$, the wave picture reduces to geometrical optics. The metamaterial regime highlights how artificial subwavelength structures can mimic and go beyond the properties of natural atoms.

the propagation of light inside the material, giving rise to a photonic band gap. The statement is remarkably strong: in a complete 3D band gap, light cannot propagate at all. Even the spontaneous emission from a quantum emitter embedded inside is suppressed at frequencies within the gap.

Structured materials, combined with modern nanofabrication, have produced many breakthroughs in the diffraction regime. Yet it is perhaps even more surprising that nanofabrication also provides a new perspective in the long-wavelength limit, where $\lambda/a \gg 1$. This conceptual advance first appeared in the microwave regime, where fabricating structures smaller than a microwave wavelength is relatively easy using printed circuit boards. It led to the emergence of metamaterials, pioneered around the year 2000 by Sir John Pendry and David Smith (Pendry *et al* 1999; Smith *et al* 2000). In this case, subwavelength structures are arranged periodically to act as 'artificial atoms,' each contributing a tailored electric or magnetic response.

The picture below the λ/a-axis in figure 1.2 shows two key building blocks: ring-like structures and vertical wires, the ingredients of the first negative-index metamaterial in the microwave regime (Shelby *et al* 2001). The microwave wavelength, in tens of centimeters, is at least five times larger than the lattice constant of the square array. To the wave, the fine details are invisible, and the array behaves as an effective medium. Vertical wires provide an electric response, while rings supply a magnetic response. Together they form artificial atoms, analogous to the role of carbon atoms in diamond, but now engineered with complete freedom (see figure 1.3). It was in this setting that the

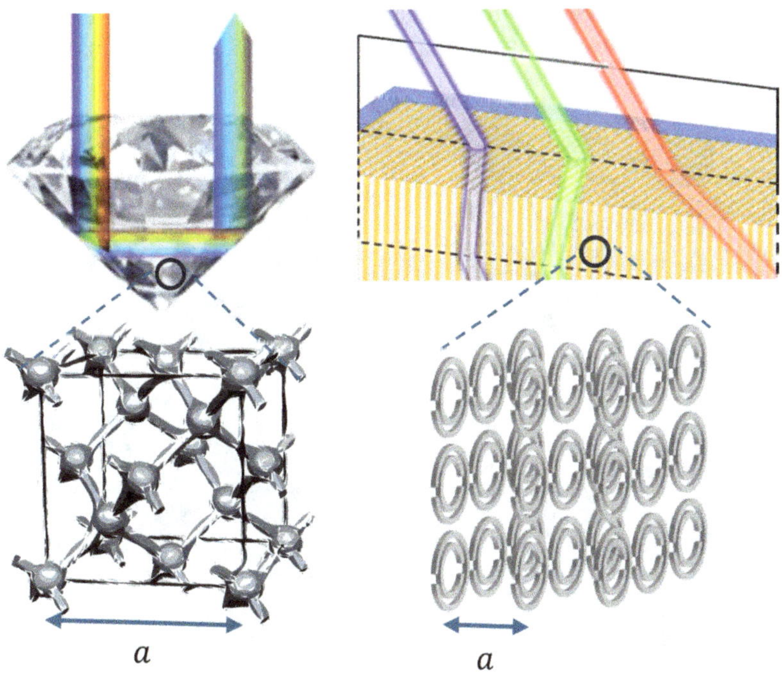

Figure 1.3. Analogy between natural and artificial materials. (a) Diamond: natural atoms collectively determine the refractive index. (b) A metamaterial: artificial atoms, such as rings and wires, collectively generate effective medium parameters that can surpass those of natural materials.

first negative refractive index was achieved—something no natural material could provide. The prefix 'meta' (meaning 'beyond') thus signifies optical properties that go beyond those found in nature. The concept of metamaterials has been extended from bulk media to surface versions called metasurfaces, which can manipulate wavefronts with subwavelength thickness. Metasurfaces have found applications in flat lenses, holograms, and beam steering devices (Yu and Capasso 2014).

The concept of λ/a is not limited to optics. In thermal metamaterials or mechanical metamaterials, one does not deal with waves in the usual sense, yet the structural route still plays the dominant role. In thermal systems, engineered structures can guide or concentrate heat flow in ways reminiscent of wave phenomena. In mechanics, patterned beams or lattices can control stiffness or Poisson's ratio, giving rise to auxetic or programmable materials. In optics, while metamaterials are mostly associated with the $\lambda/a \gg 1$ regime, there are also designs that deliberately exploit the diffraction regime, such as metagratings, which control diffraction orders to steer or split beams.

It is also worth noting that nature itself makes use of the structural route. Animals and plants often rely on subwavelength or periodic structures to tailor their appearance or functionality. Figure 1.4 shows two examples. The vivid blue color of a butterfly's wing arises not from pigments but from a photonic crystal-like structure that reflects selected wavelengths (Vukusic and Sambles 2003; Saranathan *et al* 2010). An octopus, on the other hand, adapts its skin to match its environment, using chromatophores to actively tune structural coloration for camouflage. Its distributed neural system and mechanical flexibility have inspired designs for soft robotics and adaptive metamaterials. While present-day metamaterials are still

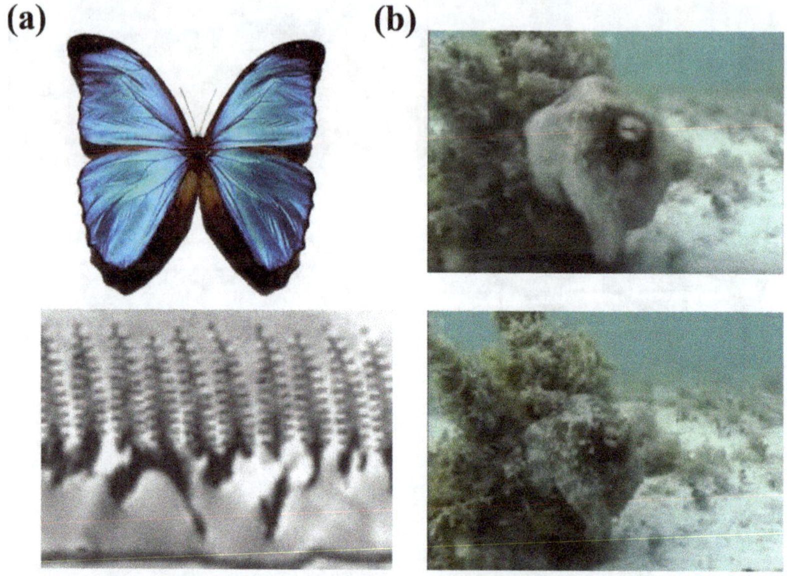

Figure 1.4. Natural examples of structural coloration and adaptation. (a) A butterfly wing with a photonic crystal-like structure that reflects blue light. (b) An octopus that adapts the color and texture of its skin for camouflage, inspiring ideas of self-adaptive metamaterials.

far from achieving the self-adaptivity of an octopus, they demonstrate how deeply the structural route pervades both nature and engineering.

When $\lambda \gg a$, the wave cannot resolve the fine structural details, and the medium can be described by only a few effective parameters. This long-wavelength regime is where metamaterials operate. The next question is whether these effective parameters can reach values not found in nature, such as a negative refractive index.

1.3 Quest for a negative refractive index

In 1968, the Russian physicist Victor Veselago posed a striking theoretical question: what would happen if the refractive index of a material were negative? Snell's law,

$$n_1 \sin \theta_1 = n_2 \sin \theta_2, \tag{1.1}$$

where n_1 and n_2 are the refractive indices of the entrance and exit media, and θ_1 and θ_2 are the incident and refracted angles, ordinarily involves positive refractive indices (e.g., 1.3 for water, 1.5 for glass, and 2.4 for diamond). More generally, the refractive index can be complex, with the imaginary part representing absorption or evanescence. For example, silver has an index of about $0.13 + 4i$ at a wavelength of 635 nm. Veselago showed that if the real part of the refractive index were negative, entirely new optical phenomena would occur, including negative refraction and reversed Cherenkov radiation (Veselago 1968). Importantly, achieving a negative index requires both the electric permittivity ϵ and the magnetic permeability μ to be negative simultaneously. At the time, however, no natural material exhibited both properties, so Veselago's idea remained a theoretical curiosity for decades.

The concept of negative refraction can be understood directly from equation (1.1). For a beam entering a medium from air, θ_1 is the incident angle and θ_2 the refracted angle. If the medium has $n_2 < 0$, then Snell's law can only be satisfied if θ_2 is negative, meaning the refracted beam bends to the 'wrong' side of the normal, as illustrated in figure 1.5.

To realize Veselago's proposal, one needs a medium with both negative permittivity and negative permeability. Natural materials do not provide this, but in the late 1990s John Pendry and collaborators proposed artificial subwavelength structures that could. Arrays of thin metallic wires behave like a plasma and yield an effective negative permittivity (Pendry *et al* 1996), while so-called split-ring resonators—metallic loops with a small gap that act like resonant LC circuits—can achieve a negative permeability (Pendry *et al* 1999). Together, these ingredients provide a design recipe: if the electric resonance from the wires and the magnetic resonance from the rings overlap in frequency, both ϵ and μ become negative, and thus the overall refractive index is negative (Smith *et al* 2000):

$$n = \sqrt{\epsilon}\sqrt{\mu}. \tag{1.2}$$

In 2001, David Smith and Sheldon Schultz demonstrated the first experimental metamaterial with a negative index in the microwave regime (Shelby *et al* 2001). By combining arrays of wires and split-ring resonators, they observed negative refraction of a microwave beam, confirming Veselago's prediction more than 30 years after it was made. This marked the birth of experimental metamaterials.

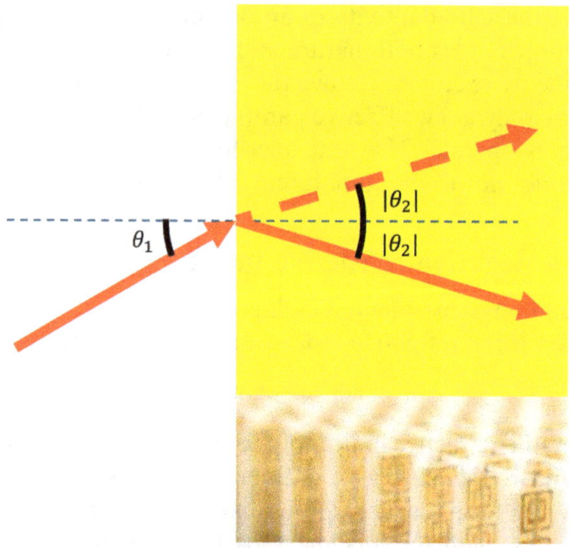

Figure 1.5. Meaning of negative refraction. If the refractive index of a medium is negative, Snell's law requires the refracted angle to lie in the opposite quadrant compared to ordinary refraction.

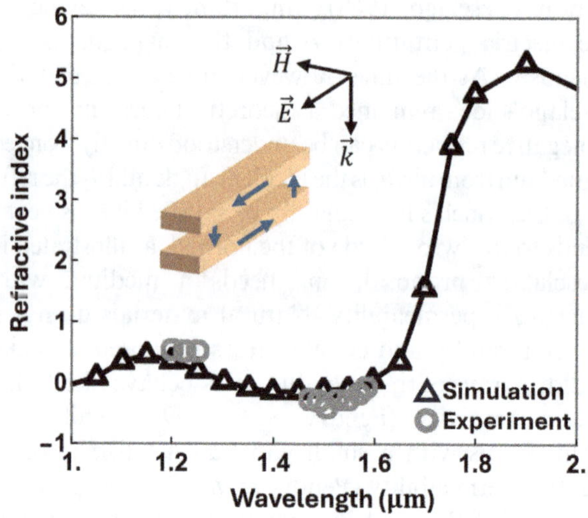

Figure 1.6. Array of paired gold nanorods used to achieve a negative index at optical frequencies. Retrieved refractive index from transmission and reflection measurements, showing negative values around 1.5 μm. Reproduced from Shalaev *et al* (2005), copyright Optical Society of America.

The successful demonstration of a negative refractive index in the microwave regime spurred efforts to realize a negative index at higher and higher frequencies, with the ultimate goal of achieving an isotropic negative index in the visible range (Valentine *et al* 2008). In the near-infrared, for example, Shalaev *et al* (2005) used a double array of parallel gold nanorods (figure 1.6). By measuring both transmission

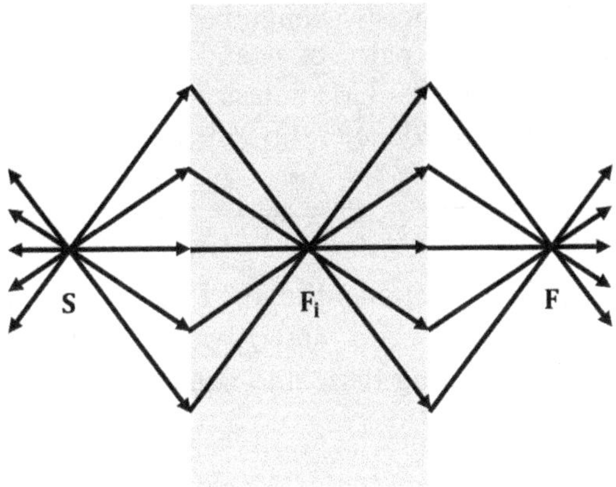

Figure 1.7. Focusing by a slab with $n = -1$. Rays from a point source S refocus inside the slab and then to a final point focus F outside. This is the principle of the perfect lens (Pendry 2000).

and reflection, they retrieved the refractive index, which was found to be negative around $\lambda \approx 1.5$ μm. The extreme capability of metamaterial in bending light later also trigger the development of transformation optics and invisibility cloaking (Pendry *et al* 2006; Schurig *et al* 2006), when the constitutive parameters of the metamaterials become inhomogeneous and tailor-made in space, although an invisibility cloak does not need to use negative indices.

Following the first realization of a metamaterial, Pendry suggested that if a negative index of -1 could be achieved by tuning both ϵ and μ to -1, a *perfect lens* could be created to resolve arbitrarily fine details (Pendry 2000). As shown in figure 1.7, rays from a point source S reconverge to a point within the slab and then to a final focus F outside it, with the distance from S to F equal to twice the slab thickness d. This occurs for all angles, not just those within the paraxial limit of geometrical optics.

This result can be understood using transfer matrix analysis, treating the slab as a one-dimensional (1D) wave propagation problem. A slab of thickness d with constant ϵ and μ has transmission and reflection coefficients t and r (equation (A.19) in appendix A.1) given by

$$\begin{aligned} t &= \frac{2}{2\cos(k_z d) - i(Z + 1/Z)\sin(k_z d)}, \\ r &= \frac{-i(Z - 1/Z)\sin(k_z d)}{2\cos(k_z d) - i(Z + 1/Z)\sin(k_z d)}. \end{aligned} \tag{1.3}$$

In this 1D problem, surface normal is along the z-direction with propagation on the xz-plane, the transverse wavevector component $k_x = k\sin(\theta)$ is conserved across the interfaces, which is equivalent to Snell's law in equation (1.1). The longitudinal component is

$$k_z = \sqrt{nk_0 - k_x}\,\sqrt{nk_0 + k_x},$$

where $k_0 = 2\pi/\lambda$ is the free-space wavenumber. The square-root branches are chosen such that the imaginary parts of n and k_z are non-negative for passive materials. For transverse magnetic (TM) polarization (with the E-field along y and H-field in the xz-plane), the impedance ratio Z between the slab and air is

$$Z = \frac{k_z}{\epsilon k_z^{(0)}} = \frac{k_z}{\epsilon\sqrt{k_0 - k_x}\,\sqrt{k_0 + k_x}},$$

where $k_z^{(0)}$ is the longitudinal component in air. Details are given in appendix A.1.

For the perfect lens with $\epsilon = -1$ and $\mu = -1$ (hence $n = -1$), we obtain $k_z = -k_z^{(0)}$ and $Z = 1$. Substituting these into equation (1.3) gives

$$t = \mathrm{e}^{-ik_z^{(0)}d}, \quad r = 0.$$

Therefore, after propagation through the slab and an additional distance d in air, the total transfer from S to F is $t = 1$ and $r = 0$. The same result arises from considering transverse-electric (TE) polarization with the E-field along y and H-field in the xz-plane) as well. The constant phase in transmission means that all rays leaving S arrive at F, consistent with the eikonal theory of geometrical optics. Importantly, this unit transmission ($t = 1$ with constant amplitude and phase) can be proved to hold for all real k_x, whether $|k_x| < k_0$ (propagating waves) or $|k_x| > k_0$ (evanescent waves). Thus, all information carried by both propagating and evanescent waves is perfectly transferred from S to F, achieving true perfect imaging.

The development of negative-index metamaterials and the proposal of the perfect lens initially generated controversy, but subsequent experiments quickly settled the debate. Since then, these ideas have inspired scientists across physics, engineering, and mathematics to explore this exciting new area. They prompt us to reconsider familiar optical laws by inventing entirely new kinds of materials.

The demonstration of a negative refractive index also raised further questions: could such materials enable devices beyond the reach of conventional optics? The following years provided many surprising answers. The path from Veselago's theoretical proposal to modern applications spans more than five decades, with several landmark discoveries along the way. Table 1.1 summarizes some of the key milestones in the development of metamaterials.

Table 1.1. Milestones of metamaterials, from theoretical proposals to practical applications.

Year	Milestone	Significance
1968	Veselago's proposal	Predicted negative refraction and reversed optical effects if ϵ and μ were both negative; remained a theoretical curiosity.
1996–99	Pendry's designs	Wires (negative ϵ) and split-ring resonators (negative μ) showed how subwavelength structures could act as artificial atoms.
2000	Perfect lens	Proposed that $n = -1$ could create a perfect lens capable of imaging beyond the diffraction limit.
2001	First experimental negative index	Negative refraction demonstrated in the microwave regime, confirming Veselago's vision.

2005	Optical negative index	Negative index realized at optical frequencies using paired gold nanorods.
2006	Transformation optics	Introduced the concept of designing materials by coordinate transformations, leading to cloaking ideas.
2006	First experimental cloak	Microwave cloak demonstrated transformation-optics principles in practice.
2007–2008	Optical fishnet structures	Negative index achieved at visible wavelengths with metal–dielectric multilayers.
2010s	Metasurfaces	Shift from bulk 3D metamaterials to 2D patterned surfaces for wavefront engineering.
2014	Flat metalenses	High-efficiency flat lenses rivaled traditional optics, accelerating applications.
2019	Global recognition	Metasurface optics highlighted as a top emerging technology by global innovation reports.
2010s–present	Flat-panel antennas	Electronically steerable metamaterial antennas adopted for civilian satellite communications.
2020s	Optical metalenses	Metasurface flat lenses deployed in smartphones and consumer electronics for 3D sensing and imaging.
2020s	Acoustic metamaterials	Engineered panels and liners used for noise reduction and soundproofing in civil construction and consumer devices.

Up to now, we have introduced the concept of structural materials in both the diffraction and long-wavelength regimes. In particular, we have introduced metamaterials, which can give rise to extraordinary optical properties that cannot be achieved using conventional means. A negative index and a perfect lens are the beginning of this era. As time goes on, we will walk through examples to show you how structures can give rise to new kinds of optical properties, such as how metamaterials can bend light in a coherent way to construct an invisibility cloak.

References

Bosman M, Keast V, Watanabe M, Maaroof A and Cortie M 2007 Mapping surface plasmons at the nanometre scale with an electron beam *Nanotechnology* **18** 165505

Bragg W L 1929 The diffraction of short electromagnetic waves by a crystal *Scientia* **23** 153

Bruggeman D A G 1935 Berechnung verschiedener physikalischer konstanten von heterogenen substanzen. i. dielektrizitätskonstanten und leitfähigkeiten der mischkörper aus isotropen substanzen *Ann. Phys., Lpz.* **416** 636–64

Cai M, Painter O and Vahala K 2000 Observation of critical coupling in a fiber taper to a silica-microsphere whispering-gallery mode system *Phys. Rev. Lett.* **85** 74–7

Garnett J 1905 Colours in metal glasses, in metallic films, and in metallic solutions. ii *Phil. Trans. R. Soc.* A **205** 237–88

Joannopoulos J, Johnson S G, Winn J and Meade R 1995 *Photonic Crystals: Molding the Flow of Light* (Princeton University Press)

John S 1987 Strong localization of photons in certain disordered dielectric superlattices *Phys. Rev. Lett.* **58** 2486–9

Pendry J 2000 Negative refraction makes a perfect lens *Phys. Rev. Lett.* **85** 3966–9

Pendry J, Holden A, Robbins D and Stewart W 1999 Magnetism from conductors and enhanced nonlinear phenomena *IEEE Trans. Microw. Theory Tech.* **47** 2075–84

Pendry J, Holden A, Stewart W and Youngs I 1996 Extremely low frequency plasmons in metallic mesostructures *Phys. Rev. Lett.* **76** 4773–6

Pendry J B, Schurig D and Smith D R 2006 Controlling electromagnetic fields *Science* **312** 1780–2

Saranathan V, Osuji C, Mochrie S, Noh H, Narayanan S, Sandy A, Dufresne E and Prum R 2010 Structure, function, and self-assembly of single network gyroid (i4132) photonic crystals in butterfly wing scales *Proc. Natl Acad. Sci.* **107** 11676–81

Schurig D, Mock J, Justice B, Cummer S, Pendry J, Starr A and Smith D R 2006 Metamaterial electromagnetic cloak at microwave frequencies *Science* **314** 977–80

Shalaev V, Cai W, Chettiar U, Yuan H K, Sarychev A, Drachev V and Kildishev A 2005 Negative index of refraction in optical metamaterials *Opt. Lett.* **30** 3356–8

Shelby R, Smith D R and Schultz S 2001 Experimental verification of a negative index of refraction *Science* **292** 77–9

Smith D R, Padilla W J, Vier D C, Nemat-Nasser S C and Schultz S 2000 Composite medium with simultaneously negative permeability and permittivity *Phys. Rev. Lett.* **84** 4184–7

Valentine J, Zhang S, Zentgraf T, Ulin-Avila E, Genov D, Bartal G and Zhang X 2008 Three-dimensional optical metamaterial with a negative refractive index *Nature* **455** 376–9

Veselago V 1968 The electrodynamics of substances with simultaneously negative values of ϵ and μ *Phys.-Usp.* **10** 509–14

Vukusic P and Sambles J R 2003 Photonic structures in biology *Nature* **424** 852–5

Yablonovitch E 1987 Inhibited spontaneous emission in solid-state physics and electronics *Phys. Rev. Lett.* **58** 2059–62

Yu N and Capasso F 2014 Flat optics with designer metasurfaces *Nat. Mater.* **13** 139–50

Chapter 2

Effective medium of metamaterials

In the first part of this chapter, we introduce a dipolar description of metamaterials to clarify their operating principles. We then present the S-parameter retrieval method, a numerical technique for extracting effective-medium parameters. In the second part, we develop a formal procedure to extract the effective medium by field-averaging the tangential fields over the boundary surfaces of a single unit cell; this both justifies conventional numerical approaches and allows connecting to classical effective medium theory. Finally, using the developed effective-medium theory and symmetry arguments, we analyze realistic metamaterials and show how to design them for prescribed functions.

2.1 Constitutive relationship

The macroscopic Maxwell's equations governing the propagation of electromagnetic waves in a medium can be written in SI units as

$$\begin{aligned} \nabla \times \mathbf{E} + \partial_t \mathbf{B} = 0, \quad \nabla \cdot \mathbf{B} = 0, \\ \nabla \times \mathbf{H} - \partial_t \mathbf{D} = \mathbf{J}_e, \nabla \cdot \mathbf{D} = \rho_e, \end{aligned} \tag{2.1}$$

where the vector fields are the electric field (**E**), magnetic field (**H**), electric flux density (**D**) and magnetic flux density (**B**). The free current density ($\mathbf{J}_e$) and the free charge density (ρ_e) are related through the charge–continuity equation:

$$\nabla \cdot \mathbf{J}_e + \partial_t \rho_e = 0.$$

For a medium without free charges and currents, we can set both quantities to zero in the above equations.

In macroscopic electromagnetism, it is important to distinguish between free charges, such as conduction electrons moving freely in a metal, and bound charges, such as the dipole moments associated with atoms or molecules. Because bound

doi:10.1088/978-0-7503-1131-1ch2

charges are always tied to the material, their effect is absorbed into the definitions of electric flux density and magnetic flux density through the constitutive relations:

$$\begin{aligned} \mathbf{D} &= \epsilon_0 \mathbf{E} + \mathbf{P} = \epsilon \mathbf{E}, \\ \mathbf{B} &= \mu_0 (\mathbf{H} + \mathbf{M}) = \mu \mathbf{H}, \end{aligned} \tag{2.2}$$

where ϵ is the electric permittivity, summarizing the electric response of the material, and μ is the magnetic permeability, summarizing the magnetic response. The electric polarization $\mathbf{P}$ is defined as the electric dipole moment per unit volume, while the magnetic polarization $\mathbf{M}$ is the magnetic dipole moment per unit volume.

These polarization fields are understood as *locally averaged* quantities: the averaging is performed over many atoms so that microscopic fluctuations are smoothed out, but the averaging region is still small enough that the fields can be treated as continuous functions of position. With this viewpoint, the fields $\mathbf{E}$, $\mathbf{H}$, $\mathbf{D}$, $\mathbf{B}$ in equations (2.1) and (2.2) should all be regarded as locally averaged, macroscopic fields. This makes Maxwell's equations much easier to apply than the microscopic ones that track individual charges.

The remaining task is to describe how the polarizations $\mathbf{P}$ and $\mathbf{M}$ depend on the applied fields $\mathbf{E}$ and $\mathbf{H}$ for different classes of materials. Once this relation is known, the effective response parameters ϵ and μ can be determined.

We begin by reviewing the classical description of the optical response of a gas or a normal dielectric. In the Lorentz model, a medium is modeled as an array of electric dipoles. Each dipole can be imagined as an electron cloud oscillating around a nucleus, much like a harmonic oscillator.

In such a case, Newton's second law for each dipole can be written as

$$m\frac{d^2\mathbf{r}}{dt^2} = -m\omega_0^2\mathbf{r} - m\gamma\frac{d\mathbf{r}}{dt} - e\mathbf{E}_{\mathrm{L}},$$

where vector $\mathbf{r}$ represents the displacement of the electron cloud from the nucleus and m is the electron mass, as shown in figure 2.1. The terms on the right-hand side have clear physical meanings. The first term is the restoring force between the electron cloud and the nucleus. It can be interpreted as a spring–mass system with natural frequency ω_0. The second term is the damping force, a phenomenological term representing material loss, for example, conversion into heat. It carries a minus sign to oppose the motion, with $\gamma > 0$. The final term is the driving force from the local electric field acting on the dipole, with the electron charge e. To study the linear response of the medium, we assume oscillations at frequency ω and set a harmonic time dependence $\exp(-\mathrm{i}\omega t)$ for all variables. This harmonic driving is responsible for the light–matter interaction.

By assuming that the charge of each nucleus is exactly balanced by the charge of its surrounding homogeneous electron cloud, the dipole moment of each atom is $\mathbf{p} = -e\mathbf{r}$. Substituting this relation into the Lorentz model, and assuming harmonic oscillations at frequency ω, we obtain the atomic polarizability α:

$$\mathbf{p} = \alpha\mathbf{E}_{\mathrm{L}} = \frac{e^2}{m}\frac{1}{\omega_0^2 - \omega^2 - \mathrm{i}\gamma\omega}\mathbf{E}_{\mathrm{L}}. \tag{2.3}$$

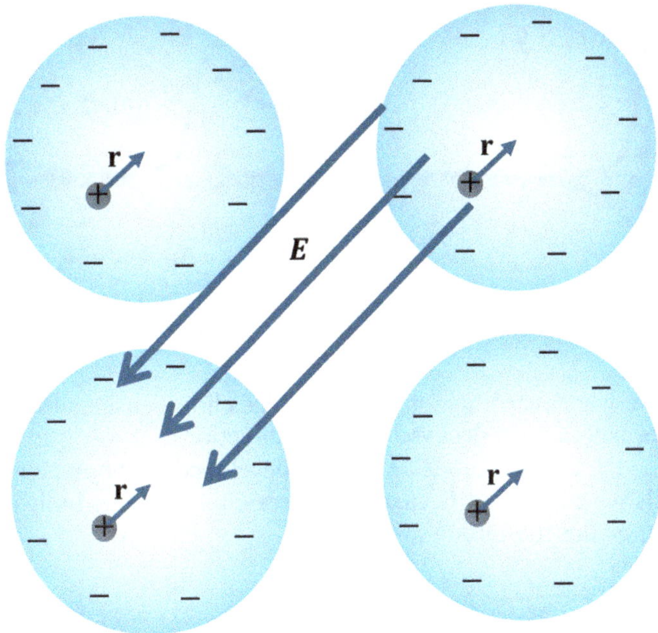

Figure 2.1. Atomic response as a Lorentz dipolar model.

The polarizability is a function of frequency. It resonates at ω_0, as expected from the spring–mass analogy, and contains the damping term γ.

As our first example, consider a dilute gas consisting of particles. Each particle has its dipole moment governed by equation (2.3). Because the number density is small, we may assume the local field is approximately the same as the macroscopic field, $\mathbf{E}_\mathrm{L} \approx \mathbf{E}$. The polarization (dipole moment per unit volume) then becomes

$$\mathbf{P} = \sum_j N_j \alpha_j \mathbf{E}, \tag{2.4}$$

where N_j is the number density of the jth particle type and α_j its atomic polarizability. Using the constitutive relation $\mathbf{D} = \epsilon_0 \mathbf{E} + \mathbf{P} = \epsilon \mathbf{E}$, we find

$$\chi = \frac{\epsilon}{\epsilon_0} - 1 = \frac{1}{\epsilon_0} \sum_j N_j \alpha_j = \sum_j \frac{\omega_{\mathrm{p}j}^2}{\omega_{0j}^2 - \omega^2 - \mathrm{i}\gamma_j \omega}, \tag{2.5}$$

where the plasma frequency of the jth particle type is

$$\omega_{\mathrm{p}j} = \sqrt{\frac{N_j e^2}{\epsilon_0 m}}.$$

Figure 2.2 shows a typical permittivity and index profile for a single Lorentz resonance. At $\omega = \omega_0$, $\mathrm{Im}(\epsilon)$ peaks and is always positive, indicating absorption loss. The refractive index $n = \sqrt{\epsilon}$ is defined using the positive square root so that

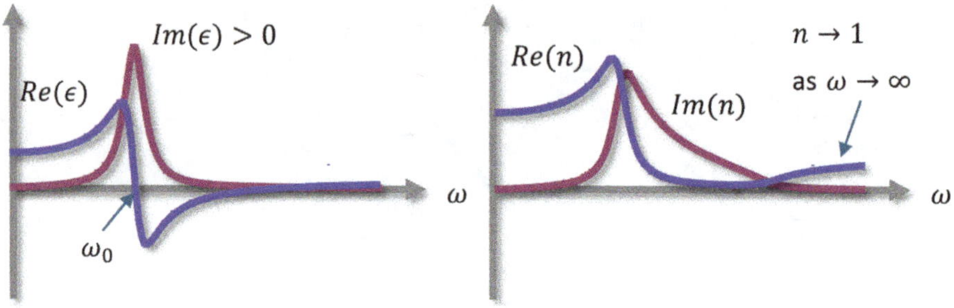

Figure 2.2. Typical permittivity ϵ and refractive index n for a Lorentz model with a single resonance.

Im(n) is positive whenever Im(ϵ) is positive. This choice ensures that a plane wave propagating as exp($in\omega z/c$) decays in the $+z$-direction.

Interestingly, the Lorentz model, although developed for bound charges, can also describe metals where current is carried by free charges. This limit is obtained by taking $\omega_0 \to 0$, giving the Drude model:

$$\frac{\epsilon}{\epsilon_0} = 1 - \frac{\omega_{\mathrm{p}}^2}{\omega^2 + \mathrm{i}\gamma\omega}. \tag{2.6}$$

Here ω_{p} is the plasma frequency, defined by the condition $\epsilon(\omega_{\mathrm{p}}) = 0$. At this frequency, Maxwell's equations admit not only transverse electromagnetic waves but also a longitudinal mode known as a bulk plasma oscillation. For example, for the ionosphere with $N = 10^{12}/\mathrm{m}^3$, we find $\omega_{\mathrm{p}}/2\pi \approx 9$ MHz. For metals such as gold, silver, and copper, the plasma frequencies are around 9 eV.

Now, let us consider dense dielectrics. Compared to a gas, the number of atoms is much larger, and each dipole feels the fields of its neighbors. The local field $\mathbf{E}_{\mathrm{L}}$ differs from the macroscopic field $\mathbf{E}$. A simple approximation is the Lorentz local-field correction:

$$\mathbf{E}_{\mathrm{L}} = \mathbf{E} - \mathbf{E}_{\mathrm{P}}, \tag{2.7}$$

where the polarization contribution from a homogeneous distribution of dipoles inside a small sphere is

$$\mathbf{E}_{\mathrm{P}} = -\frac{\mathbf{P}}{3\epsilon_0}. \tag{2.8}$$

This picture arises when we smear out dipoles to form continuous macroscopic fields. In principle, there should also be a correction from the discrete lattice (a lattice sum), but for many regular or random ensembles, this term vanishes or is negligible. Substituting equation (2.7) into the polarization relation, we obtain

$$\mathbf{P} = \frac{\sum_j N_j \alpha_j}{1 - \frac{1}{3\epsilon_0}\sum_j N_j \alpha_j}\mathbf{E}, \tag{2.9}$$

which gives the Clausius–Mossotti (CM) relation:

$$\frac{\epsilon - \epsilon_0}{\epsilon + 2\epsilon_0} = \frac{1}{3\epsilon_0}\sum_j N_j \alpha_j. \tag{2.10}$$

The CM relation is more general than the dilute-gas case. In the low-concentration limit where $\epsilon \approx \epsilon_0$, it reduces back to equation (2.5).

In this section, we have discussed the Lorentz model and the corresponding permittivity function in explaining the optical response function of natural materials such as gases, dielectric materials, and metals.

Figure 2.3 shows a medieval stained-glass window in Exeter Cathedral. The red color comes from a very thin flashed layer on the surface that contains copper nanoparticles embedded in a glass matrix. These nanoparticles produce a plasmonic extinction band in the green and therefore appear red in transmission when the window is viewed from the darker interior toward daylight. The schematic in the same figure depicts the glass matrix doped with Cu nanoparticles. In this case, we can use the CM equation (equation (2.10)) to estimate an absorption peak in the green (around 580 nm). Here, the effective dipole response arises from the collective free-electron oscillations on the metallic nanoparticle surfaces under the applied electric field, rather than from bound atomic dipoles. For small spherical particle with permittivity ϵ_m (a function of frequency), the polarizability is given by

$$\alpha = 3\epsilon_b V_{sph}\frac{\epsilon_m - \epsilon_b}{\epsilon_m + 2\epsilon_b},$$

where V_{sph} is the volume of a particle. Substituting it into the CM equation, with a replacement of background from ϵ_0 to ϵ_b, gives us

$$\frac{\epsilon - \epsilon_b}{\epsilon + 2\epsilon_b} = f\frac{\epsilon_m - \epsilon_b}{\epsilon_m + 2\epsilon_b}, \tag{2.11}$$

where $f = NV_{sph}$ is the volume filling fraction of the particles. This is called the Maxwell–Garnett formula. The derived ϵ can now be referred to as the effective permittivity of the composite, which is also a function of frequency. Experimental observations show an absorption spectrum with a peak at around 580 nm,

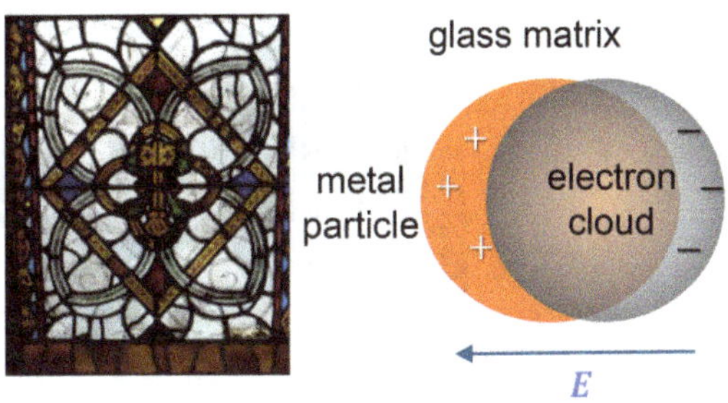

Figure 2.3. Medieval glass window in Exeter Cathedral and a schematic representing the glass matrix with embedded metal nanoparticles, in making the glass window.

corresponding to a similar peak if we plot the imaginary part of the effective refractive index through $n = \sqrt{\epsilon}$. Absorption at such a wavelength results in a red color in transmission. The absorption peak also indicates the existence of a resonance mode at such a wavelength. The resonance mode confined to the nanoparticle is called a localized surface plasmon, while the resonance mode at the plasma frequency, where $\epsilon \approx 0$, is called the bulk plasma. The frequency of such a surface plasmon mode ω_{SP} can be seen from equation (2.11) as

$$\mathrm{Re}(\epsilon) \approx -2\epsilon_{\mathrm{b}} \Leftrightarrow \mathrm{Re}(\epsilon_{\mathrm{m}}(\omega_{\mathrm{SP}})) \approx -2\epsilon_{\mathrm{b}}.$$

By plugging in the Drude dispersion, equation (2.6), of a metal into the above equation, we obtain

$$\omega_{\mathrm{SP}} = \frac{\omega_{\mathrm{p}}}{\sqrt{1 + 2\epsilon_{\mathrm{b}}/\epsilon_0}}.$$

2.2 Understanding metamaterials with a dipolar model

With the CM equation established, we now develop an analytical effective-medium model for realistic metamaterial structures to understand how they operate. To this end, we first analyze the classical negative-index metamaterial using this model and then verify the results numerically.

The first metamaterial in the world with a negative index, shown in figure 2.4 from Shelby *et al* (2001), is made from a two-dimensional (2D) array of wires and split-ring resonators. It is designed for the E_z-polarization in 2D wave propagation. The vertical wires provide the necessary electric response, while the split rings provide the magnetic response.

Let us look at the vertical wires first. Here, we employ the RLC model to describe the response of an individual metamaterial atom. Without going into numerical details, this allows us to focus on the essential elements. Because the wires are assumed to be very long, effectively forming a 2D system, and the electric field is oriented along the wires, the local field $\mathbf{E}_0$ and the effective (macroscopic) field $\mathbf{E}_{\mathrm{eff}}$ are approximately the same. The current I induced by the field is

$$I = \frac{E_0 l}{R - \mathrm{i}\omega L} \approx \frac{E_{\mathrm{eff}} l}{R - \mathrm{i}\omega L},$$

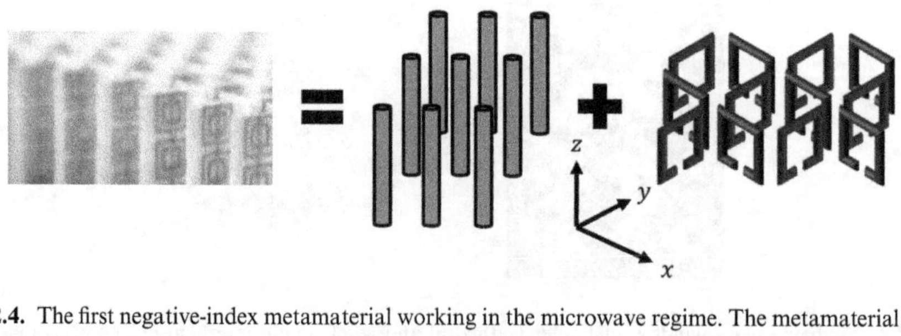

Figure 2.4. The first negative-index metamaterial working in the microwave regime. The metamaterial consists of an array of vertical wires and an array of split-rings (the square rings with a splitting on each ring). Image taken from Shelby *et al* (2001), reprinted with permission from AAAS.

where l is the wire length, R is the resistance representing ohmic losses, and L is the inductance accounting for the inertia of electrons in the wire.

The polarization can then be obtained using the relation $J = -\mathrm{i}\omega P$:

$$P_{\text{eff}} = \frac{I}{-\mathrm{i}\omega A} = -\frac{1}{\mathrm{i}\omega A}\frac{E_{\text{eff}}l}{R - \mathrm{i}\omega L},$$

where A is the area of each unit cell. Since $P_{\text{eff}} = (\epsilon_{\text{eff}} - \epsilon_0)E_{\text{eff}}$, the effective permittivity is

$$\frac{\epsilon_{\text{eff}}}{\epsilon_0} = 1 - \frac{l}{\mathrm{i}\omega\epsilon_0 A}\frac{1}{R - \mathrm{i}\omega L} = 1 - \frac{\omega_{\text{ep}}^2}{\omega^2 + \mathrm{i}\omega R/L}, \tag{2.12}$$

where the effective electric plasma frequency is $\omega_{\text{ep}} = \sqrt{l/(\epsilon_0 LA)}$. This expression has the same form as the Drude model of a metal, but here the electric plasma frequency can be engineered by the geometry (Pendry *et al* 1996).

Now let us turn to the artificial magnetic response from the split rings. A similar engineering approach can be used to derive the magnetic polarizability of a single split ring, very much like a transformer with additional capacitance (Pendry *et al* 1999). A single split ring responds to the local magnetic field through Faraday's law. The electromotive force (e.m.f.) generated is

$$\text{e.m.f.} = \mathrm{i}\omega\mu_0 H_0 A_{\text{ring}} = I\left(R - \mathrm{i}\omega L - \frac{1}{\mathrm{i}\omega C}\right),$$

where A_{ring} is the area of the ring. Here R again represents loss, L the inductance, and the additional capacitance C arises from the gap between the arms of the split ring, which interrupts the circuit path. From this, the magnetic dipole of a single ring and its magnetic polarizability can be written as

$$m = IA_{\text{ring}} = \alpha_{\text{m}} H_0,$$

with

$$\alpha_{\text{m}} = -\frac{aV\omega^2}{\omega^2 - \omega_{\text{m}0}^2 + \mathrm{i}\omega R/L}.$$

Here V is the unit cell volume, while the magnetic resonance frequency and strength coefficient are

$$\omega_{\text{m}0} = \frac{1}{\sqrt{LC}}, \qquad a = \frac{\mu_0 A_{\text{ring}}^2}{LV}.$$

Finally, invoking the local-field formula $H_0 = H_{\text{eff}} + M/3$ leads to the CM relation:

$$\frac{\mu_{\text{eff}} - \mu_0}{\mu_{\text{eff}} + 2\mu_0} = \frac{1}{3}\frac{\alpha_{\text{m}}}{V} = -\frac{1}{3}\frac{a\omega^2}{\omega^2 - \omega_{\text{m}0}^2 + \mathrm{i}\omega R/L}.$$

In the dilute limit, this reduces to

$$\frac{\mu_{\text{eff}}}{\mu_0} \approx 1 - \frac{a\omega^2}{\omega^2 - \omega_{\text{m0}}^2 + \mathrm{i}\omega R/L}. \tag{2.13}$$

As expected, if the frequency ranges where $\epsilon_{\text{eff}} < 0$ and $\mu_{\text{eff}} < 0$ overlap, a negative refractive index is achieved. This was the situation for the first microwave metamaterial. There are, however, subtleties. First, the frequency dispersion here resembles that of a Lorentz model, but with the numerator proportional to ω^2. Thus the resonant magnetic response vanishes at the DC limit, leaving only weak diamagnetism from the metal excluding magnetic flux. Second, the above arguments apply only for an electric field aligned with the long wires and a magnetic field perpendicular to the rings. This highlights that permittivity and permeability are not scalars but second-rank tensors:

$$\begin{pmatrix} D_x \\ D_y \\ D_z \end{pmatrix} = \begin{pmatrix} \epsilon_{xx} & \epsilon_{xy} & \epsilon_{xz} \\ \epsilon_{yx} & \epsilon_{yy} & \epsilon_{yz} \\ \epsilon_{zx} & \epsilon_{zy} & \epsilon_{zz} \end{pmatrix} \begin{pmatrix} E_x \\ E_y \\ E_z \end{pmatrix},$$
$$\begin{pmatrix} B_x \\ B_y \\ B_z \end{pmatrix} = \begin{pmatrix} \mu_{xx} & \mu_{xy} & \mu_{xz} \\ \mu_{yx} & \mu_{yy} & \mu_{yz} \\ \mu_{zx} & \mu_{zy} & \mu_{zz} \end{pmatrix} \begin{pmatrix} H_x \\ H_y \\ H_z \end{pmatrix}. \tag{2.14}$$

For the orientation of the metamaterial structures in figure 2.4, with principal axes along x, y, and z, only the diagonal terms are non-zero. The off-diagonal terms such as ϵ_{xy} or μ_{xy} vanish. The resonant responses derived above correspond to ϵ_{zz}, μ_{xx}, and μ_{yy}.

Achieving a negative index at optical frequencies (infrared and visible) has been much more challenging since the first microwave demonstration. One difficulty is the need for nanofabrication, since the metamaterial unit cells must be much smaller than the wavelength to justify an effective medium description. Another is the higher absorption loss at optical frequencies, which leads to a large imaginary part of the refractive index n.

To overcome these challenges, various designs have been proposed and implemented. One example is the double nano-rod structure shown in figure 1.6 from Shalaev *et al* (2005), which exhibits a negative index at infrared wavelengths with $\mathrm{Re}(n) = -0.3$ at $\lambda = 1.5\ \mu\text{m}$. Its advantage is that a single element supports both electric and magnetic resonances: the electric mode corresponds to currents in the same direction along the two rods, while the magnetic mode corresponds to opposite currents along the rods with a perpendicular magnetic field. Figure 2.5 shows the 'fishnet' structure (Zhang *et al* 2005; Valentine *et al* 2008), consisting of metal patterned into a fishnet with two layers separated by a dielectric spacer. This structure achieves relatively lower losses, with a figure of merit $|\mathrm{Re}(n)/\mathrm{Im}(n)|$ of about 3–4 at the same frequency.

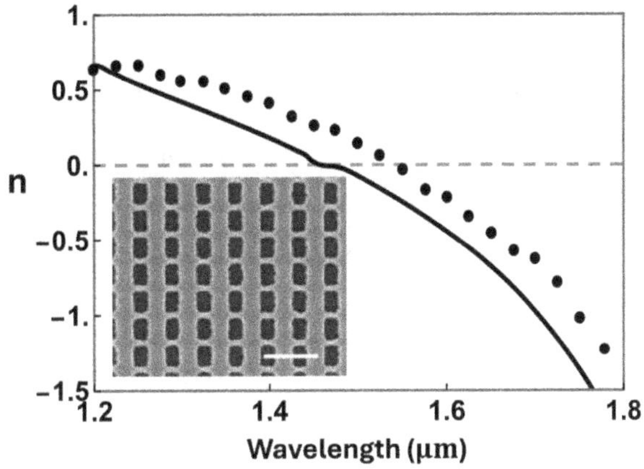

Figure 2.5. Fishnet metamaterial with a negative refractive index at optical frequencies. The inset shows an SEM image of the sample with a 1 μm scale bar. Experimental results (symbols) and simulations (line) are shown. Reproduced from Valentine *et al* (2008), with permission from Springer Nature.

The CM equation, also known as the dipolar model, provides useful insight for understanding and designing metamaterial structures. However, it lacks numerical rigor and cannot be easily applied to connected structures such as the fishnet.

2.3 S-parameter retrieval method

A widely used approach for extracting effective medium parameters is the S-parameter retrieval method (Smith *et al* 2005) for metamaterials with permittivity and permeability. It has been extended to chiral or bianisotropic metamaterials later (Zhao *et al* 2010; Chen *et al* 2005). This technique (here we focus on just the metamaterials with permittivity and permeability) is particularly appealing because it works directly with experimentally measurable quantities: the scattering (S)-parameters, or equivalently, the complex transmission and reflection amplitudes of a metamaterial slab.

Let us consider the normal incidence of a plane wave on a homogeneous slab of metamaterial with effective permittivity ϵ_{eff} and permeability μ_{eff} in vacuum. From now on, we switch from SI to Heaviside–Lorentz units, in which the permittivity and permeability become dimensionless quantities, leading to simpler formulas. For simplicity, we treat these as scalar quantities, assuming a single TE or TM polarization. The methodology is straightforward: (i) *assume* the slab behaves as a spatially homogeneous medium; (ii) obtain the complex transmission and reflection coefficients t and r under normal incidence; and (iii) *solve* the two unknown effective parameters ϵ_{eff} and μ_{eff} from the two equations that relate (t, r) to the slab's dispersion and impedance. Under these assumptions, by using the results of transfer matrix method on a slab of constant material parameters (equation (A.19)):

$$
\begin{aligned}
t &= \frac{2}{2\cos(nka) - \mathrm{i}(Z + 1/Z)\sin(nka)}, \\
r &= \frac{-\mathrm{i}(Z - 1/Z)\sin(nka)}{2\cos(nka) - \mathrm{i}(Z + 1/Z)\sin(nka)},
\end{aligned}
$$

one can prove that t and r satisfy the following:

$$\mathrm{i}\frac{1-r-t}{1+r+t} = \frac{1}{Z}\tan\left(\frac{nka}{2}\right),$$
$$\mathrm{i}\frac{1+r-t}{1-r+t} = Z\tan\left(\frac{nka}{2}\right),$$

where $Z = \sqrt{\mu_{\text{eff}}}/\sqrt{\epsilon_{\text{eff}}}$ is the effective surface impedance, $n = \sqrt{\mu_{\text{eff}}}\sqrt{\epsilon_{\text{eff}}}$ is the effective refractive index, k is the wavenumber in freespace and a is the slab thickness.

In practice, t and r are obtained frequency by frequency from experiments or full-wave simulations. These equations can then be solved for the two unknowns, Z and nka. However, the solution for nka is not unique: it is subject to an additive constant $2m\pi$, where m is an integer. To select the correct branch, one can use physical constraints. In particular, the response functions on the left-hand side of the above formulas always lie in the upper half of the complex plane (this will be proved in a later problem).

By taking the ratio of the two equations, we obtain a unique expression for Z:

$$Z = \frac{\sqrt{\dfrac{1+r-t}{1-r+t}}}{\sqrt{\dfrac{1-r-t}{1+r+t}}},$$

which is guaranteed to have a positive real part, consistent with causality. Substituting this into the equations then gives

$$nka = 2\tan^{-1}\left(i\sqrt{\frac{1+r-t}{1-r+t}}\sqrt{\frac{1-r-t}{1+r+t}}\right),$$

which has a positive imaginary part, as required by passivity. Here, we have chosen the branch that is analytically continued from the thin-slab limit, where the solution is unique because $\mathrm{e}^{inka} \approx 1 + inka$. For thicker slabs, uniqueness is lost because a phase and the same phase plus 2π cannot be distinguished from a single measurement. Nevertheless, analytic continuation from the thin-slab limit provides a smooth, physically consistent solution across frequency.

Finally, the effective permittivity and permeability are obtained from $\epsilon_{\text{eff}} = n/Z$ and $\mu_{\text{eff}} = nZ$, leading to

$$\begin{aligned}\epsilon_{\text{eff}} &= \frac{2}{ka}\tan^{-1}\left(\mathrm{i}\sqrt{\frac{1+r-t}{1-r+t}}\sqrt{\frac{1-r-t}{1+r+t}}\right)\frac{\sqrt{\dfrac{1-r-t}{1+r+t}}}{\sqrt{\dfrac{1+r-t}{1-r+t}}},\\ \mu_{\text{eff}} &= \frac{2}{ka}\tan^{-1}\left(\mathrm{i}\sqrt{\frac{1+r-t}{1-r+t}}\sqrt{\frac{1-r-t}{1+r+t}}\right)\frac{\sqrt{\dfrac{1+r-t}{1-r+t}}}{\sqrt{\dfrac{1-r-t}{1+r+t}}}.\end{aligned} \tag{2.15}$$

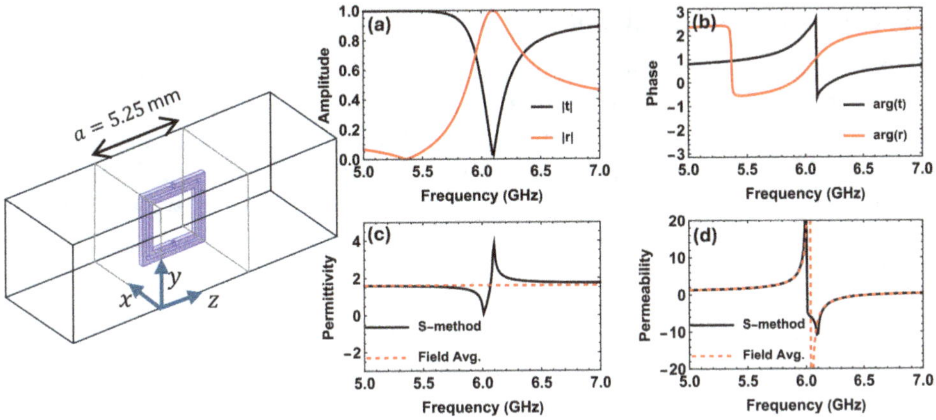

Figure 2.6. (a, b) Transmission and reflection for a metamaterial with split-ring resonators. (c, d) Retrieved effective medium using the S-parameter retrieval method (equation (2.15)). Results using field-averaging method are shown in red lines. The inset shows the cubic unit cell of the metamaterial.

The quantities t and r are readily obtained from full-wave simulations or experiments on a thin slab, for example, a sample only one unit cell thick. This makes the S-parameter retrieval method particularly convenient, and explains why it has become one of the most widely used techniques in metamaterials research.

Figure 2.6 shows a case where the metamaterial consists of a cubic array of split rings, without the vertical wires needed for a negative index. Panels (a) and (b) show t and r obtained from a full-wave simulation for one layer of the unit cell along the z-direction at normal incidence. In this case, the electric resonance lies at a much higher frequency, outside the plotted range, while the magnetic resonance produces a resonant response in the retrieved permeability as expected but also permittivity (panels (c) and (d) using black lines). This apparent resonance in ϵ_{eff} is not physical: it is an artifact called an *antiresonance*. The dispersion curve bends in the opposite direction (downwards) from what is expected near the magnetic resonance.

Such artifacts are not uncommon in the literature when applying the S-parameter retrieval method to metamaterials. They arise because the method does not filter out non-local contributions near resonance, when either ϵ_{eff} or μ_{eff} becomes very large. As a result, a resonance in μ_{eff} can induce a spurious anti-resonant feature in ϵ_{eff}. Although here we used a metamaterial with only a magnetic resonance as an example, the same artifact can appear in other structures, including negative-index metamaterials.

Nevertheless, the S-parameter retrieval method remains accurate once we move away from resonance. In fact, local effective medium theory itself breaks down very close to resonance, so we do not expect the retrieved effective parameters to remain valid there. If we insist that the numerically extracted parameters should correspond exactly to a local effective medium, we must turn to the field-averaging method developed in the next section.

2.4 Effective and local fields

The *S*-parameter retrieval method is accurate only in the very-thin limit, i.e., when the slab thickness is much smaller than the wavelength. For typical metamaterials, however, the wavelength is only about 3 to 5 times the lattice constant or the slab thickness. Moreover, to capture interlayer coupling, we may need to consider metamaterial slabs that are much thicker than a wavelength. Here, we will develop a field-averaging method to extract effective-medium parameters. In this section, we first establish the concepts of effective fields.

Field-averaging formulas

We now discuss how the microscopic fields inside a metamaterial unit cell can be averaged to give their macroscopic versions. Again, we adopt Heaviside–Lorentz units, which emphasize the symmetry between electric and magnetic fields. We begin directly from the macroscopic Maxwell equations, but now the relative permittivity $\epsilon(\mathbf{r})$ is a function of position, representing, for example, a split-ring or other artificial atoms. Figure 2.7(a) illustrates a hypothetical unit cell with such a permittivity distribution. The lattice constant is assumed to be much smaller than the wavelength, so that homogenization is meaningful. In this way, we are effectively using the microscopic permittivity of each inclusion, as introduced in the last section, to build up a higher-level description of the material. The wave equations are

$$\begin{aligned} \nabla \times \mathbf{E} &= ik\,\mathbf{B} = ik(\mathbf{H} + \mathbf{M}), \\ \nabla \times \mathbf{H} &= -ik\,\mathbf{D} = -ik(\mathbf{E} + \mathbf{P}), \end{aligned}$$

where $k = \omega/c$ is the freespace wavenumber. These equations describe how the fields propagate in the presence of a non-uniform permittivity profile. Within the unit cell, the fields fluctuate strongly due to local resonances. Our goal is to replace these fluctuating

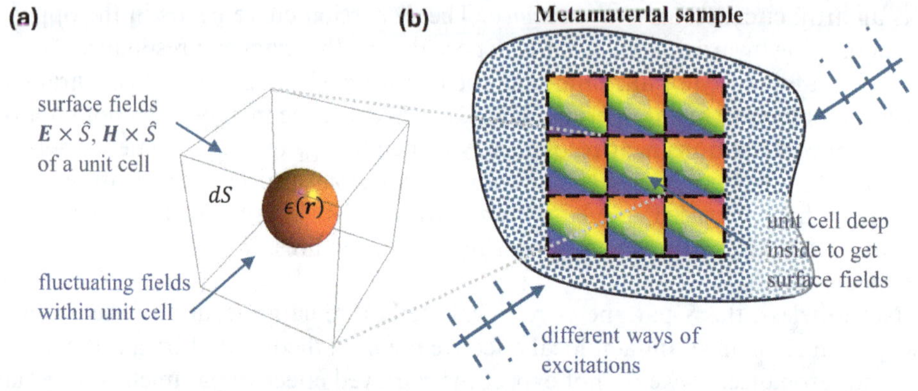

Figure 2.7. Formal procedure to extract an effective medium using dynamic field averaging. (a) A cubic unit cell of a metamaterial, in which surface fields on the six enclosing faces are used to evaluate the effective fields ($\mathbf{D}_{\text{eff}}$, $\mathbf{B}_{\text{eff}}$) and ($\mathbf{E}_{\text{eff}}$, $\mathbf{H}_{\text{eff}}$). (b) Different excitations on a metamaterial sample of finite size are used to generate different local fields arriving at the unit cell deep inside the bulk region. Effective fields extracted from the unit cell are then used to solve the constitutive matrix.

fields with effective fields ($\mathbf{E}_{\text{eff}}$, $\mathbf{H}_{\text{eff}}$, $\mathbf{D}_{\text{eff}}$, $\mathbf{B}_{\text{eff}}$) that vary smoothly. Once this is done, we can define effective constitutive parameters (such as ϵ_{eff}) which remain constant inside a unit cell. Such a replacement is useful whenever the rapidly varying microscopic fields cannot be directly observed. For example, an observer located many wavelengths away from a sample consisting of many unit cells cannot resolve the detailed polarization distribution $\mathbf{P}$ within a unit cell, but only its averaged effect.

To proceed, a usual starting point is to consider integral version of the Maxwell equations, i.e., Stokes' theorem to see the importance of line integrals of the E and H fields and the area integrals of the D and B fields (Smith and Pendry 2006; Ward and Pendry 1996). Here, we take a slightly different approach by taking advantage of the whole enclosed area of unit cell to increase numerical accuracy on defining effective fields. We volume-average the wave equations and define $\mathbf{D}_{\text{eff}}$ and $\mathbf{B}_{\text{eff}}$ as

$$\begin{aligned} \mathbf{D}_{\text{eff}} &\triangleq \frac{1}{ikV}\oint \mathbf{H}\times \mathrm{d}\mathbf{S} = \frac{1}{V}\int \mathbf{D}\,\mathrm{d}V, \\ \mathbf{B}_{\text{eff}} &\triangleq -\frac{1}{ikV}\oint \mathbf{E}\times \mathrm{d}\mathbf{S} = \frac{1}{V}\int \mathbf{B}\,\mathrm{d}V, \end{aligned} \tag{2.16}$$

where V is the unit-cell volume. The use of surface integrals here is significant: it allows us to treat the unit cell as a 'black box,' avoiding the need to track the rapidly fluctuating fields inside.

In a similar way, we require expressions for $\mathbf{E}_{\text{eff}}$ and $\mathbf{H}_{\text{eff}}$. For this purpose, we employ the following 'inner-product' identity that connects volume and surface integrals:

$$\frac{1}{ik}\oint\left(\mathbf{E}^*_{\text{loc}}\cdot \mathbf{H}\times \mathrm{d}\mathbf{S} - \mathbf{H}^*_{\text{loc}}\cdot \mathbf{E}\times \mathrm{d}\mathbf{S}\right) = \int\left(\mathbf{E}^*_{\text{loc}}\cdot \mathbf{P} + \mathbf{H}^*_{\text{loc}}\cdot \mathbf{M}\right)\mathrm{d}V, \tag{2.17}$$

for an arbitrary solution ($\mathbf{E}$, $\mathbf{H}$) of the medium and a chosen vacuum solution ($\mathbf{E}_{\text{loc}}$, $\mathbf{H}_{\text{loc}}$). The vacuum fields $\mathbf{E}_{\text{loc}}$, $\mathbf{H}_{\text{loc}}$ are the slowly varying fields incident upon the unit cell. They can be expanded about the cell center using the dipole local-field formula (equation (A.26)):

$$\begin{aligned} \mathbf{E}_{\text{loc}}(\mathbf{r}) &\approx \mathbf{E}_{\text{loc}}(\mathbf{0}) - \frac{ik}{2}\mathbf{r}\times \mathbf{H}_{\text{loc}}(\mathbf{0}), \\ \mathbf{H}_{\text{loc}}(\mathbf{r}) &\approx \mathbf{H}_{\text{loc}}(\mathbf{0}) + \frac{ik}{2}\mathbf{r}\times \mathbf{E}_{\text{loc}}(\mathbf{0}). \end{aligned}$$

By substituting these expansions into equation (2.17) and collecting coefficients of the basis vectors $\mathbf{E}_{\text{loc}}(\mathbf{0})$ and $\mathbf{H}_{\text{loc}}(\mathbf{0})$, we are led to the following definitions:

$$\begin{aligned} \mathbf{P}_{\text{eff}} &= \frac{\mathbf{p}_{\text{eff}}}{V} = \frac{1}{V}\int\left(\mathbf{P} + \frac{1}{2}ik\,\mathbf{r}\times \mathbf{M}\right)\mathrm{d}V, \\ \mathbf{M}_{\text{eff}} &= \frac{\mathbf{m}_{\text{eff}}}{V} = \frac{1}{V}\int\left(\mathbf{M} - \frac{1}{2}ik\,\mathbf{r}\times \mathbf{P}\right)\mathrm{d}V, \\ \mathbf{E}_{\text{eff}} &= \mathbf{D}_{\text{eff}} - \mathbf{P}_{\text{eff}}, \\ \mathbf{H}_{\text{eff}} &= \mathbf{B}_{\text{eff}} - \mathbf{M}_{\text{eff}}, \end{aligned} \tag{2.18}$$

so that equation (2.17) takes the form

$$\mathbf{E}_{\text{eff}} \approx \frac{1}{2V} \oint \mathbf{r} \times (\mathbf{E} \times d\mathbf{S}),$$
$$\mathbf{H}_{\text{eff}} \approx \frac{1}{2V} \oint \mathbf{r} \times (\mathbf{H} \times d\mathbf{S}). \tag{2.19}$$

These surface-field integrals give us $\mathbf{E}_{\text{eff}}$ and $\mathbf{H}_{\text{eff}}$ (see figure 2.7(a)). Together with equation (2.16), they form a complete recipe for obtaining effective fields. In essence, equation (2.19) works like an interpolation formula, estimating the field value at the cell center from the surrounding surface fields.

Effective medium extraction procedure

At this point, we can see how the field-averaging formulas connect to the effective medium description. The method provides a systematic way to obtain effective fields and, from them, the effective constitutive relations. This approach is general: it combines numerical accuracy with applicability to a wide range of metamaterial structures.

The procedure is illustrated in figure 2.7(b). Consider a metamaterial, and focus on a unit cell located deep inside its bulk region. By exciting the metamaterial from outside with different incident directions and polarizations, we generate different local fields $\mathbf{E}_{\text{loc}}(\mathbf{0})$ and $\mathbf{H}_{\text{loc}}(\mathbf{0})$ arriving at the chosen unit cell. In fact, there are six linearly independent local-field profiles, so we need to excite the metamaterial in six independent ways.

The constitutive relation is expressed by a 6×6 matrix that relates $(\mathbf{D}_{\text{eff}}, \mathbf{B}_{\text{eff}})$ to $(\mathbf{E}_{\text{eff}}, \mathbf{H}_{\text{eff}})$. This matrix can be determined by evaluating the effective fields for each excitation using full-wave simulations. After each simulation, the surface-field integrals in equations (2.16) and (2.19) are applied on the boundary surfaces of the unit cell to extract the effective fields. Once six sets of effective fields are obtained, the 6×6 constitutive matrix can be solved by matrix inversion.

In fact, such a procedure is rather formal and appears demanding. For practical usage, it is better to have a simplified version. Fortunately, many metamaterials possess symmetries that simplify the process. For instance, if we are only concerned with normal incidence on a metamaterial with mirror symmetry along the propagation direction, two simulations for forward incidence with orthogonal polarizations are sufficient, and only a 2×2 matrix inversion is required. If the metamaterial has additional mirror symmetry in the transverse plane, TE and TM polarizations decouple. In that case, a single simulation per polarization is enough, with no need for matrix inversion.

In developing this method, we have assumed that the medium can be described as a *local* effective medium. This means the constitutive matrix does not depend on the excitation wavevector $\mathbf{k}$, i.e., it is valid in the limit $\mathbf{k} \to 0$. This is reasonable for metamaterials with a lattice constant much smaller than the wavelength, so that non-local effects are negligible. The extracted effective parameters should also vary only slightly with the choice of unit cell, provided it is located deep inside the structure, where bulk properties dominate. Finally, this approach is able to account for interactions between unit cells if needed: by starting with one cell and gradually adding surrounding cells, one can check for convergence of the effective parameters extracted from the central unit cell.

Relationship between local and effective fields

Before we apply the field-averaging method to numerically determine the effective medium of a metamaterial, let us first take a detour to understand the meaning of the field-averaging formulas. From an analytical point of view, equation (2.19) provides a direct way to calculate the contribution of dipole moments to the effective fields. By writing the total field as the sum of dipole radiation from $\mathbf{p}_{\text{eff}}$ and $\mathbf{m}_{\text{eff}}$ (using equation (A.24)), together with the local fields $\mathbf{E}_{\text{loc}}$ and $\mathbf{H}_{\text{loc}}$, and integrating over the closed surface of a sphere or cube of volume V, we can expand in powers of k. Keeping only the leading-order terms (real and imaginary) gives

$$\begin{aligned} \mathbf{E}_{\text{loc}}(\mathbf{0}) &= \mathbf{E}_{\text{eff}} + \frac{\mathbf{p}_{\text{eff}}}{3V} - \frac{ik^3\,\mathbf{p}_{\text{eff}}}{6\pi}, \\ \mathbf{H}_{\text{loc}}(\mathbf{0}) &= \mathbf{H}_{\text{eff}} + \frac{\mathbf{m}_{\text{eff}}}{3V} - \frac{ik^3\,\mathbf{m}_{\text{eff}}}{6\pi}. \end{aligned} \tag{2.20}$$

This result provides a solid foundation for the familiar local-field formulas, and connects back to the CM equation:

$$\frac{1}{\alpha} = \frac{1}{3V}\frac{\epsilon + 2}{\epsilon - 1} - \frac{ik^3}{6\pi}.$$

Note that we are in Heaviside–Lorentz units here and only have one kind of 'atoms' (cf equation (2.10)). The additional term on the right-hand side with an imaginary coefficient is the so-called radiative correction (lowest order term for the imaginary part as the equation actually a complex-valued equation). A similar correction also appears in the polarizability of a small particle (permittivity ϵ_s and volume V_s), giving it an imaginary part at finite frequency even when the particle is lossless:

$$\frac{1}{\alpha} = \frac{1}{3V_s}\frac{\epsilon_s + 2}{\epsilon_s - 1} - \frac{ik^3}{6\pi}.$$

These two equations have exactly the same form. In terms of the current language, the CM equation links the effective permittivity ϵ to the polarizability α of the artificial atoms in freespace. Suppose the artificial atom is simply a small sphere with permittivity ϵ_s and volume V_s, we can choose the surface integration in equation (2.16) and equation (2.19) just to the surface of the small sphere. Then, the permittivity ϵ in CM equation simply becomes the permittivity ϵ_s of the small sphere. Then, the two corrections cancel each other, ensuring that the Maxwell–Garnett equation (equation (2.11)) remains valid. Interestingly, the Maxwell–Garnett equation survives the radiation correction but the CM equation does not. A deeper reason is that the wave equation has real-valued permittivity in the lossless case but polarizability is

Table 2.1. Effective field formulas including contributions from both local fields and polarizations generated from various objects in freespace. (a) a point dipole; (b) and (c): a line dipole with dipole orientation along or perpendicular to the line with field invariant along the line; (d) a planar sheet dipole with dipole orientation on-plane with field invariant on the plane. P and M are the electric and magnetic polarizations volume densities. V, A or d represents the volume, area or thickness of the object. k is the wavenumber in freespace.

(a)		$E_{\text{eff}} = E_{\text{loc}} - \frac{P}{3} + \frac{ik^3PV}{6\pi}$ $H_{\text{eff}} = H_{\text{loc}} - \frac{M}{3} + \frac{ik^3MV}{6\pi}$
(b)		$E_{\text{eff}} = E_{\text{loc}} + \frac{ik^2PA}{4}$ $H_{\text{eff}} = H_{\text{loc}} + \frac{ik^2MA}{4}$
(c)	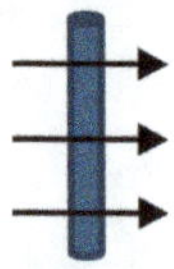	$E_{\text{eff}} = E_{\text{loc}} - \frac{P}{2} + \frac{ik^2PA}{8}$ $H_{\text{eff}} = H_{\text{loc}} - \frac{M}{2} + \frac{ik^2MA}{8}$
(d)		$E_{\text{eff}} = E_{\text{loc}} + \frac{ikPd}{2}$ $H_{\text{eff}} = H_{\text{loc}} + \frac{ikMd}{2}$

complex-valued due to the radiation loss, hence requiring the radiation correction in order to remove the imaginary part in the CM equation. For completeness, table 2.1 summarizes the effective field formulas with contribution from local fields and polarizations for various geometries and polarization orientations.

Field-averaging in two dimensions

So far we have considered field averaging in a 3-dimensional unit cell. The same idea applies to 2D problems as well. In this case, all material parameters and fields are assumed to be invariant along the z-direction. Equation (2.16) remains valid and is applied to the enclosing boundary (a closed line) of the 2D unit cell on the x–y plane. If we imagine the cell as a cylinder extended along z, the contributions from the top and bottom surfaces cancel out. However, the definitions of $\mathbf{E}_{\text{eff}}$ and $\mathbf{H}_{\text{eff}}$ must be modified. At the quasistatic limit, the local-field expression becomes (equation (A.29))

$$\begin{aligned} \mathbf{E}_{\text{loc}}(\mathbf{r}) &\approx \mathbf{E}_{\text{loc}}(\mathbf{0}) + ik\ \mathbf{H}_{\text{loc}}(\mathbf{0}) \cdot \left(\bar{\bar{I}} - \frac{\hat{z}\hat{z}}{2}\right) \times \boldsymbol{\rho}, \\ \mathbf{H}_{\text{loc}}(\mathbf{r}) &\approx \mathbf{H}_{\text{loc}}(\mathbf{0}) - ik\ \mathbf{E}_{\text{loc}}(\mathbf{0}) \cdot \left(\bar{\bar{I}} - \frac{\hat{z}\hat{z}}{2}\right) \times \boldsymbol{\rho}, \end{aligned} \tag{2.21}$$

where $\boldsymbol{\rho} = x\hat{x} + y\hat{y}$ is the in-plane position vector. Substituting this into the inner-product identity (equation (2.17)) and defining $\mathbf{E}_{\text{eff}} = \mathbf{D}_{\text{eff}} - \mathbf{P}_{\text{eff}}$, $\mathbf{H}_{\text{eff}} = \mathbf{B}_{\text{eff}} - \mathbf{M}_{\text{eff}}$, we obtain

$$\begin{aligned} \mathbf{E}_{\text{eff}} &= \frac{1}{V}\oint\left(\bar{\bar{I}} - \frac{\hat{z}\hat{z}}{2}\right)\cdot[\boldsymbol{\rho} \times (\mathbf{E} \times d\mathbf{S})], \\ \mathbf{H}_{\text{eff}} &= \frac{1}{V}\oint\left(\bar{\bar{I}} - \frac{\hat{z}\hat{z}}{2}\right)\cdot[\boldsymbol{\rho} \times (\mathbf{H} \times d\mathbf{S})], \end{aligned} \tag{2.22}$$

where the integration is carried out only on the boundary of the 2D area. Alternatively, one may view equation (2.19) as still valid, with the top and bottom contributions effectively absorbed into the additional factor $\bar{\bar{I}} - \hat{z}\hat{z}/2$.

2.5 Field-averaging method

Now, we choose the one-dimensional (1D) model for the field-averaging technique so that we can see its connection to the *S*-parameter retrieval method, which assumes plane wave excitation along a fixed direction, such as normal incidence on a metamaterial slab.

1D application

To illustrate the field-averaging method in a simple setting, we now apply it to a 1D case. As an example, consider the negative-index metamaterial formed by split-rings and wires. Its response can be modeled as the secondary radiation from a planar sheet of electric and magnetic polarizations, as shown in figure 2.8, for waves propagating along the $+z$- or $-z$-directions.

We assume a single polarization: the electric field along the x-direction, the magnetic field along the y-direction, and propagation along the z-direction. Application of the surface-integral formulas in equation (2.16) gives

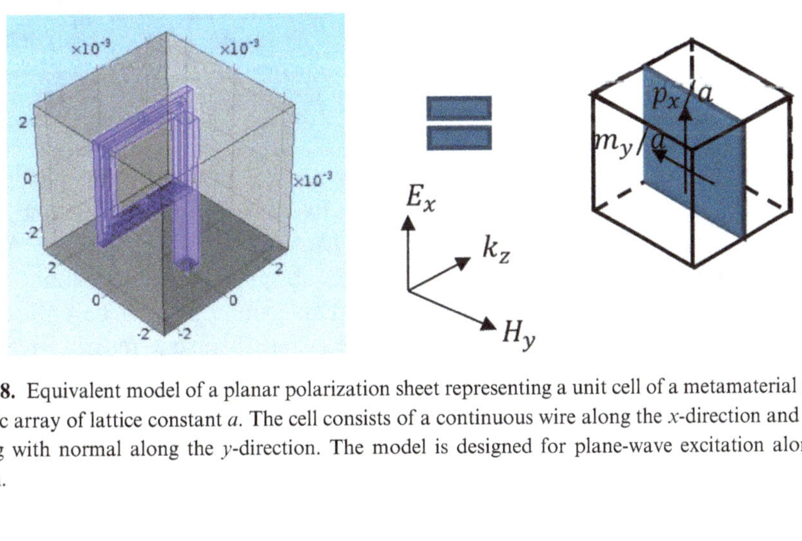

Figure 2.8. Equivalent model of a planar polarization sheet representing a unit cell of a metamaterial arranged in a cubic array of lattice constant a. The cell consists of a continuous wire along the x-direction and a double split ring with normal along the y-direction. The model is designed for plane-wave excitation along the z-direction.

$$
\begin{aligned}
D_{\text{eff},x} &\triangleq \frac{\bar{H}_y(a/2) - \bar{H}_y(-a/2)}{\text{i}ka}, \\
B_{\text{eff},y} &\triangleq \frac{\bar{E}_x(a/2) - \bar{E}_x(-a/2)}{\text{i}ka},
\end{aligned}
\tag{2.23}
$$

where a is the lattice constant of the cubic lattice, and the overbar denotes averaging over the transverse plane. The unit cell is centered at $z = 0$. This result may look 'trivial,' as it is essentially the finite-difference version of $\nabla \times \mathbf{H} = -\text{i}k\mathbf{D}$ and $\nabla \times \mathbf{E} = \text{i}k\mathbf{B}$. The expression depends only on the surface-averaged fields at the front ($z = a/2$) and back ($z = -a/2$) of the cell, because contributions from the other four surfaces cancel under the periodic boundary conditions and the invariance of the fields along x and y.

The local fields can be written in closed form as

$$
\begin{aligned}
E_{\text{loc},x}(z) &= E_{\text{loc},x}(0)\cos(kz) + \text{i}H_{\text{loc},y}(0)\sin(kz), \\
H_{\text{loc},y}(z) &= H_{\text{loc},y}(0)\cos(kz) + \text{i}E_{\text{loc},x}(0)\sin(kz),
\end{aligned}
$$

where k is the freespace wavenumber. In this 1D example, we do not perform a Taylor expansion of the local fields; the exact trigonometric form is still manageable, hoping for higher accuracy (often the lattice constant is not many times smaller than the freespace wavelength). Substituting these expressions into equation (2.17), and requiring validity for arbitrary input amplitudes $E_{\text{loc},x}(0)$ and $H_{\text{loc},y}(0)$, together with the relations

$$
\begin{aligned}
\frac{p_{\text{eff},x}}{a} &= D_{\text{eff},x} - E_{\text{eff},x}, \\
\frac{m_{\text{eff},y}}{a} &= B_{\text{eff},y} - H_{\text{eff},y},
\end{aligned}
$$

we obtain the definitions of the effective fields:

$$
\begin{aligned}
E_{\text{eff},x} &\triangleq \frac{\bar{H}_y(a/2) - \bar{H}_y(-a/2)}{\text{i}ka}\left(1 - \cos\frac{ka}{2}\right) + \frac{\bar{E}_x(a/2) + \bar{E}_x(-a/2)}{ka}\sin\frac{ka}{2}, \\
H_{\text{eff},y} &\triangleq \frac{\bar{E}_x(a/2) - \bar{E}_x(-a/2)}{\text{i}ka}\left(1 - \cos\frac{ka}{2}\right) + \frac{\bar{H}_y(a/2) + \bar{H}_y(-a/2)}{ka}\sin\frac{ka}{2}.
\end{aligned}
\tag{2.24}
$$

Because the electric and magnetic responses are decoupled in this structure, we can determine them using only two simulations: one with $E_{\text{loc},x}(0) = 1$, $H_{\text{loc},y}(0) = 0$ and another with $E_{\text{loc},x}(0) = 0$, $H_{\text{loc},y}(0) = 1$. We also assume the metamaterial is mirror-symmetric about the $z = 0$ plane, so that bianisotropy (cross-coupling terms) vanishes. This symmetry can be enforced, for instance, by symmetrizing the split-ring geometry. In this case, matrix inversion is unnecessary: D relates directly to E, and B directly to H.

The effective permittivity and permeability follow from combining the above formulas:

$$\frac{1}{\epsilon_{\text{eff}}} = \frac{E_{\text{eff}}}{D_{\text{eff}}} = 1 - \cos\frac{ka}{2} + \mathrm{i}\,\frac{\bar{E}(a/2) + \bar{E}(-a/2)}{\bar{H}(a/2) - \bar{H}(-a/2)}\sin\frac{ka}{2},$$
$$\frac{1}{\mu_{\text{eff}}} = \frac{H_{\text{eff}}}{B_{\text{eff}}} = 1 - \cos\frac{ka}{2} + \mathrm{i}\,\frac{\bar{H}(a/2) + \bar{H}(-a/2)}{\bar{E}(a/2) - \bar{E}(-a/2)}\sin\frac{ka}{2}. \tag{2.25}$$

The above formulas can be evaluated after performing full-wave simulations of the metamaterial. Figures 2.9(c) and (d) show the numerical results for the structure in figure 2.6 with vertical wires added. The red curves represent the effective medium obtained using the field-averaging method. The electric response displays Drude-like frequency dispersion, while the magnetic response near 6 GHz shows Lorentz-type dispersion, consistent with analytic expectations. In contrast, the results obtained from the *S*-parameter retrieval method (equation (2.15)) exhibit an antiresonance feature in ϵ_{eff} around the magnetic resonance (Koschny *et al* 2003; Smith *et al* 2005), which should not appear in a local effective-medium theory.

It is important to note that the field-averaging method and the *S*-parameter retrieval method must agree in the thin-slab limit, where the slab thickness is much smaller than a wavelength. To see this, let us apply equation (2.25) to the case of transmission and reflection from a metamaterial slab. If the slab is illuminated symmetrically from both sides, we can extract ϵ_{eff} using

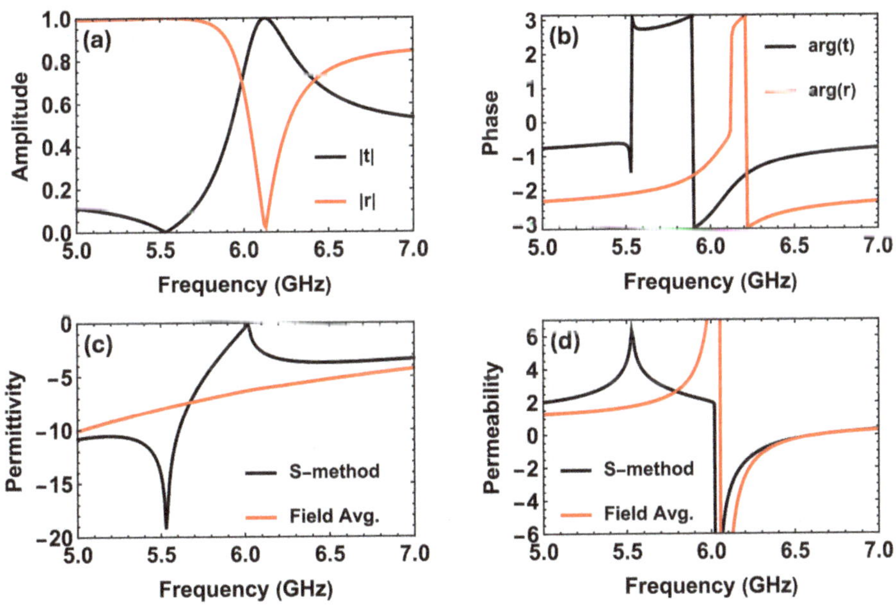

Figure 2.9. Effective medium of the negative-index metamaterial constructed from split-rings and continuous wires. (a, b) Transmission and reflection coefficients. (c, d) Effective permittivity and permeability. Red: obtained by the field-averaging method with planar polarization equivalence. Blue: obtained by the *S*-parameter retrieval method (equation (2.15)).

$$\bar{E}(a/2) = \bar{E}(-a/2) = 1 + r + t, \qquad \bar{H}(a/2) = -\bar{H}(-a/2) = -1 + r + t.$$

For antisymmetric illumination, we can extract μ_{eff} using

$$\bar{E}(a/2) = -\bar{E}(-a/2) = -1 - r + t, \qquad \bar{H}(a/2) = \bar{H}(-a/2) = 1 - r + t.$$

Substituting into equation (2.25) yields

$$\begin{aligned} \frac{1}{\epsilon_{\text{eff}}} &= 1 - \cos\frac{ka}{2} - \mathrm{i}\frac{1 + r + t}{1 - r - t}\sin\frac{ka}{2}, \\ \frac{1}{\mu_{\text{eff}}} &= 1 - \cos\frac{ka}{2} - \mathrm{i}\frac{1 - r + t}{1 + r - t}\sin\frac{ka}{2}. \end{aligned} \tag{2.26}$$

These are the formulas to get permittivity and permeability in the field-averaging method in figure 2.9. If we expand only to leading order in ka, we obtain

$$\begin{aligned} \epsilon_{\text{eff}} &= \frac{2\mathrm{i}}{ka}\frac{1 - r - t}{1 + r + t}, \\ \mu_{\text{eff}} &= \frac{2\mathrm{i}}{ka}\frac{1 + r - t}{1 - r + t}, \end{aligned} \tag{2.27}$$

which are exactly the Taylor-expanded versions of the S-parameter retrieval formulas in equation (2.15).

We still refer to equation (2.26) as the *field-averaging method.* It is simply the application of the field-averaging technique to the reflection (r) and transmission (t) coefficients of a thin slab. As long as the slab is thin compared to a wavelength, and only far-field coupling between neighboring layers is significant, this method can be used reliably.

2.6 Spatial symmetries

A negative index metamaterial is only the first example to show the new power in designing our own metamaterial atoms. Negative refraction in an effectively homogeneous medium is beyond what natural materials can give us. On the other hand, if we think about many of the existing materials or composites for polarization control, absorption, and so on, the metamaterials route can already be very useful if we would like to have our material with a stronger response, optimal control, specific working frequency and bandwidth, etc. We may also start from a specification of functionality and reversely work out the required material parameters and metamaterial designs. In such cases, a simple description using a scalar ϵ and a scalar μ is not enough, and we would need to break certain kinds of spatial symmetries of our metamaterial atoms to generate the desired functionality.

2.6.1 General S-parameter retrieval

To know the relationship between spatial symmetries of metamaterials and the associated response, the effective medium description serves as a bridge. We need to develop the effective medium formulas for metamaterials of low spatial symmetry.

We now generalize the field-averaging method developed in one dimension to metamaterials with *low spatial symmetry*. In contrast to the symmetric cases treated earlier, here we assume no polarization decoupling and allow for bianisotropy. Even under normal incidence on a slab, the constitutive relation must therefore capture cross-coupling between electric and magnetic responses.

Let us begin with a metamaterial without assuming any symmetries. The constitutive relationship can go even lower than the one in equation (2.14) with possible cross-coupling between electric and magnetic fields (bianisotropy). Even if we are only discussing normal incidence on a metamaterial slab, it has to be specified by a 4×4 $\mathcal{C}$ matrix defined by

$$\begin{pmatrix} E_{\text{eff},x} \\ E_{\text{eff},y} \\ H_{\text{eff},x} \\ H_{\text{eff},y} \end{pmatrix} = \mathcal{C} \begin{pmatrix} D_{\text{eff},x} \\ D_{\text{eff},y} \\ B_{\text{eff},x} \\ B_{\text{eff},y} \end{pmatrix}.$$

We have assumed that the slab, with a thickness of a, has its surface normal in the z-direction. The constitutive $\mathcal{C}$ matrix is defined to relate **D**, **B** to **E**, **H** instead of the other way around for mathematical convenience, which will become clear very soon. On the other hand, we have more S-parameters than the scalar t and r, as we cannot assume polarization decoupling, and the metamaterial can look different in the forward and backward directions. In the current case, we need a 4×4 S-matrix:

$$\begin{pmatrix} E_{2x}^{+} \\ E_{2y}^{+} \\ E_{1x}^{-} \\ E_{1y}^{-} \end{pmatrix} = S \begin{pmatrix} E_{1x}^{+} \\ E_{1y}^{+} \\ E_{2x}^{-} \\ E_{2y}^{-} \end{pmatrix} = \begin{pmatrix} S_{xx}^{++} & S_{xy}^{++} & S_{xx}^{+-} & S_{xy}^{+-} \\ S_{yx}^{++} & S_{yy}^{++} & S_{yx}^{+-} & S_{yy}^{+-} \\ S_{xx}^{-+} & S_{xy}^{-+} & S_{xx}^{--} & S_{xy}^{--} \\ S_{yx}^{-+} & S_{yy}^{-+} & S_{yx}^{--} & S_{yy}^{--} \end{pmatrix} \begin{pmatrix} E_{1x}^{+} \\ E_{1y}^{+} \\ E_{2x}^{-} \\ E_{2y}^{-} \end{pmatrix},$$

where the superscript denotes forward (+) or backward (-) traveling plane waves along z, and subscript 1 (2) denotes the reference plane of the **E**-field at the back (front) interface of the slab at $z = -a/2$ ($z = +a/2$). The fields on the two boundary surfaces can be written as a superposition of forward and backward waves:

$$\begin{pmatrix} E_{ix} \\ E_{iy} \\ H_{ix} \\ H_{iy} \end{pmatrix} = B \begin{pmatrix} E_{ix}^{+} \\ E_{iy}^{+} \\ E_{ix}^{-} \\ E_{iy}^{-} \end{pmatrix} = \begin{pmatrix} 1 & 0 & 1 & 0 \\ 0 & 1 & 0 & 1 \\ 0 & -1 & 0 & 1 \\ 1 & 0 & -1 & 0 \end{pmatrix} \begin{pmatrix} E_{ix}^{+} \\ E_{iy}^{+} \\ E_{ix}^{-} \\ E_{iy}^{-} \end{pmatrix}$$

for $i = 1, 2$. Then, the same procedure about field-averaging can be applied to the fields on the boundary surfaces to define the effective fields. There are a total of four different sets of incident fields (x/y-polarizations and forward/backward propagation) and hence four sets of effective fields, from which we can extract the full constitutive $\mathcal{C}$ matrix. While this process may seem complicated, the result is surprisingly simple (Feng *et al.* (2013)), taking the form of a matrix generalization of equation (2.26):

$$\mathcal{C} = \left(1 - \cos\frac{ka}{2}\right)I + \mathrm{i}\left(\sin\frac{ka}{2}\right) B\,(S + I)\,(S - I)^{-1}\,B^{-1}, \tag{2.28}$$

where I is the 4×4 identity matrix.

Again, this has the convenience of the conventional S-parameter retrieval method in that the input is the measured (or simulated) S-matrix, with the artificial antiresonance potentially removed. The trigonometric functions are not Taylor-expanded in k, which improves accuracy for electrically thin but not infinitesimal slabs. The derived effective-medium formula is, in fact, similar to a Cayley transform in a scattering problem (Hörner *et al* 2023). For a passive system, the scattering matrix S has all its eigenvalues lying within or on the unit circle. These eigenvalues are all mapped to the lower half-plane for the eigenvalues of the $\mathcal{C}$-matrix, as expected for a passive local effective medium.

A further remark is that if the metamaterial is thicker than a wavelength, we can bypass the above formula and instead extract the effective fields directly from the boundary of a central unit cell. These fields, obtained from four independent simulations, are then used to invert the $\mathcal{C}$ matrix. In practice, this approach is limited to simulations, since experiments cannot access the fields inside the slab.

2.6.2 Metamaterials with low spatial symmetries

To illustrate how metamaterials with reduced spatial symmetries look, figure 2.10 shows representative samples operating from microwave to optical frequencies. These structures are deliberately designed without certain mirror or rotational symmetries. As a result, they can achieve remarkable effects such as efficient polarization conversion (Pfeiffer and Grbic 2014), strong circular dichroism or optical activity (Decker *et al* 2010; Gansel *et al* 2009), and even chirality-induced negative index (Zhang 2009; Plum *et al* 2009).

The effective medium description follows from the S-parameters through equation (2.28), and the role of symmetry becomes clear from the constitutive matrix. For instance, inserting $S = \mathrm{e}^{\mathrm{i}ka}I$ gives the trivial vacuum result $\mathcal{C} = I$. An interesting case is a perfect polarization rotator, where forward x-polarization is converted entirely into y-polarization (and vice versa). Choosing $S_{xy}^{++} = S_{yx}^{--} = -\mathrm{e}^{\mathrm{i}\phi}$ and $S_{yx}^{++} = S_{xy}^{--} = \mathrm{e}^{\mathrm{i}\phi}$ leads, for $\phi = 3ka/4$, to

$$\mathcal{C}^{-1} = \begin{pmatrix} 1 & 0 & -\mathrm{i}\csc(ka/4)/2 & 0 \\ 0 & 1 & 0 & -\mathrm{i}\csc(ka/4)/2 \\ \mathrm{i}\csc(ka/4)/2 & 0 & 1 & 0 \\ 0 & \mathrm{i}\csc(ka/4)/2 & 0 & 1 \end{pmatrix}.$$

This medium is isotropic in $\epsilon_{xx} = \epsilon_{yy}$, $\mu_{xx} = \mu_{yy}$, but strongly chiral. Figure 2.10(a) shows a corresponding design, which indeed has no mirror symmetries and produces nearly ideal polarization rotation around 10 GHz.

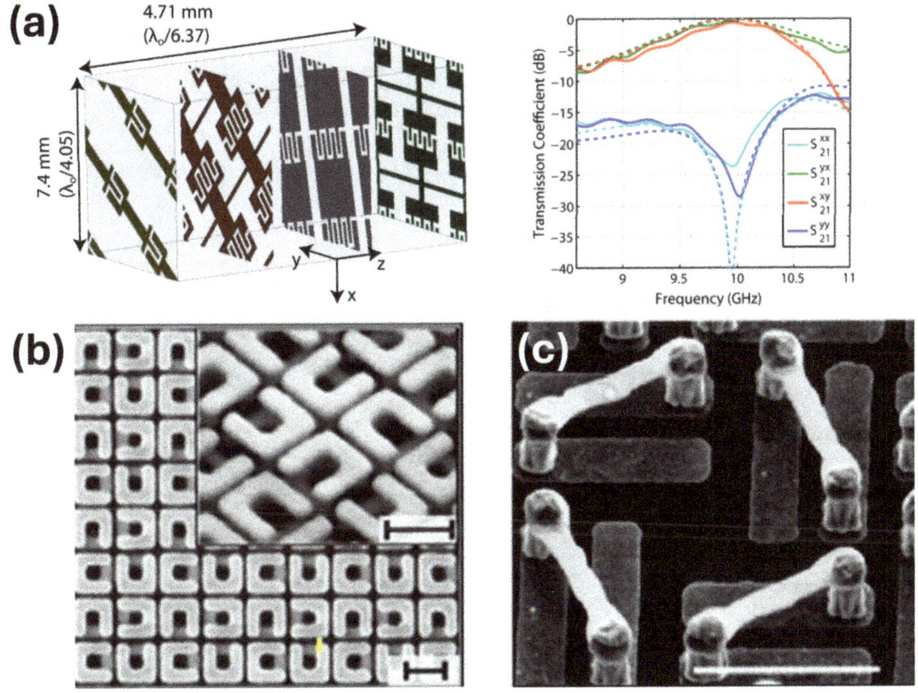

Figure 2.10. Metamaterials with lower spatial symmetries. (a) Linear polarization rotator with nearly 100% transmission. Reprinted with permission from Pfeiffer and Grbic (2014). Copyright (2014) by the American Physical Society. (b) Bilayer split-ring design showing strong optical activity and dichroism. Reprinted with permission from Decker *et al* (2010). © Optical Society of America. (c) A negative index metamaterial for one circular polarization, working at around 1 THz. Reprinted with permission from Zhang (2009). Copyright (2009) by the American Physical Society.

Mirror symmetries. A mirror operation M_i ($i = x, y, z$) acts differently on electric (polar) and magnetic (axial) fields. A metamaterial possessing M_i symmetry must satisfy

$$M_i\, \mathcal{C}^{-1}\, M_i^{-1} = \mathcal{C}^{-1}.$$

From this condition, the allowed form of $\mathcal{C}^{-1}$ can be worked out (see table 2.2). Symmetry forces certain tensor components to vanish. For example, all chiral parameters (ξ_{xx}, ξ_{yy}, ζ_{xx}, ζ_{yy}) vanish if any mirror symmetry is present. Thus, to realize a chiral medium, all mirror symmetries must be broken. Similarly, magneto-electric coupling (ξ_{xy}, ζ_{xy}) requires breaking M_z, while off-diagonal permittivity or permeability requires breaking both M_x and M_y.

Rotational symmetries. A similar argument applies to an n-fold rotation C_n about the z-axis. Requiring invariance,

$$C_n \mathcal{C}^{-1} C_n^{-1} = \mathcal{C}^{-1},$$

Table 2.2. Examples of unit cells with given mirror symmetries and the resulting form of $\mathcal{C}^{-1}$.

<table>
<tr><th>Mirror symmetry</th><th>Example structure</th><th>Form of $\mathcal{C}^{-1}$</th></tr>
<tr><td>$\mathbf{M}_x$</td><td>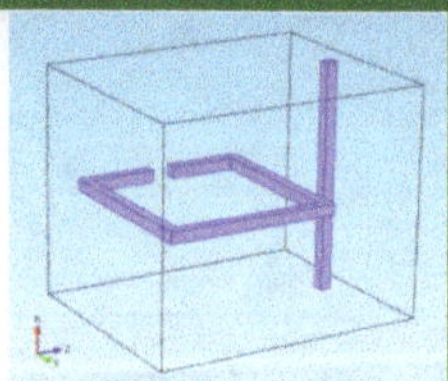</td><td rowspan="2">$\begin{pmatrix} \epsilon_{xx} & 0 & 0 & \xi_{xy} \\ 0 & \epsilon_{yy} & \xi_{yx} & 0 \\ 0 & \zeta_{xy} & \mu_{xx} & 0 \\ \zeta_{yx} & 0 & 0 & \mu_{yy} \end{pmatrix}$</td></tr>
<tr><td>$\mathbf{M}_y$</td><td>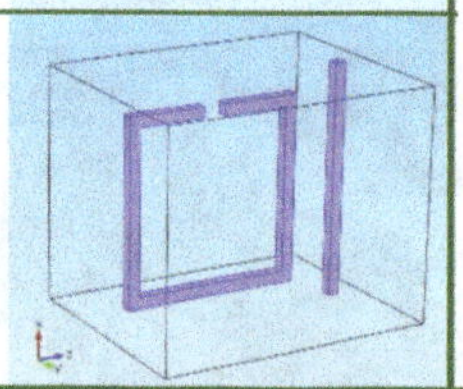</td></tr>
<tr><td>$\mathbf{M}_z$</td><td>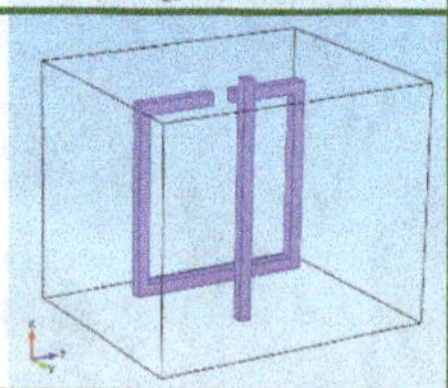</td><td>$\begin{pmatrix} \epsilon_{xx} & \epsilon_{xy} & 0 & 0 \\ \epsilon_{yx} & \epsilon_{yy} & 0 & 0 \\ 0 & 0 & \mu_{xx} & \mu_{xy} \\ 0 & 0 & \mu_{yx} & \mu_{yy} \end{pmatrix}$</td></tr>
</table>

yields the forms summarized in table 2.3. For $n \geqslant 3$, the medium is isotropic in the x–y plane (with reciprocity assumed). Hence, if off-diagonal terms such as ϵ_{xy}, μ_{xy} are desired, rotational symmetry must not exceed C_2 (reciprocity requests symmetric ϵ and μ tensors). Moreover, two-fold rotational symmetry has no constraints on the transverse components of the constitutive matrix while the constraints are on the other matrix elements not shown here.

Asymmetric transmission. With the symmetry rules in place, we can target specific effects such as asymmetric transmission, where conversion between polarizations differs for forward and backward incidence. For circular polarization, the figure of merit simplifies (for $\mu = 1$) to

$$\left| \frac{t_{\mathrm{RL}}^{+}}{t_{\mathrm{RL}}^{-}} \right| \approx \left| \frac{(\epsilon_{xx} - \epsilon_{yy}) + 2\mathrm{i}\epsilon_{xy}}{(\epsilon_{xx} - \epsilon_{yy}) - 2\mathrm{i}\epsilon_{xy}} \right|. \tag{2.29}$$

Asymmetry disappears if either $\epsilon_{xy} = 0$ or $\epsilon_{xx} = \epsilon_{yy}$. Thus, the unit cell must be anisotropic, lack in-plane mirror symmetries, and exhibit loss. Planar chiral metamaterials (e.g., inset of figure 2.11), which lack any in-plane mirror symmetries, satisfy these conditions (Fedotov *et al* 2006; Zhukovsky *et al* 2009; Drezet *et al* 2008;

Table 2.3. Forms of $\mathcal{C}^{-1}$ allowed by C_2, C_3, and C_4 rotational symmetry. Higher orders are equivalent to C_3/C_4.

<table>
<tr><th>Rotational Symmetry (about $\hat{z}$)</th><th>Example structure</th><th>Form of $\mathcal{C}^{-1}$</th></tr>
<tr><td>C_2</td><td></td><td>$$\begin{pmatrix} \epsilon_{xx} & \epsilon_{xy} & \xi_{xx} & \xi_{xy} \\ \epsilon_{yx} & \epsilon_{yy} & \xi_{yx} & \xi_{yy} \\ \zeta_{xx} & \zeta_{xy} & \mu_{xx} & \mu_{xy} \\ \zeta_{yx} & \zeta_{yy} & \mu_{yx} & \mu_{yy} \end{pmatrix}$$</td></tr>
<tr><td>C_3</td><td>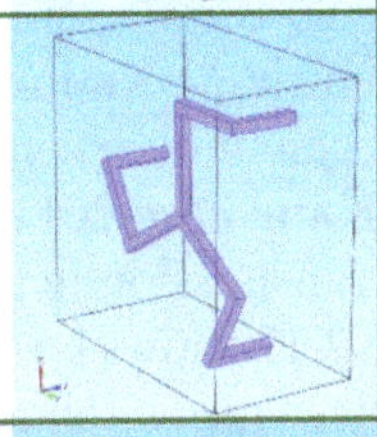</td><td rowspan="2">$$\begin{pmatrix} \epsilon_{xx} & \epsilon_{xy} & \xi_{xx} & \xi_{xy} \\ -\epsilon_{xy} & \epsilon_{xx} & -\xi_{xy} & \xi_{xx} \\ \zeta_{xx} & \zeta_{xy} & \mu_{xx} & \mu_{xy} \\ -\zeta_{xy} & \zeta_{xx} & -\mu_{xy} & \mu_{xx} \end{pmatrix}$$</td></tr>
<tr><td>C_4</td><td>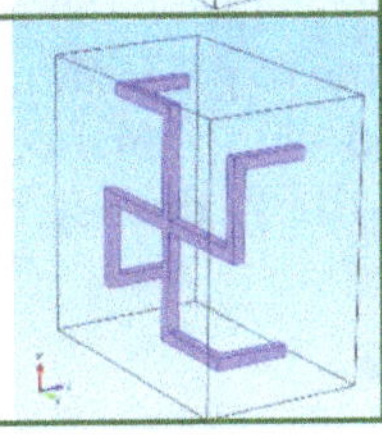</td></tr>
</table>

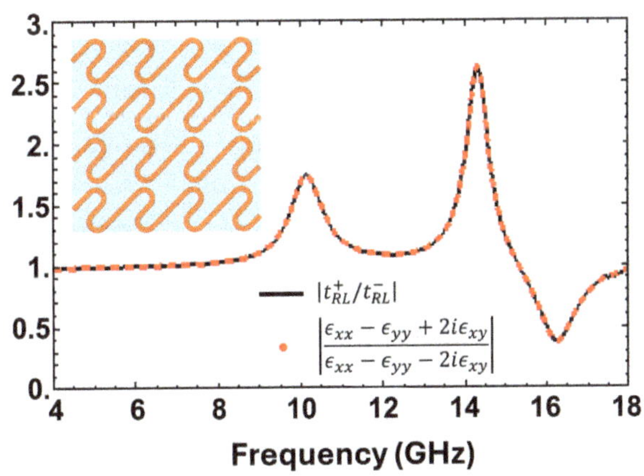

Figure 2.11. Planar chiral metamaterial using structural parameters reported in Fedotov *et al* (2006). Asymmetric transmission compared with equation (2.29) using simulation data. Adapted figure with permission from Fedotov *et al* (2006), copyright (2006) American Physical Society.

Singh *et al* 2009). Figure 2.11 shows an example with configuration from Fedotov *et al* (2006): a copper film patterned into a fish-scale structure (copper stripes on dielectric substrate). Effective parameters are retrieved from the simulated S-parameters, and the asymmetric transmission computed from equation (2.29) agrees well with the simulated results.

For linear polarization, the corresponding expression is

$$\left|\frac{t_{yx}^{+}}{t_{yx}^{-}}\right| = \left|\frac{(\epsilon_{xy} - \mu_{xy}) + (\xi_{xx} + \xi_{yy})}{(\epsilon_{xy} - \mu_{xy}) - (\xi_{xx} + \xi_{yy})}\right|. \tag{2.30}$$

From this equation, we can see that we need both anisotropy and chirality, and the anisotropy from electric and magnetic responses should be different. Here, chirality means that the unit cell must break all mirror symmetries, both in-plane or out-of-plane. Such anisotropic and chiral metamaterials can induce asymmetric transmission for linear polarization (Menzel *et al* 2010; Kang *et al* 2011). Furthermore, we also require loss to have prominent asymmetric transmission in amplitude as the chiral terms are imaginary for lossless media. Figure 2.12 shows an example of such unit cell composed of both chiral and achiral units with simulated transmission amplitudes showing asymmetric transmission for linear polarization (Zhang *et al* 2013). In the same figure, equation (2.30) is verified with the effective parameters retrieved from the simulated S-parameters around the frequency obtaining asymmetric transmission.

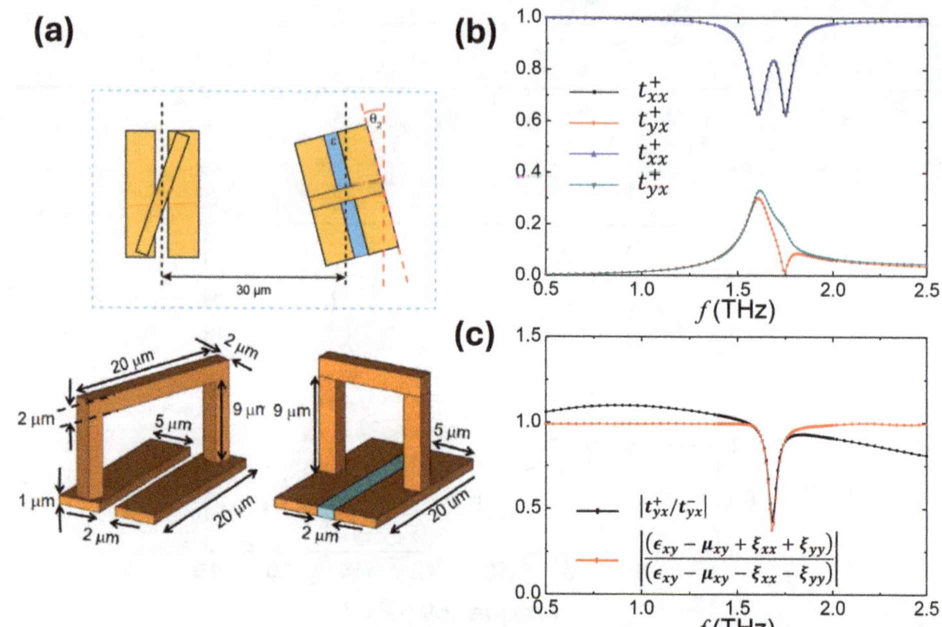

Figure 2.12. A chiral metamaterial showing asymmetric transmission for linear polarization. Reprinted with permission from Zhang *et al* (2013). Copyright (2013) by the American Physical Society.

2.7 Problems

Problem 2.1 For two kinds of metallic particles embedded in a glass matrix, prove that the Maxwell–Garnett formula is modified to

$$\frac{\epsilon - \epsilon_b}{\epsilon + 2\epsilon_b} = f_1 \frac{\epsilon_{m1} - \epsilon_b}{\epsilon_{m1} + 2\epsilon_b} + f_2 \frac{\epsilon_{m2} - \epsilon_b}{\epsilon_{m2} + 2\epsilon_b},$$

where the subscripts 1 and 2 are added to the volume filling fraction and metal permittivity for the two kinds of metallic particles.

Problem 2.2 For a slowly varying electric field approximated by $\mathbf{E} = \mathbf{E}_0 + \bar{\bar{a}} \cdot \mathbf{r}$, with constant antisymmetric second-rank tensor $\bar{\bar{a}}$, verify that applying equation (2.19) over a cube of dimensions $L_1 \times L_2 \times L_3$ centered at the origin returns $\mathbf{E}_0$. Repeat the same calculation by integrating over a spherical surface centered at the origin.

Problem 2.3 Show that $\nabla \times \mathbf{E}_{\rm loc}(\mathbf{r}) = ik\, \mathbf{H}_{\rm loc}$ and $\nabla \times \mathbf{H}_{\rm loc}(\mathbf{r}) = -ik\, \mathbf{E}_{\rm loc}$ for the local-field solution in equation (2.21), assuming the fields are invariant along the z-direction.

Problem 2.4 A 2D rectangular unit cell of dimensions $a \times b$ in the x–y plane is shown below, with line-averaged fields defined on its four edges:

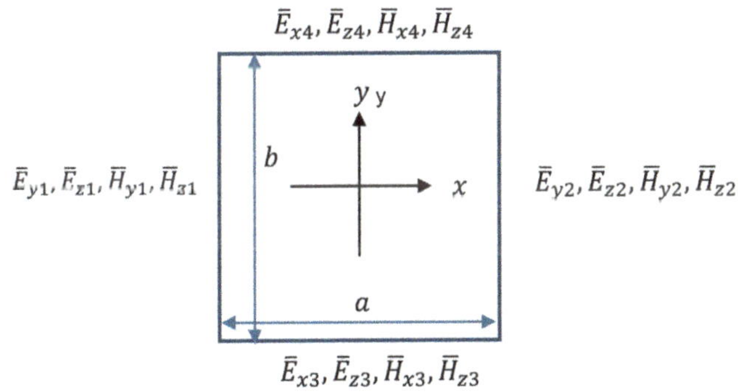

By applying equations (2.16) and (2.20), prove that the effective fields are

$$
\begin{aligned}
\mathbf{D}_{\text{eff}} &= \frac{1}{\mathrm{i}\omega ab}[\hat{x}\ a\ (\bar{H}_{z3} - \bar{H}_{z4}) + \hat{y}\ b\ (\bar{H}_{z2} - \bar{H}_{z1})] \\
&+ \frac{1}{\mathrm{i}\omega ab}\ \hat{z}[b\ (\bar{H}_{y1} - \bar{H}_{y2}) + a\ (\bar{H}_{x4} - \bar{H}_{x3})], \\
\mathbf{B}_{\text{eff}} &= \frac{1}{\mathrm{i}\omega ab}[\hat{x}\ a\ (\bar{E}_{z4} - \bar{E}_{z3}) + \hat{y}\ b\ (\bar{E}_{z1} - \bar{E}_{z2})] \\
&+ \frac{1}{\mathrm{i}\omega ab}\ \hat{z}[b\ (\bar{E}_{y2} - \bar{E}_{y1}) + a\ (\bar{E}_{x4} - \bar{E}_{x3})], \\
\mathbf{E}_{\text{eff}} &= \frac{\hat{x}}{2}(\bar{E}_{x3} + \bar{E}_{x4}) + \frac{\hat{y}}{2}(\bar{E}_{y1} + \bar{E}_{y2}) + \frac{\hat{z}}{4}(\bar{E}_{z1} + \bar{E}_{z2} + \bar{E}_{z3} + \bar{E}_{z4}), \\
\mathbf{H}_{\text{eff}} &= \frac{\hat{x}}{2}(\bar{H}_{x3} + \bar{H}_{x4}) + \frac{\hat{y}}{2}(\bar{H}_{y1} + \bar{H}_{y2}) + \frac{\hat{z}}{4}(\bar{H}_{z1} + \bar{H}_{z2} + \bar{H}_{z3} + \bar{H}_{z4}),
\end{aligned}
$$

where the overbar denotes the edge-averaged field on each boundary segment.

Problem 2.5 (a) By considering illumination of a metamaterial slab with ϵ_{eff} and μ_{eff} using symmetric and antisymmetric incident waves, establish that $|t + r| \leqslant 1$ and $|t - r| \leqslant 1$ for a passive slab, so that the outgoing power never exceeds the incoming power. Show that $\mathrm{i}\frac{1-r-t}{1+r+t}$ and $\mathrm{i}\frac{1+r-t}{1-r+t}$ always lie in the upper half of the complex plane.

(b) Conversely, using the result from part (a), argue that the effective permittivity and permeability obtained from the field-averaging method (equation (2.26)) have non-negative imaginary parts for a passive metamaterial.

References

Chen X, Wu B-I, Kong J and Grzegorczyk T 2005 Retrieval of the effective constitutive parameters of bianisotropic metamaterials *Phys. Rev.* **71** 046610

Decker M, Zhao R, Soukoulis C, Linden S and Wegener M 2010 Twisted split-ring-resonator photonic metamaterial with huge optical activity *Opt. Lett.* **35** 1593–5

Drezet A, Genet C, Laluet J and Ebbesen T 2008 Optical chirality without optical activity: How surface plasmons give a twist to light *Opt. Express* **16** 12559–70

Fedotov V, Mladyonov P, Prosvirnin S, Rogacheva A V, Chen Y and Zheludev N 2006 Asymmetric propagation of electromagnetic waves through a planar chiral structure *Phys. Rev. Lett.* **97** 167401

Feng T, Liu F, Tam W and Li J 2013 Effective parameters retrieval for complex metamaterials with low symmetries *Europhys. Lett.* **102** 18003

Gansel J K, Thiel M, Rill M, Decker M, Bade K, Saile V, von Freymann G, Linden S and Wegener M 2009 Gold helix photonic metamaterial as broadband circular polarizer *Science* **325** 1513–5

Hörner H, Rachbauer L M, Rotter S and Leonhardt U 2023 Van der waals chain: A simple model for casimir forces in dielectrics *Phys. Rev.* **B 108** 235430

Kang M, Chen J, Cui H, Li Y and Wang H 2011 Asymmetric transmission for linearly polarized electromagnetic radiation *Opt. Express* **19** 8347–56

Koschny T, Markoš P, Markoš P, Smith D R and Soukoulis C 2003 Resonant and antiresonant frequency dependence of the effective parameters of metamaterials *Phys. Rev.* **68** 065602

Menzel C, Helgert C, Rockstuhl C, Kley E, Tünnermann A, Pertsch T and Lederer F 2010 Asymmetric transmission of linearly polarized light at optical metamaterials *Phys. Rev. Lett.* **104** 253902

Pendry J, Holden A, Robbins D and Stewart W 1999 Magnetism from conductors and enhanced nonlinear phenomena *IEEE Trans. Microw. Theory Tech.* **47** 2075–84

Pendry J, Holden A, Stewart W and Youngs I 1996 Extremely low frequency plasmons in metallic mesostructures *Phys. Rev. Lett.* **76** 4773–6

Pfeiffer C and Grbic A 2014 Bianisotropic metasurfaces for optimal polarization control: Analysis and synthesis *Phys. Rev. Appl.* **2** 044011

Plum E, Zhou J, Dong J, Fedotov V, Koschny T, Soukoulis C and Zheludev N 2009 Metamaterial with negative index due to chirality *Phys. Rev.* B **79** 035407

Shalaev V, Cai W, Chettiar U, Yuan H-K, Sarychev A, Drachev V and Kildishev A 2005 Negative index of refraction in optical metamaterials *Opt. Lett.* **30** 3356–8

Shelby R, Smith D R and Schultz S 2001 Experimental verification of a negative index of refraction *Science* **292** 77–9

Singh R, Plum E, Menzel C, Rockstuhl C, Azad A K, Cheville R, Lederer F, Zhang W and Zheludev N 2009 Terahertz metamaterial with asymmetric transmission *Phys. Rev.* B **80** 153104

Smith D R and Pendry J 2006 Homogenization of metamaterials by field averaging (invited paper) *J. Opt. Soc. Am.* B **23** 391–403

Smith D R, Smith D R, Vier D, Koschny T, Koschny T, Soukoulis C and Soukoulis C 2005 Electromagnetic parameter retrieval from inhomogeneous metamaterials *Phys. Rev.* **71** 036617

Valentine J, Zhang S, Zentgraf T, Ulin-Avila E, Genov D, Bartal G and Zhang X 2008 Three-dimensional optical metamaterial with a negative refractive index *Nature* **455** 376–9

Ward A and Pendry J 1996 Refraction and geometry in Maxwellas equations *J. Mod. Opt.* **43** 773–93

Zhang S, Fan W, Malloy K, Brueck S, Panoiu N and Osgood R 2005 Near-infrared double negative metamaterials *Opt. Express* **13** 4922–30

Zhang S, Liu F, Zentgraf T and Li J 2013 Interference-induced asymmetric transmission through a monolayer of anisotropic chiral metamolecules *Phys. Rev.* A **88** 023823

Zhang S, shik Park Y, Li J, Lu X, Zhang W and Zhang X 2009 Negative refractive index in chiral metamaterials *Phys. Rev. Lett.* **102** 023901

Zhao R, Koschny T and Soukoulis C 2010 Chiral metamaterials: retrieval of the effective parameters with and without substrate *Opt. Express* **18** 14553–67

Zhukovsky S, Novitsky A and Galynsky V 2009 Elliptical dichroism: operating principle of planar chiral metamaterials *Opt. Lett.* **34** 1988–90

IOP Publishing

Essential Concepts in Metamaterials

Jensen Li

Chapter 3

Principles of transformation optics

In this chapter, we walk through the basic design procedures of transformation optics step by step. We begin by discussing the form-invariant property of Maxwell's equations under coordinate transformations. We then introduce the concepts of physical space and virtual space and establish how material parameters transform between them. This forms the theoretical framework for the applications developed in the following chapter.

3.1 Heuristic derivation

In this section, we will first take a 'microscopic' look at what we mean by transformation from the perspective of a full-wave simulation. Through this, we can discover some of the invariant quantities that arise when we stretch and transform the space through which light propagates. These insights will help us as we gradually introduce the formal developments using tensors and coordinate transformation (Pendry *et al* 2006) in the following sections.

We imagine that we divide the space into small parallelepiped pieces. One small piece is spanned by three basis vectors: $\mathbf{a}$, $\mathbf{b}$, and $\mathbf{c}$, as shown in figure 3.1. Each parallelepiped can be considered as a finite element for the purpose of full-wave simulation, with tangential fields specified on the six faces. The center of the element is set as the origin. The surface electric fields are represented by the projection of the fields along the three basis vectors: $\mathbf{a} \cdot \mathbf{E}$, $\mathbf{b} \cdot \mathbf{E}$, and $\mathbf{c} \cdot \mathbf{E}$. They are evaluated at positions $\pm\mathbf{a}/2$, $\pm\mathbf{b}/2$, and $\pm\mathbf{c}/2$ at the six faces. Therefore, there are a total of 12 complex degrees of freedom (DOF) in specifying the surface fields. They are more than enough to represent the electromagnetic fields within the small parallelepiped. Alternatively, there are 12 complex numbers to specify the tangential $\mathbf{H}$-fields using the same kind of labeling at the same locations, and they have to be compatible to the tangential $\mathbf{E}$-fields through the response of the unspecified ϵ and μ of each element. Based on this response of a single element,

doi:10.1088/978-0-7503-1131-1ch3

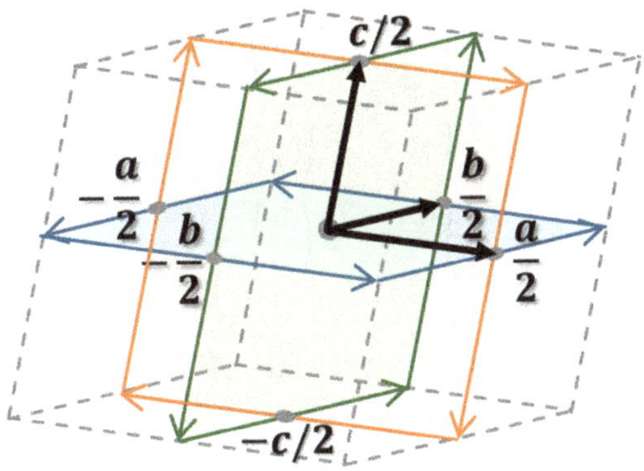

Figure 3.1. A finite element spanned by three vectors **a**, **b**, and **c** with surface fields specified at the grey nodes on the six faces and bulk fields specified at the centre node of the element.

full-wave simulation, often using tetrahedron instead of parallelepiped elements, can be performed by joining neighboring elements through the continuity of tangential electric and magnetic fields on the common interfaces.

In fact, the single element response is exactly what we discussed in the last chapter. We only need to shrink the unit cell from the last chapter further down to be much smaller than a wavelength so that no resonance is allowed within a single element. In this case, the fields inside vary smoothly and can be approximated by linear polynomials in spatial coordinates (we did that for local fields previously) using a first-order ansatz that is consistent with the curl equations. Writing a linear field as $\mathbf{E}(\mathbf{r}) = \mathbf{E}(\mathbf{0}) + \overline{\overline{A}} \cdot \mathbf{r}$ with constant matrix $\overline{\overline{A}}$, the curl condition $\nabla \times \mathbf{E} = ik\ \mathbf{B}(\mathbf{0})$ fixes the antisymmetric part of $\overline{\overline{A}}$. The symmetric part does not enter the local curl condition, so that it has a higher order than the dipolar contribution. Therefore, we can write the electromagnetic fields within the element as

$$\begin{aligned} \mathbf{E}(\mathbf{r}) &\approx \mathbf{E}(\mathbf{0}) - \frac{ik}{2}\mathbf{r} \times \mathbf{B}(\mathbf{0}), \\ \mathbf{H}(\mathbf{r}) &\approx \mathbf{H}(\mathbf{0}) + \frac{ik}{2}\mathbf{r} \times \mathbf{D}(\mathbf{0}). \end{aligned} \tag{3.1}$$

By substituting equation (3.1) into equations (2.16) and (2.19), the effective fields obtained from surface integrals now have to be the fields evaluated at the center of the finite element filled with a homogeneous medium:

$$\begin{aligned} \mathbf{D}(\mathbf{0}) &= \frac{1}{ikV}\oint \mathbf{H} \times d\mathbf{S}, & \mathbf{B}(\mathbf{0}) &= -\frac{1}{ikV}\oint \mathbf{E} \times d\mathbf{S}, \\ \mathbf{E}(\mathbf{0}) &= \frac{1}{2V}\oint \mathbf{r} \times (\mathbf{E} \times d\mathbf{S}), & \mathbf{H}(\mathbf{0}) &= \frac{1}{2V}\oint \mathbf{r} \times (\mathbf{H} \times d\mathbf{S}). \end{aligned} \tag{3.2}$$

The surface integrals can then be evaluated in terms of the fields at the surface nodes. It results in the discrete version of curl that recovers **B(0)** from surface **E**,

$$
\begin{aligned}
ik\mathbf{b}\times\mathbf{c}\cdot\mathbf{B(0)} &= \mathbf{b}\cdot\mathbf{E}\left(-\frac{\mathbf{c}}{2}\right)-\mathbf{b}\cdot\mathbf{E}\left(\frac{\mathbf{c}}{2}\right)-\mathbf{c}\cdot\mathbf{E}\left(-\frac{\mathbf{b}}{2}\right)+\mathbf{c}\cdot\mathbf{E}\left(\frac{\mathbf{b}}{2}\right),\\
ik\mathbf{c}\times\mathbf{a}\cdot\mathbf{B(0)} &= \mathbf{c}\cdot\mathbf{E}\left(-\frac{\mathbf{a}}{2}\right)-\mathbf{c}\cdot\mathbf{E}\left(\frac{\mathbf{a}}{2}\right)-\mathbf{a}\cdot\mathbf{E}\left(-\frac{\mathbf{c}}{2}\right)+\mathbf{a}\cdot\mathbf{E}\left(\frac{\mathbf{c}}{2}\right),\\
ik\mathbf{a}\times\mathbf{b}\cdot\mathbf{B(0)} &= \mathbf{a}\cdot\mathbf{E}\left(-\frac{\mathbf{b}}{2}\right)-\mathbf{a}\cdot\mathbf{E}\left(\frac{\mathbf{b}}{2}\right)-\mathbf{b}\cdot\mathbf{E}\left(-\frac{\mathbf{a}}{2}\right)+\mathbf{b}\cdot\mathbf{E}\left(\frac{\mathbf{a}}{2}\right).
\end{aligned}
\tag{3.3}
$$

For example, taking the curl of **E** on the green plane in figure 3.1 obtains one component of the center **B(0)** from the surface **E**. The approximated evaluation of equation (3.2) also results in interpolating the surface **E** to the center **E(0)**

$$
\begin{aligned}
\mathbf{a}\cdot\mathbf{E(0)} &= \frac{1}{4}\left[\mathbf{a}\cdot\mathbf{E}\left(-\frac{\mathbf{b}}{2}\right)+\mathbf{a}\cdot\mathbf{E}\left(\frac{\mathbf{b}}{2}\right)+\mathbf{a}\cdot\mathbf{E}\left(-\frac{\mathbf{c}}{2}\right)+\mathbf{a}\cdot\mathbf{E}\left(\frac{\mathbf{c}}{2}\right)\right],\\
\mathbf{b}\cdot\mathbf{E(0)} &= \frac{1}{4}\left[\mathbf{b}\cdot\mathbf{E}\left(-\frac{\mathbf{c}}{2}\right)+\mathbf{b}\cdot\mathbf{E}\left(\frac{\mathbf{c}}{2}\right)+\mathbf{b}\cdot\mathbf{E}\left(-\frac{\mathbf{a}}{2}\right)+\mathbf{b}\cdot\mathbf{E}\left(\frac{\mathbf{a}}{2}\right)\right],\\
\mathbf{c}\cdot\mathbf{E(0)} &= \frac{1}{4}\left[\mathbf{c}\cdot\mathbf{E}\left(-\frac{\mathbf{a}}{2}\right)+\mathbf{c}\cdot\mathbf{E}\left(\frac{\mathbf{a}}{2}\right)+\mathbf{c}\cdot\mathbf{E}\left(-\frac{\mathbf{b}}{2}\right)+\mathbf{c}\cdot\mathbf{E}\left(\frac{\mathbf{b}}{2}\right)\right].
\end{aligned}
\tag{3.4}
$$

On the other hand, the actual homogeneous medium filling the bulk of the element satisfies

$$
\begin{aligned}
\mathbf{D(0)} &= \overline{\overline{\epsilon}}\cdot\mathbf{E(0)},\\
\mathbf{B(0)} &= \overline{\overline{\mu}}\cdot\mathbf{H(0)},
\end{aligned}
\tag{3.5}
$$

from which we can now obtain the center **D(0)** and **H(0)** from **E(0)** and **B(0)**. Since **E(0)** and **B(0)** have been reconstructed directly from surface **E**-data via equations (3.3) and (3.4), the consistent computational pipeline is to determine the 'missing' quantities **D(0)** and **H(0)** from **E(0)** and **B(0)** as

$$
\mathbf{D(0)} = \overline{\overline{\epsilon}}\cdot\mathbf{E(0)},\qquad \mathbf{H(0)} = \overline{\overline{\mu}}^{-1}\cdot\mathbf{B(0)}\ . \tag{3.6}
$$

Finally, the surface **H** can be obtained through projection to the surface boundary again using equation (3.1). Evaluating $\mathbf{H(r)} = \mathbf{H(0)} + \frac{ik}{2}\mathbf{r}\times\mathbf{D(0)}$ at $\mathbf{r} = \pm\frac{\mathbf{a}}{2}$, $\pm\frac{\mathbf{b}}{2}$, $\pm\frac{\mathbf{c}}{2}$ and projecting onto the local tangents yields, for example,

$$
\begin{aligned}
\mathbf{b}\cdot\mathbf{H}\left(-\frac{\mathbf{a}}{2}\right) &= \mathbf{b}\cdot\mathbf{H(0)} + \frac{ik}{4}\mathbf{a}\times\mathbf{b}\cdot\mathbf{D(0)},\\
\mathbf{c}\cdot\mathbf{H}\left(-\frac{\mathbf{a}}{2}\right) &= \mathbf{c}\cdot\mathbf{H(0)} - \frac{ik}{4}\mathbf{c}\times\mathbf{a}\cdot\mathbf{D(0)},\text{etc.}
\end{aligned}
\tag{3.7}
$$

Alternatively, we can also start from surface *H*-fields instead of surface *E*-fields. We also note that the total number of intrinsic DOFs of all the surface fields is actually six: 12 from equation (3.1) constrained by 6 equations from equation (3.5). The full-wave simulation is essentially dictated by the response of a single element now. Equations (3.3) to (3.7) are equivalent to the Maxwell's equations with equation (3.5) being the constitutive relationship in addition. Therefore, by projecting the **E** field using a line element (e.g., $\mathbf{a} \cdot \mathbf{E}$) and the **B** field using an area element (e.g., $\mathbf{b} \times \mathbf{c} \cdot \mathbf{B}$), and using them to formulate the simulation, the wave equation is completely free of the explicit coordinates of the vertices of the finite element. This makes explicit the computational invariants under element distortion: edge-projected **E** and **H** and face-projected flux densities **B**, **D**.

The implication is: suppose we have two configurations where the vertices of the finite elements can be at different coordinates. Let's say one is spanned by **a**, **b**, **c** and another is spanned by $\mathbf{a}'$, $\mathbf{b}'$, $\mathbf{c}'$. These two configurations are said to be computationally equivalent if the two 3×3 matrix representations of $\overline{\overline{\epsilon}}$ and $\overline{\overline{\mu}}$ from equation (3.4), obtained by projecting **D**, **B** into area elements and **E**, **H** into line elements, are completely the same for the two configurations. In such a case, if this equivalence is true for all elements in the simulation domain, we only need to simulate either one; the simulation result of the other is automatically obtained by transforming one set of **E** or **H** fields (on the surface) to another by conserving $\mathbf{a} \cdot \mathbf{E} = \mathbf{a}' \cdot \mathbf{E}'$, $\mathbf{a} \cdot \mathbf{H} = \mathbf{a}' \cdot \mathbf{H}'$, etc. Scanning through all the three different line elements, we can define the transformation matrix as

$$[\mathbf{E}' \to \mathbf{E}] = \begin{pmatrix} a_x & a_y & a_z \\ b_x & b_y & b_z \\ c_x & c_y & c_z \end{pmatrix}^{-1} \begin{pmatrix} a_x' & a_y' & a_z' \\ b_x' & b_y' & b_z' \\ c_x' & c_y' & c_z' \end{pmatrix}.$$

For the **B** and **D** fields, they are conserved on the flux level, e.g. $\mathbf{b} \times \mathbf{c} \cdot \mathbf{D} = \mathbf{b}' \times \mathbf{c}' \cdot \mathbf{D}'$. Introducing the 3×3 'face' matrices whose rows are the face normals/areas $(\mathbf{b} \times \mathbf{c})^T$, $(\mathbf{c} \times \mathbf{a})^T$, $(\mathbf{a} \times \mathbf{b})^T$, one obtains

$$\begin{aligned}
&[\mathbf{D} \to \mathbf{D}'] \\
&= \begin{pmatrix} (\mathbf{b}' \times \mathbf{c}')_x & (\mathbf{b}' \times \mathbf{c}')_y & (\mathbf{b}' \times \mathbf{c}')_z \\ (\mathbf{c}' \times \mathbf{a}')_x & (\mathbf{c}' \times \mathbf{a}')_y & (\mathbf{c}' \times \mathbf{a}')_z \\ (\mathbf{a}' \times \mathbf{b}')_x & (\mathbf{a}' \times \mathbf{b}')_y & (\mathbf{a}' \times \mathbf{b}')_z \end{pmatrix}^{-1} \begin{pmatrix} (\mathbf{b} \times \mathbf{c})_x & (\mathbf{b} \times \mathbf{c})_y & (\mathbf{b} \times \mathbf{c})_z \\ (\mathbf{c} \times \mathbf{a})_x & (\mathbf{c} \times \mathbf{a})_y & (\mathbf{c} \times \mathbf{a})_z \\ (\mathbf{a} \times \mathbf{b})_x & (\mathbf{a} \times \mathbf{b})_y & (\mathbf{a} \times \mathbf{b})_z \end{pmatrix} \\
&= \frac{\mathbf{a} \cdot (\mathbf{b} \times \mathbf{c})}{\mathbf{a}' \cdot (\mathbf{b}' \times \mathbf{c}')} [\mathbf{E}' \to \mathbf{E}]^T.
\end{aligned}$$

$$[\epsilon'] = [\mathbf{D} \to \mathbf{D}'][\epsilon][\mathbf{E}' \to \mathbf{E}]. \tag{3.8}$$

In fact, the matrix $[\mathbf{E}' \to \mathbf{E}]$ is actually the transpose of what we call the Jacobian matrix $[\Lambda]$, which will be formulated properly in a later section, in transforming between the two elements if we imagine the transformation between elements are performed differentially in coordinates. Then, equation (3.8) is turned into the standard transformation rule

$$[\epsilon'] = \frac{1}{\det[\Lambda]}[\Lambda][\epsilon][\Lambda]^T. \tag{3.9}$$

The transformation for the permeability tensor is the same. The discussion here is purely from the perspective of computational equivalence, involving a passive transformation in coordinates if you wish to think about it that way (Ward and

Pendry 1996). This is actually a precursor for transformation optics. In transformation optics, we turn the coordinate transformation into an 'active' one. In this case, we purposely fill up the space using the transformed permittivity to ensure that the new configuration is equivalent to the original one before transformation.

3.2 Curvilinear coordinates

Transformation optics is based on writing Maxwell's equations in curvilinear coordinates and exploiting the invariance of the form of the equations as a design tool for optical devices. To begin with, we must first introduce the contravariant and covariant index notations briefly used to represent vectors and tensors. These are powerful notations that use upper and lower indices to label the components of vectors and tensors when we talk about coordinate transformations later. If you are familiar with using these indices to write vector components, which are often used in courses on special or general relativity, you are ready to proceed to section 3.3.

We usually describe three-dimensional (3D) Euclidean space with Cartesian coordinates (x, y, z), or in index notation (x^1, x^2, x^3). By convention we have already been using an upper index to label the coordinates. The same space, however, can also be described by other coordinates. A very general choice is to use curvilinear right-handed coordinates (q^1, q^2, q^3). When we draw lines of constant spacing for these q-coordinates, the grid becomes curved, as illustrated in figure 3.2 for a two-dimensional (2D) case.

To work in such coordinates, we need new basis vectors. A natural choice is to differentiate the position vector $\mathbf{r}$ with respect to the coordinates. The tangent basis vectors are therefore

$$\frac{\partial \mathbf{r}}{\partial q^i},$$

which point along the coordinate lines. Any vector $\mathbf{E}$, such as the electric field in Maxwell's equations, can then be expressed as a combination of these basis vectors,

$$\mathbf{E} = E^i \frac{\partial \mathbf{r}}{\partial q^i}. \tag{3.10}$$

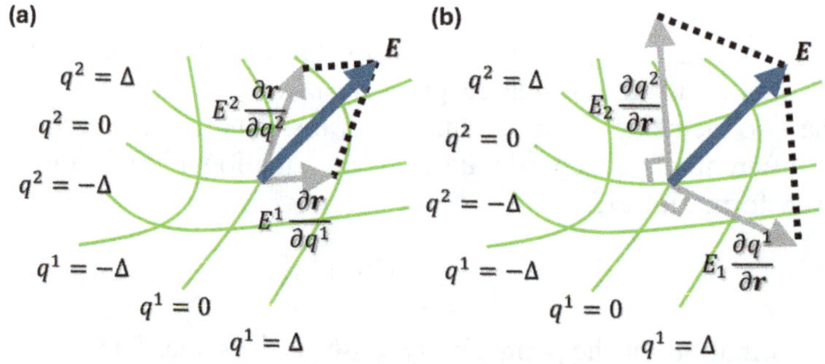

Figure 3.2. Vector $\mathbf{E}$ in (a) contravariant E^i and (b) covariant E_i components. Basis vectors $\partial\mathbf{r}/\partial q^i$ are tangential to coordinate lines. The dual basis vectors $\partial q^i/\partial\mathbf{r}$ are perpendicular to the coordinate lines.

The coefficients E^i are called contravariant components. Here and throughout we use Einstein notation: whenever an index is repeated, we sum over all its values ($i = 1, 2, 3$).

There is also a dual set of basis vectors, obtained by taking the gradient of the coordinates,

$$\frac{\partial q^i}{\partial \mathbf{r}},$$

which are perpendicular to the surfaces of constant q^i. The same vector can then be written as

$$\mathbf{E} = E_i \frac{\partial q^i}{\partial \mathbf{r}}. \tag{3.11}$$

The coefficients E_i are called covariant components. Both descriptions refer to the same vector but decompose it in different ways. Figure 3.2 illustrates this difference for two dimensions. If you have encountered lattice vectors and reciprocal lattice vectors in solid-state physics, the analogy is close: one set of components can be probed by dotting with the dual basis vectors,

$$E^i = \frac{\partial q^i}{\partial \mathbf{r}} \cdot \mathbf{E}, \qquad E_i = \frac{\partial \mathbf{r}}{\partial q^i} \cdot \mathbf{E}.$$

To compare contravariant and covariant components, we need the metric tensor. This is defined by

$$g_{ij} = \frac{\partial \mathbf{r}}{\partial q^i} \cdot \frac{\partial \mathbf{r}}{\partial q^j}, \qquad g^{ij} = \frac{\partial q^i}{\partial \mathbf{r}} \cdot \frac{\partial q^j}{\partial \mathbf{r}}.$$

The metric acts as a 'machine' for taking dot products between vectors:

$$g(\mathbf{a}, \mathbf{b}) = \mathbf{a} \cdot \mathbf{b} = a_i g^{ij} b_j = a^i g_{ij} b^j.$$

With the metric, the two sets of components are related by the rule for raising and lowering indices,

$$E^i = g^{ij} E_j, \qquad E_i = g_{ij} E^j. \tag{3.12}$$

An important special case is the diagonal entries of the metric. These define the scale factors

$$h_1 = \sqrt{g_{11}}, \qquad h_2 = \sqrt{g_{22}}, \qquad h_3 = \sqrt{g_{33}},$$

so that line elements along the coordinate directions are

$$|\mathbf{dr}| = h_1 dq^1, \quad |\mathbf{dr}| = h_2 dq^2, \quad |\mathbf{dr}| = h_3 dq^3.$$

A concrete example helps at this stage. In cylindrical coordinates, $(q^1, q^2, q^3) = (r, \theta, z)$, the Cartesian coordinates are

$$x = r\cos\theta, \quad y = r\sin\theta, \quad z = z.$$

The tangent basis vectors are

$$\frac{\partial \mathbf{r}}{\partial r} = (\cos\theta, \sin\theta, 0), \quad \frac{\partial \mathbf{r}}{\partial \theta} = (-r\sin\theta, r\cos\theta, 0), \quad \frac{\partial \mathbf{r}}{\partial z} = (0, 0, 1).$$

The corresponding metric entries are

$$g_{rr} = 1, \quad g_{\theta\theta} = r^2, \quad g_{zz} = 1, \quad g_{ij} = 0 \quad (i \neq j).$$

Hence, the scale factors are $h_r = 1$, $h_\theta = r$, $h_z = 1$. From these we see that the line elements are $\mathrm{d}r$, $r\,\mathrm{d}\theta$, and $\mathrm{d}z$. These scale factors also explain the familiar appearance of r in area and volume elements. For example, the area element in the r–θ plane is $\mathrm{d}A = r\,\mathrm{d}r\,\mathrm{d}\theta$, and the volume element is $\mathrm{d}V = r\,\mathrm{d}r\,\mathrm{d}\theta\,\mathrm{d}z$.

In Maxwell's equations we also encounter second-rank tensors, such as the permeability tensor $\overline{\overline{\mu}}$. Its components can be written using these basis vectors,

$$\overline{\overline{\mu}} = \mu^{ij}\frac{\partial \mathbf{r}}{\partial q^i}\frac{\partial \mathbf{r}}{\partial q^j} = \mu^i{}_j\frac{\partial \mathbf{r}}{\partial q^i}\frac{\partial q^j}{\partial \mathbf{r}}. \tag{3.13}$$

To handle cross products in general coordinates we introduce the Levi-Civita symbol. In Cartesian coordinates you may know it as the object encoding the orientation of indices in the cross product. In curvilinear coordinates it is rescaled by the geometry. We define

$$\varepsilon^{ijk} = \frac{1}{\Omega}[i, j, k], \qquad \varepsilon_{ijk} = \Omega[i, j, k],$$

where $[i, j, k]$ is the usual symbol taking values ± 1 for even or odd permutations, and 0 if indices repeat. The factor Ω will be explained below when we discuss the Jacobian. With this definition, the dot and cross products can be written in component form.

We are now ready to explain the geometric factor Ω used above. It is defined through the Jacobian of the coordinate transformation,

$$\Lambda^i{}_j = \frac{\partial x^i}{\partial q^j}, \tag{3.14}$$

which relates directly to the metric,

$$g_{ij} = \Lambda^k_i \Lambda^k_j. \tag{3.15}$$

The determinant of the Jacobian is

$$\Omega = \det\Lambda^i_j = \frac{\partial(x, y, z)}{\partial(q^1, q^2, q^3)} = \sqrt{\det g_{ij}}, \tag{3.16}$$

which gives the volume element,

$$\mathrm{d}V = \Omega\ \mathrm{d}q^1\mathrm{d}q^2\mathrm{d}q^3. \tag{3.17}$$

For completeness, an area element spanned by varying q^2, q^3 can be written as

$$\mathrm{d}A = \left| \frac{\partial \mathbf{r}}{\partial q^2} \times \frac{\partial \mathbf{r}}{\partial q^3} \right| \mathrm{d}q^2\mathrm{d}q^3 = \Omega\sqrt{g^{11}}\ \mathrm{d}q^2\mathrm{d}q^3.$$

As a check, let us revisit cylindrical coordinates. From the metric we already found that $\mathrm{d}V = r\ \mathrm{d}r\ \mathrm{d}\theta\ \mathrm{d}z$. Computing the Jacobian determinant directly also gives $\Omega = r$. The two approaches are consistent.

It is helpful to summarize the general procedure. First write $x^k = x^k(q^1, q^2, q^3)$. Then compute $\Lambda^k{}_i = \partial x^k/\partial q^i$. From this you can construct the tangent basis vectors $\partial \mathbf{r}/\partial q^i$ and the metric $g_{ij} = \Lambda^k{}_i\Lambda^k{}_j$. The scale factors follow as $h_i = \sqrt{g_{ii}}$. The Jacobian determinant is $\Omega = \sqrt{\det g}$, from which the differential elements are read off: $\mathrm{d}\ell = h_i\ \mathrm{d}q^i$, $\mathrm{d}A = \Omega\sqrt{g^{ii}}\ \mathrm{d}q^j\mathrm{d}q^k$ for the 2D elements, and $\mathrm{d}V= \Omega\ \mathrm{d}q^1\mathrm{d}q^2\mathrm{d}q^3$. Index conversions are made with $E^i = g^{ij}E_j$ and $E_i = g_{ij}E^j$.

A few common pitfalls are worth noting. The tangent and dual bases are not the same unless the coordinates are orthonormal. Even in orthogonal coordinates they are not unit vectors, so the scale factors must always be included. When dealing with cross products, remember that the Levi-Civita symbol comes with the volume factor Ω.

3.3 Form invariance

The Maxwell's equations, in differential form and written in Heaviside–Lorentz units, are

$$\begin{aligned} &\nabla \times \mathbf{E} + \frac{1}{c}\partial_t \mathbf{B} = 0,\ \mathbf{D} = \overline{\overline{\epsilon}} \cdot \mathbf{E}, \\ &\nabla \times \mathbf{H} - \frac{1}{c}\partial_t \mathbf{D} = \frac{1}{c}\mathbf{J},\ \mathbf{B} = \overline{\overline{\mu}} \cdot \mathbf{H}, \end{aligned} \tag{3.18}$$

where c is the speed of light in vacuum. In a general curvilinear coordinate system q^i (with metric g_{ij} and $\Omega = \sqrt{\det g}$ as introduced previously), the differential operators act on scalars and vectors as

$$\begin{aligned} \nabla V &\triangleq \frac{\partial q^i}{\partial \mathbf{r}}\frac{\partial V}{\partial q^i}, \\ \nabla \cdot \mathbf{D} &\triangleq \frac{\partial q^i}{\partial \mathbf{r}} \cdot \frac{\partial \mathbf{D}}{\partial q^i} = \frac{1}{\Omega}\frac{\partial}{\partial q^i}(\Omega D^i), \\ \nabla \times \mathbf{E} &\triangleq \frac{\partial q^j}{\partial \mathbf{r}} \times \frac{\partial \mathbf{E}}{\partial q^j} = \frac{\partial \mathbf{r}}{\partial q^i}\varepsilon^{ijk}\frac{\partial E_k}{\partial q^j}, \end{aligned}$$

where $\varepsilon^{ijk} = \frac{1}{\Omega}[i, j, k]$ is the (contravariant) permutation tensor. Proof (sketch for the two identities above using results in problem 3.5):

$$
\begin{aligned}
\nabla \cdot \mathbf{D} &= \frac{\partial q^i}{\partial \mathbf{r}} \cdot \frac{\partial \mathbf{D}}{\partial q^i} = \frac{1}{\Omega}\frac{\partial}{\partial q^i}\left(\Omega \frac{\partial q^i}{\partial \mathbf{r}} \cdot \mathbf{D}\right) = \frac{1}{\Omega}\frac{\partial}{\partial q^i}(\Omega D^i), \\
(\nabla \times \mathbf{E})^i &= \frac{\partial q^i}{\partial \mathbf{r}} \cdot \left(\frac{\partial q^j}{\partial \mathbf{r}} \times \frac{\partial \mathbf{E}}{\partial q^j}\right) = \varepsilon^{ijk}\frac{\partial \mathbf{r}}{\partial q^k} \cdot \frac{\partial \mathbf{E}}{\partial q^j} \\
&= \varepsilon^{ijk}\frac{\partial}{\partial q^j}\left(\frac{\partial \mathbf{r}}{\partial q^k} \cdot \mathbf{E}\right) - \varepsilon^{ijk}\frac{\partial^2 \mathbf{r}}{\partial q^j \partial q^k} \cdot \mathbf{E} = \varepsilon^{ijk}\frac{\partial E_k}{\partial q^j}.
\end{aligned}
$$

With these, Maxwell's equations in component form in q-coordinates become

$$
\begin{aligned}
&\varepsilon^{ijk}\frac{\partial E_k}{\partial q^j} + \frac{1}{c}\partial_t B^i = 0, \; D^i = \epsilon^{ij}E_j, \\
&\varepsilon^{ijk}\frac{\partial H_k}{\partial q^j} - \frac{1}{c}\partial_t D^i = \frac{1}{c}J^i, \; B^i = \mu^{ij}H_j.
\end{aligned}
\tag{3.19}
$$

In the Cartesian limit ($\varepsilon^{ijk} \to [i, j, k]$, $g_{ij} \to \delta_{ij}$, $\Omega \to 1$), we recover the familiar matrix form

$$
\begin{aligned}
&[\nabla\times][E] + \frac{1}{c}\partial_t[B] = 0, \; [D] = [\epsilon][E], \\
&[\nabla\times][H] - \frac{1}{c}\partial_t[D] = \frac{1}{c}[J], \; [B] = [\mu][H],
\end{aligned}
\tag{3.20}
$$

with the abbreviations

$$
\begin{aligned}
&[D] = \begin{bmatrix} D^x \\ D^y \\ D^z \end{bmatrix}, \quad [B] = \begin{bmatrix} B^x \\ B^y \\ B^z \end{bmatrix}, \quad [E] = \begin{bmatrix} E_x \\ E_y \\ E_z \end{bmatrix}, \quad [H] = \begin{bmatrix} H_x \\ H_y \\ H_z \end{bmatrix}, \\
&[\epsilon] = \begin{bmatrix} \epsilon^{xx} & \epsilon^{xy} & \epsilon^{xz} \\ \epsilon^{yx} & \epsilon^{yy} & \epsilon^{yz} \\ \epsilon^{zx} & \epsilon^{zy} & \epsilon^{zz} \end{bmatrix}, \quad [\mu] = \begin{bmatrix} \mu^{xx} & \mu^{xy} & \mu^{xz} \\ \mu^{yx} & \mu^{yy} & \mu^{yz} \\ \mu^{zx} & \mu^{zy} & \mu^{zz} \end{bmatrix}, \\
&[\nabla\times] = \begin{bmatrix} 0 & -\partial_z & \partial_y \\ \partial_z & 0 & -\partial_x \\ -\partial_y & \partial_x & 0 \end{bmatrix}, \quad [J] = \begin{bmatrix} J^x \\ J^y \\ J^z \end{bmatrix}.
\end{aligned}
\tag{3.21}
$$

For Cartesian coordinates, it is not necessary to distinguish superscripts and subscripts since $\partial\mathbf{r}/\partial x^i = \partial x^i/\partial\mathbf{r} = \hat{x}^i = \hat{x}_i$, but we keep the notation to align with the curvilinear case.

Up to this point, we know how to write Maxwell's equations, equation (3.18), in general coordinates for the physical space (where our device lives). We now introduce a virtual space with primed coordinates (x', y', z') (index form $x'^{i'}$). Given a coordinate transformation between (x, y, z) and (x', y', z'), we ask: can we find transformation rules for fields and material parameters so that Maxwell's equations hold in the virtual space in the same form,

$$\nabla' \times \mathbf{E}' + \frac{1}{c}\partial_t \mathbf{B}' = 0, \quad \mathbf{D}' = \overline{\overline{\epsilon}}' \cdot \mathbf{E}', \qquad (3.22)$$
$$\nabla' \times \mathbf{H}' - \frac{1}{c}\partial_t \mathbf{D}' = \frac{1}{c}\mathbf{J}', \quad \mathbf{B}' = \overline{\overline{\mu}}' \cdot \mathbf{H}'.$$

In Cartesian components this reads

$$[\nabla'\times][E'] + \frac{1}{c}\partial_t[B'] = 0, \quad [D'] = [\epsilon'][E'], \qquad (3.23)$$
$$[\nabla'\times][H'] - \frac{1}{c}\partial_t[D'] = \frac{1}{c}[J'], \quad [B'] = [\mu'][H'],$$

with

$$[D'] = \begin{bmatrix} D'^{x'} \\ D'^{y'} \\ D'^{z'} \end{bmatrix}, \quad [B'] = \begin{bmatrix} B'^{x'} \\ B'^{y'} \\ B'^{z'} \end{bmatrix}, \quad [E'] = \begin{bmatrix} E'_{x'} \\ E'_{y'} \\ E'_{z'} \end{bmatrix}, \quad [H'] = \begin{bmatrix} H'_{x'} \\ H'_{y'} \\ H'_{z'} \end{bmatrix},$$

$$[\epsilon'] = \begin{bmatrix} \epsilon'^{x'x'} & \epsilon'^{x'y'} & \epsilon'^{x'z'} \\ \epsilon'^{y'x'} & \epsilon'^{y'y'} & \epsilon'^{y'z'} \\ \epsilon'^{z'x'} & \epsilon'^{z'y'} & \epsilon'^{z'z'} \end{bmatrix}, \quad [\mu'] = \begin{bmatrix} \mu'^{x'x'} & \mu'^{x'y'} & \mu'^{x'z'} \\ \mu'^{y'x'} & \mu'^{y'y'} & \mu'^{y'z'} \\ \mu'^{z'x'} & \mu'^{z'y'} & \mu'^{z'z'} \end{bmatrix}, \qquad (3.24)$$

$$[\nabla'\times] = \begin{bmatrix} 0 & -\partial_{z'} & \partial_{y'} \\ \partial_{z'} & 0 & -\partial_{x'} \\ -\partial_{y'} & \partial_{x'} & 0 \end{bmatrix}, \quad [J'] = \begin{bmatrix} J'^{x'} \\ J'^{y'} \\ J'^{z'} \end{bmatrix}.$$

This step underlies transformation optics: we interpret the virtual space as Euclidean (Cartesian calculus) even if (x', y', z') are curvilinear when viewed from the physical space.

To connect the two, we write fields in the physical space using either the Cartesian basis or the primed (curvilinear) basis at the same point:

$$\mathbf{E} = E_i \frac{\partial x^i}{\partial \mathbf{r}} = E_{i'} \frac{\partial x'^{i'}}{\partial \mathbf{r}}, \qquad \mathbf{D} = D^i \frac{\partial \mathbf{r}}{\partial x^i} = D^{i'} \frac{\partial \mathbf{r}}{\partial x'^{i'}}.$$

The two representations for $\mathbf{E}$ at the origin of the physical space are sketched in figure 3.3. In the physical space, (x, y, z) are Cartesian while (x', y', z') appear curvilinear.

The change of basis between these representations is

$$\frac{\partial x'^{i'}}{\partial x^i} E_{i'} = E_i, \qquad D^{i'} = \frac{\partial x'^{i'}}{\partial x^i} D^i, \qquad \epsilon^{i'j'} = \frac{\partial x'^{i'}}{\partial x^i}\frac{\partial x'^{j'}}{\partial x^j} \epsilon^{ij},$$

and similarly for $\mu^{i'j'}$. Note the distinct rules for covariant quantities (like E_i) and contravariant ones (like D^i).

Using these relations, Maxwell's equations for the physical space written in primed coordinates become

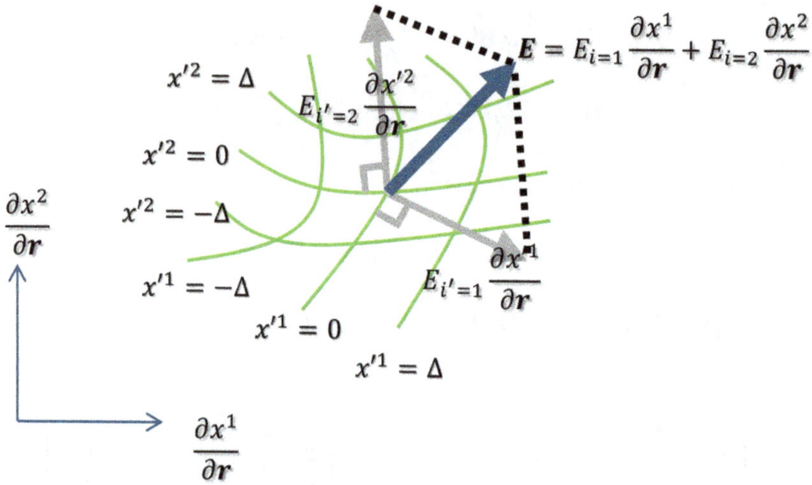

Figure 3.3. Vector **E** at the origin of the physical space in Cartesian components $E_i\ \partial x^i/\partial \mathbf{r}$ or in curvilinear components $E_{i'}\ \partial x'^{i'}/\partial \mathbf{r}$.

$$\frac{\partial(x', y', z')}{\partial(x, y, z)}\,[i', j', k']\,\frac{\partial E_{k'}}{\partial x'^{j'}} + \frac{1}{c}\partial_t B^{i'} = 0, \quad D^{i'} = \epsilon^{i'j'}E_{j'},$$
$$\frac{\partial(x', y', z')}{\partial(x, y, z)}\,[i', j', k']\,\frac{\partial H_{k'}}{\partial x'^{j'}} - \frac{1}{c}\partial_t D^{i'} = \frac{1}{c}J^{i'}, \quad B^{i'} = \mu^{i'j'}H_{j'}.$$

If we now define the primed (virtual-space) fields and materials by

$$\begin{aligned}
&E'_{i'} = E_{i'}, \quad H'_{i'} = H_{i'},\\
&D'^{i'} = \frac{\partial(x, y, z)}{\partial(x', y', z')}\,D^{i'}, \quad B'^{i'} = \frac{\partial(x, y, z)}{\partial(x', y', z')}\,B^{i'},\\
&\epsilon'^{i'j'} = \frac{\partial(x, y, z)}{\partial(x', y', z')}\,\epsilon^{i'j'}, \quad \mu'^{i'j'} = \frac{\partial(x, y, z)}{\partial(x', y', z')}\,\mu^{i'j'},\\
&J'^{i'} = \frac{\partial(x, y, z)}{\partial(x', y', z')}\,J^{i'},
\end{aligned} \tag{3.25}$$

then the equations take the form

$$[i', j', k']\,\frac{\partial E'_{k'}}{\partial x'^{j'}} + \frac{1}{c}\partial_t B'^{i'} = 0, \qquad D'^{i'} = \epsilon'^{i'j'}E'_{j'},$$
$$[i', j', k']\,\frac{\partial H'_{k'}}{\partial x'^{j'}} - \frac{1}{c}\partial_t D'^{i'} = \frac{1}{c}J'^{i'}, \qquad B'^{i'} = \mu'^{i'j'}H'_{j'},$$

which are exactly the desired form-invariant Maxwell equations in the virtual space, equation (3.22).

Finally, equation (3.25) can be written in matrix form (Cartesian components) as the standard transformation-optics rules:

$$\begin{aligned}
&[E] = [\Lambda]^T[E'], \qquad [H] = [\Lambda]^T[H'],\\
&[D'] = \frac{\partial(x, y, z)}{\partial(x', y', z')}[\Lambda]\,[D], \qquad [B'] = \frac{\partial(x, y, z)}{\partial(x', y', z')}[\Lambda]\,[B],\\
&[J'] = \frac{\partial(x, y, z)}{\partial(x', y', z')}[\Lambda]\,[J],
\end{aligned} \tag{3.26}$$

and

$$[\epsilon'] = \frac{\partial(x, y, z)}{\partial(x', y', z')}[\Lambda]\,[\epsilon]\,[\Lambda]^T, \qquad [\mu'] = \frac{\partial(x, y, z)}{\partial(x', y', z')}[\Lambda]\,[\mu]\,[\Lambda]^T, \tag{3.27}$$

where the Jacobian matrix $[\Lambda]$ is defined with row index i' and column index i by

$$\Lambda_i^{i'} = \frac{\partial x'^{i'}}{\partial x^i}.$$

We also have the Jacobian determinant in this case satisfying

$$\frac{\partial(x, y, z)}{\partial(x', y', z')} = \frac{1}{\det[\Lambda]}.$$

A direct substitution of equations (3.26)–(3.27) into equation (3.23) verifies the equivalences

$$\nabla \times \mathbf{E} + \frac{1}{c}\partial_t \mathbf{B} = 0 \;\Leftrightarrow\; \nabla' \times \mathbf{E}' + \frac{1}{c}\partial_t \mathbf{B}' = 0,$$

$$\nabla \times \mathbf{H} - \frac{1}{c}\partial_t \mathbf{D} = \frac{1}{c}\mathbf{J} \;\Leftrightarrow\; \nabla' \times \mathbf{H}' - \frac{1}{c}\partial_t \mathbf{D}' = \frac{1}{c}\mathbf{J}',$$

$$\mathbf{D} = \overline{\overline{\epsilon}} \cdot \mathbf{E} \;\Leftrightarrow\; \mathbf{D}' = \overline{\overline{\epsilon}}' \cdot \mathbf{E}', \qquad \mathbf{B} = \overline{\overline{\mu}} \cdot \mathbf{H} \;\Leftrightarrow\; \mathbf{B}' = \overline{\overline{\mu}}' \cdot \mathbf{H}'.$$

In transformation optics, we finally reinterpret the transformed space (with primed coordinates in this case) as also Euclidean. For the notations, there are subtleties. For example, the matrix notations $[D']$ and $[\epsilon']$ include an additional factor beyond the usual tensor transformation. The tensors built from these matrices D' and $\overline{\overline{\epsilon}}'$ in the reinterpreted Euclidean space are written with primes to distinguish them from the usual convention of using the same symbols for tensors in different coordinate systems.

Conservation laws under transformation

We can also express integral conservation under transformation. For line integrals, since

$$E_i\,\mathrm{d}x^i = E_i\frac{\partial x^i}{\partial x'^{i'}}\,\mathrm{d}x'^{i'} = E_{i'}\,\mathrm{d}x'^{i'} = E'_{i'}\,\mathrm{d}x'^{i'},$$

we have the circulation invariants

$$\int \mathbf{E} \cdot \mathrm{d}\mathbf{r} = \int \mathbf{E}' \cdot \mathrm{d}\mathbf{r}', \qquad \int \mathbf{H} \cdot \mathrm{d}\mathbf{r} = \int \mathbf{H}' \cdot \mathrm{d}\mathbf{r}'. \tag{3.28}$$

These integrals are performed over corresponding *mapped paths*: each point of the path in the physical space is carried to its image path in the virtual space by the coordinate transformation, while the fields are transformed by equations (3.26) and (3.27).

Likewise for fluxes, the surface integrals are taken over *mapped surfaces* with their induced orientations and area elements:

$$\int \mathbf{D} \cdot \mathrm{d}\mathbf{A} = \int \mathbf{D}' \cdot \mathrm{d}\mathbf{A}', \qquad \int \mathbf{B} \cdot \mathrm{d}\mathbf{A} = \int \mathbf{B}' \cdot \mathrm{d}\mathbf{A}'. \tag{3.29}$$

Here $\mathrm{d}\mathbf{A}$ and $\mathrm{d}\mathbf{A}'$ denote the oriented area elements in the physical and virtual spaces, respectively, related by the Jacobian (including the $\det\Lambda$ factor for area scaling and the mapping of normals). These equalities encapsulate the conservation of electric displacement and magnetic flux under the transformation when fields and media obey equations (3.26) and (3.27). These statements complete the formal development of transformation optics under general coordinate transformations.

In this section, we have established that the form-invariance of Maxwell's equations under general coordinate transformations (Post 1962) leads to the transformation rules for fields and materials, and the integral conservation laws. These results are the foundation of transformation optics. The same form-invariance also holds for acoustic waves and no wonder there is a parallel theory of transformation acoustics. Interestingly, the form invariance breaks down for elastodynamics as pointed out by Milton *et al* (2006) without introducing unconventional constitutive parameters, which is why transformation elastodynamics is more complicated and less developed.

3.4 Transformation media

While the transformation rules of the fields and material parameters can be expressed in terms of the Cartesian components, between (x, y, z) and (x', y', z'), it is often more convenient to work with other coordinates in both the physical and virtual spaces.

Figure 3.4 shows this general situation. The transformation is described between two sets of coordinates: the q-coordinates in the physical space and the q'-coordinates in the virtual space. For example, for a spherical cloak both sides use spherical polar coordinates, and the mapping is simply a rescaling of the radial coordinate r while the two angular coordinates remain unchanged.

While equation (3.25) is still valid in terms of Cartesian components, we now rewrite it in terms of basis vectors attached to the q- and q'-coordinates. In this way we obtain

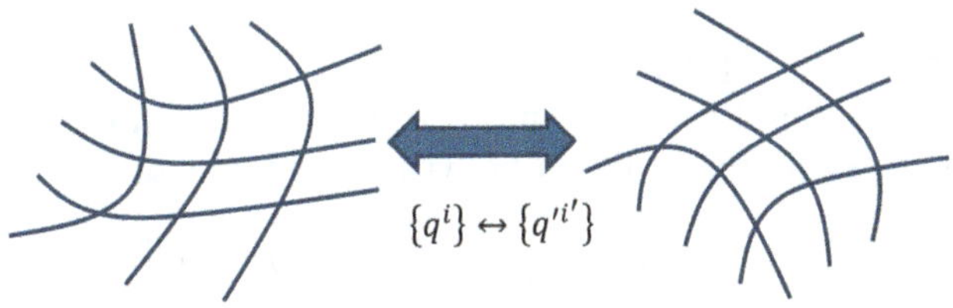

Figure 3.4. The general situation in transformation optics. Physical space on the left with general $(q^{i=1}, q^{i=2}, q^{i=3})$ coordinates visualized by the grid, is transformed to the virtual space on the right with $(q'^{i'=1}, q'^{i'=2}, q'^{i'=3})$ coordinates. The transformation is described through formulas between the two sets of q-coordinates.

$$\begin{aligned} &[E] = [\Lambda]^T[E'], \quad [H] = [\Lambda]^T[H'], \\ &[D'] = \frac{\partial(x, y, z)}{\partial(x', y', z')}[\Lambda][D], \quad [B'] = \frac{\partial(x, y, z)}{\partial(x', y', z')}[\Lambda][B], \\ &[J'] = \frac{\partial(x, y, z)}{\partial(x', y', z')}[\Lambda][J], \end{aligned} \tag{3.30}$$

and

$$[\epsilon'] = \frac{\partial(x, y, z)}{\partial(x', y', z')}[\Lambda][\epsilon][\Lambda]^T, \qquad [\mu'] = \frac{\partial(x, y, z)}{\partial(x', y', z')}[\Lambda][\mu][\Lambda]^T. \tag{3.31}$$

These formulas look the same as equations (3.26) and (3.27), except that the Jacobian matrix $[\Lambda]$ is now defined between the q- and q'-coordinates, with row index i' and column index i:

$$\Lambda_i^{i'} = \frac{\partial q'^{i'}}{\partial q^i}. \tag{3.32}$$

The matrices $[E]$ and $[E']$ contain the covariant components E_i and $E'_{i'}$ in q- and q'-coordinates:

$$\mathbf{E} = E_i \frac{\partial q^i}{\partial \mathbf{r}}, \quad \mathbf{E}' = E'_{i'} \frac{\partial q'^{i'}}{\partial \mathbf{r}'}. \tag{3.33}$$

The matrices $[D]$ and $[D']$ contain the contravariant components D^i and $D'^{i'}$:

$$\mathbf{D} = D^i \frac{\partial \mathbf{r}}{\partial q^i}, \quad \mathbf{D}' = D'^{i'} \frac{\partial \mathbf{r}'}{\partial q'^{i'}}. \tag{3.34}$$

The permittivity tensors $[\epsilon]$ and $[\epsilon']$ are expressed as

$$\overline{\overline{\epsilon}} = \epsilon^{ij} \frac{\partial \mathbf{r}}{\partial q^i} \frac{\partial \mathbf{r}}{\partial q^j}, \quad \overline{\overline{\epsilon}}' = \epsilon'^{i'j'} \frac{\partial \mathbf{r}'}{\partial q'^{i'}} \frac{\partial \mathbf{r}'}{\partial q'^{j'}}. \tag{3.35}$$

Similar definitions apply to $[H]$, $[B]$, $[\mu]$.

The prefactor involving the Jacobian determinant in equations (3.30) and (3.31) comes from volume scaling, and it can be written in terms of the q-coordinates as

$$\frac{\partial(x, y, z)}{\partial(x', y', z')} = \frac{\partial(x, y, z)}{\partial(q^1, q^2, q^3)} \frac{\partial(q^1, q^2, q^3)}{\partial(q'^1, q'^2, q'^3)} \frac{\partial(q'^1, q'^2, q'^3)}{\partial(x', y', z')}.$$

An important observation is that the physical and virtual spaces are actually on an equal footing. Equations (3.30) and (3.31) can be derived either way: by writing Maxwell's equations in the physical space using the q'-coordinates, or by writing them in the virtual space using the q-coordinates. The outcome is the same. In this sense, the permittivity and permeability tensors $\overline{\overline{\epsilon}}$, $\overline{\overline{\mu}}$ belong to the physical space, while $\overline{\overline{\epsilon}}'$, $\overline{\overline{\mu}}'$ belong to the virtual space. Both simply describe how light perceives the geometry.

We next focus on the special case when both q and q' are orthogonal coordinate systems. This means the coordinate grid lines are perpendicular, though they may be curved, as in cylindrical or spherical coordinates. In this case the contravariant and covariant basis vectors point in the same directions. We can define unit vectors $\hat{q}_i = \hat{q}^i$ and scale factors h_i as

$$\frac{\partial \mathbf{r}}{\partial q^i} = h_i \hat{q}_i, \qquad \frac{\partial q^i}{\partial \mathbf{r}} = \frac{1}{h_i} \hat{q}_i.$$

The repeated index i is not summed in the above equation. Tensors are then expressed using their principal values (with tilde notation), for example

$$\overline{\overline{\epsilon}} = \tilde{\epsilon}^{ij} \hat{q}_i \hat{q}_j, \qquad \mathbf{E} = \tilde{E}_i \hat{q}^i, \qquad \mathbf{D} = \tilde{D}^i \hat{q}_i. \tag{3.36}$$

Substituting into equation (3.31) gives

$$[\tilde{\epsilon}'] = \frac{1}{\det[\tilde{\Lambda}]} [\tilde{\Lambda}][\tilde{\epsilon}][\Lambda]^T, \qquad [\tilde{\mu}'] = \frac{1}{\det[\tilde{\Lambda}]} [\tilde{\Lambda}][\tilde{\mu}][\tilde{\Lambda}]^T, \tag{3.37}$$

where

$$[\tilde{\Lambda}] = \begin{pmatrix} h_1' & 0 & 0 \\ 0 & h_2' & 0 \\ 0 & 0 & h_3' \end{pmatrix} \begin{pmatrix} \frac{\partial q'^1}{\partial q^1} & \frac{\partial q'^1}{\partial q^2} & \frac{\partial q'^1}{\partial q^3} \\ \frac{\partial q'^2}{\partial q^1} & \frac{\partial q'^2}{\partial q^2} & \frac{\partial q'^2}{\partial q^3} \\ \frac{\partial q'^3}{\partial q^1} & \frac{\partial q'^3}{\partial q^2} & \frac{\partial q'^3}{\partial q^3} \end{pmatrix} \begin{pmatrix} \frac{1}{h_1} & 0 & 0 \\ 0 & \frac{1}{h_2} & 0 \\ 0 & 0 & \frac{1}{h_3} \end{pmatrix}. \tag{3.38}$$

By the symmetry mentioned earlier, we can also write the reverse transformation as

$$[\tilde{\epsilon}] = \frac{1}{\det[\tilde{\Lambda}]} [\tilde{\Lambda}][\tilde{\epsilon}'][\tilde{\Lambda}]^T, \qquad [\tilde{\mu}] = \frac{1}{\det[\tilde{\Lambda}]} [\tilde{\Lambda}][\tilde{\mu}'][\tilde{\Lambda}]^T, \tag{3.39}$$

with

$$[\tilde{\Lambda}] = \begin{pmatrix} h_1 & 0 & 0 \\ 0 & h_2 & 0 \\ 0 & 0 & h_3 \end{pmatrix} \begin{pmatrix} \frac{\partial q^1}{\partial q'^1} & \frac{\partial q^1}{\partial q'^2} & \frac{\partial q^1}{\partial q'^3} \\ \frac{\partial q^2}{\partial q'^1} & \frac{\partial q^2}{\partial q'^2} & \frac{\partial q^2}{\partial q'^3} \\ \frac{\partial q^3}{\partial q'^1} & \frac{\partial q^3}{\partial q'^2} & \frac{\partial q^3}{\partial q'^3} \end{pmatrix} \begin{pmatrix} \frac{1}{h_1'} & 0 & 0 \\ 0 & \frac{1}{h_2'} & 0 \\ 0 & 0 & \frac{1}{h_3'} \end{pmatrix}. \tag{3.40}$$

From this point onwards, if we are dealing with orthogonal coordinates q and q', we can omit the tildes for simplicity, and express permittivity and permeability tensors directly in their principal values.

Example: Field rotator A field rotator is a device that twists the electromagnetic field inside a region by a chosen angle (Chen and Chan 2007; Chen *et al* 2009). In other words, a wavefront entering the device will emerge rotated, as if the space inside has been turned. To illustrate this idea, let us consider a circular field rotator that rotates the wavefront by an angle of θ_0 (figure 3.5). We specify the transformation in polar coordinates. The mapping is

$$r = r', \theta = \theta(r', \theta') = \begin{cases} \theta' + \theta_0, & r' \leqslant a, \\ \theta' + \theta_0\left(1 - \dfrac{r' - a}{b - a}\right), & a < r' \leqslant b, \\ \theta', & r' > b, \end{cases} \qquad z = z'.$$

The meaning of this formula is straightforward. In the inner core $r' < a$, the whole region is rotated by a constant angle θ_0. In the outer region $r' > b$, no rotation is applied and the space is unchanged. In the annular shell between a and b, the rotation gradually reduces from θ_0 down to zero so that the device matches smoothly to free space.

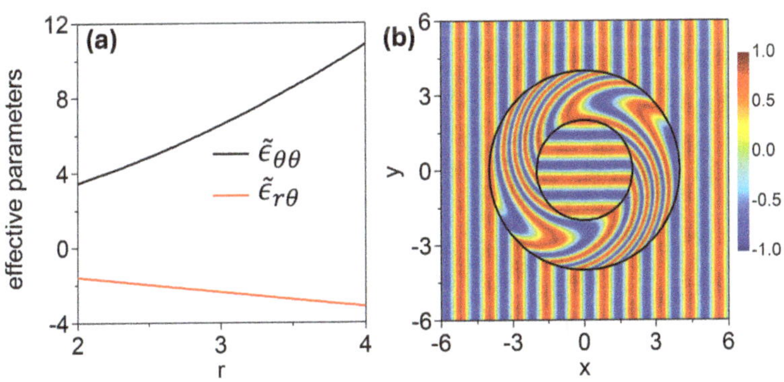

Figure 3.5. Field rotator. (a) Induced material parameters inside the device. (b) Wavefront rotation effect.

Using such a mapping, the outer region with $r > b$ and the inner region with $r < a$ remain air as the Jacobian matrix is the identity matrix. For the middle ring, the induced Jacobian and material parameters can be obtained using equations (3.39) and (3.40).

Step-by-step derivation in the annulus $a < r \leqslant b$. We work with orthogonal polar coordinates on both sides, so we can use the orthogonal-coordinate formulas in equations (3.39) and (3.40) (symbols with tildes). The scale factors are

$$h_r = 1, \quad h_\theta = r, \quad h_z = 1; \qquad h'_{r'} = 1, \quad h'_{\theta'} = r' = r, \quad h'_{z'} = 1.$$

Thus, the coordinate-derivative matrix $\partial q/\partial q'$ (with $q = (r, \theta, z), q' = (r', \theta', z')$) is

$$\frac{\partial(q^1, q^2, q^3)}{\partial(q'^1, q'^2, q'^3)} = \begin{pmatrix} \frac{\partial r}{\partial r'} & \frac{\partial r}{\partial \theta'} & \frac{\partial r}{\partial z'} \\ \frac{\partial \theta}{\partial r'} & \frac{\partial \theta}{\partial \theta'} & \frac{\partial \theta}{\partial z'} \\ \frac{\partial z}{\partial r'} & \frac{\partial z}{\partial \theta'} & \frac{\partial z}{\partial z'} \end{pmatrix} = \begin{pmatrix} 1 & 0 & 0 \\ -\alpha & 1 & 0 \\ 0 & 0 & 1 \end{pmatrix},$$

where $\alpha = -\frac{\partial \theta}{\partial r'} = \frac{\theta_0}{b-a}$ is a constant. Next, we assemble $[\tilde{\Lambda}]$ from equation (3.40):

$$[\tilde{\Lambda}] = \underbrace{\begin{pmatrix} h_r & 0 & 0 \\ 0 & h_\theta & 0 \\ 0 & 0 & h_z \end{pmatrix}}_{\mathrm{diag}(1,r,1)} \underbrace{\begin{pmatrix} 1 & 0 & 0 \\ -\alpha & 1 & 0 \\ 0 & 0 & 1 \end{pmatrix}}_{\partial q/\partial q'} \underbrace{\begin{pmatrix} \frac{1}{h'_{r'}} & 0 & 0 \\ 0 & \frac{1}{h'_{\theta'}} & 0 \\ 0 & 0 & \frac{1}{h'_{z'}} \end{pmatrix}}_{\mathrm{diag}(1,1/r,1)} = \begin{pmatrix} 1 & 0 & 0 \\ -\alpha r & 1 & 0 \\ 0 & 0 & 1 \end{pmatrix}.$$

This matrix is lower-triangular with unit diagonal, so $\det[\tilde{\Lambda}] = 1$. Taking the virtual space to be vacuum, $[\tilde{\epsilon}'] = [\tilde{\mu}'] = [I]$. From equation (3.39),

$$[\tilde{\epsilon}] = \frac{1}{\det[\tilde{\Lambda}]}[\tilde{\Lambda}]\,[\tilde{\epsilon}']\,[\tilde{\Lambda}]^T = [\tilde{\Lambda}][\tilde{\Lambda}]^T, \qquad [\tilde{\mu}] = [\tilde{\Lambda}][\tilde{\Lambda}]^T.$$

Then, we obtain

$$[\tilde{\Lambda}][\tilde{\Lambda}]^T = \begin{pmatrix} 1 & 0 & 0 \\ -\alpha r & 1 & 0 \\ 0 & 0 & 1 \end{pmatrix}\begin{pmatrix} 1 & -\alpha r & 0 \\ 0 & 1 & 0 \\ 0 & 0 & 1 \end{pmatrix} = \begin{pmatrix} 1 & -\alpha r & 0 \\ -\alpha r & 1+\alpha^2 r^2 & 0 \\ 0 & 0 & 1 \end{pmatrix}.$$

Therefore, in the annulus,

$$\tilde{\epsilon}_{rr} = \tilde{\mu}_{rr} = 1, \qquad \tilde{\epsilon}_{\theta\theta} = \tilde{\mu}_{\theta\theta} = 1 + \left(\frac{\theta_0 r}{b-a}\right)^2,$$

$$\tilde{\epsilon}_{r\theta} = \tilde{\epsilon}_{\theta r} = \tilde{\mu}_{r\theta} = \tilde{\mu}_{\theta r} = -\frac{\theta_0 r}{b-a}, \qquad \tilde{\epsilon}_{zz} = \tilde{\mu}_{zz} = 1,$$

which are plotted in figure 3.5(a) for a specific example.

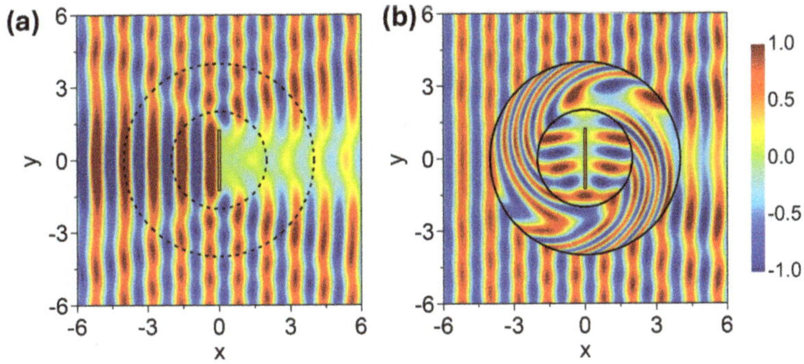

Figure 3.6. Illusion from the wavefront rotator. (a) A vertical PEC rectangle creates a large shadow behind it under plane wave incidence. (b) When the PEC is embedded inside the wavefront rotator, the shadow becomes much smaller.

Two important points are worth highlighting. First, the permittivity and permeability take exactly the same form, which is typical for transformation media. Second, the off-diagonal terms ($r\theta$) indicate coupling between radial and angular directions—this is the mathematical signature of a rotation. Figure 3.5(a) shows the simulation results for a plane-wave incidence from the left. The field inside the rotator is rotated by 90°.

Such a circular field rotator can be immediately used to create an effect of illusions (figure 3.6). When a vertical PEC (perfect electric conductor) slab is placed inside the device, it appears as if the slab is placed horizontally due to the 90° wavefront rotation inside the device.

3.5 Approaching electrostatics

The established rule of transformation optics on the material parameters, equation (3.25), is determined solely by the property of the coordinate transformation, i.e., the Jacobian matrix. The formulas themselves do not explicitly involve frequency. Nevertheless, there is always an implicit frequency dependence. To see this, note that realizing large deviations of the required material parameters from vacuum becomes increasingly difficult without introducing strong frequency dispersion in the metamaterial. In practice, therefore, frequency dependence is unavoidable. Conceptually, however, the design principle itself is frequency-independent: once the transformed material parameters are obtained in the physical space, they are, in principle, valid for all frequencies. The transformation optics framework therefore already covers the zero-frequency limit. What is special at this limit is that the general rules reduce to simpler equations such as Laplace's equation, which allows us to make useful connections to other physical systems.

Let us now consider explicitly the electrostatic regime as an example. In this case, only the permittivity, but not the magnetic permeability, is required from the transformation method. The general rule becomes

$$[\epsilon] = \frac{\partial(x', y', z')}{\partial(x, y, z)}[\Lambda][\epsilon'][\Lambda]^T, \tag{3.41}$$

where the Jacobian of the coordinate transformation is

$$\Lambda^i_{i'} = \frac{\partial x^i}{\partial x'^{i'}}.$$

To see this more directly, we can start from the electrostatic Laplace's equation in the virtual space:

$$\nabla' \cdot (\overline{\overline{\epsilon}}' \cdot \nabla ' V') = 0,$$

where V' is the potential in the virtual coordinates. Rewriting the same equation in the physical coordinates x^i, which are curvilinear in the virtual space, gives

$$\frac{\partial}{\partial x^i}\left(\frac{\partial(x', y', z')}{\partial(x, y, z)}\epsilon'^{ij}\frac{\partial V'}{\partial x^j}\right) = 0.$$

This can be interpreted as Laplace's equation in the physical space,

$$\nabla \cdot (\overline{\overline{\epsilon}} \cdot \nabla V) = \frac{\partial}{\partial x^i}\left(\epsilon^{ij}\frac{\partial V}{\partial x^j}\right) = 0,$$

if we define

$$V = V', \qquad \epsilon^{ij} = \frac{\partial(x', y', z')}{\partial(x, y, z)}\epsilon'^{ij} = \frac{\partial(x', y', z')}{\partial(x, y, z)}\frac{\partial x^i}{\partial x'^{i'}}\epsilon'^{i'j'}\frac{\partial x^j}{\partial x'^{j'}}. \tag{3.42}$$

This is precisely equivalent to equation (3.41).

Although auxiliary fields are not essential in the electrostatic case, it is instructive to see how they transform. In the virtual space, the constitutive relation is

$$D'^i = \epsilon'^{ij}E'_j, \qquad E'_i = -\frac{\partial V'}{\partial x^i},$$

while in the physical space we have

$$D^i = \epsilon^{ij}E_j, \qquad E_i = -\frac{\partial V}{\partial x^i},$$

when all fields are expressed in terms of the physical coordinates x^i.

The zero-frequency limit is particularly illuminating because it reduces the formalism to Laplace's equation, the simplest case of the transformation method. Many other steady-state processes share exactly this structure. For instance, heat conduction in the steady state satisfies

$$\nabla \cdot (\overline{\overline{\kappa}} \cdot \nabla T) = 0,$$

with κ the thermal conductivity and T the temperature. The same form governs particle diffusion and direct-current electrical conduction. In each of these situations, the material parameter (κ, a diffusivity, or a conductivity) plays the same role as the permittivity in electrostatics. This means that the transformation rules carry over directly: one can design cloaks, concentrators, or other devices for static or transport problems in just the same way as for waves. Entire research directions, such as transformation thermotics and transformation diffusion, have grown from this realization.

An interesting connection of the electrostatic limit is to electrical impedance tomography (EIT), a technique that reconstructs the internal conductivity σ profile of an object from voltage and current measurements on its boundary. The governing equation

$$\nabla \cdot (\overline{\overline{\sigma}} \cdot \nabla V) = 0$$

is indistinguishable in form from the electrostatic equation for a permittivity profile. In the strict dc limit, where displacement currents vanish, σ, ϵ, and κ are simply interchangeable coefficients in the same Laplace-type equation. This mathematical equivalence implies that anisotropic conductivities related by a coordinate transformation cannot be distinguished from boundary measurements (Greenleaf *et al* 2003). In other words, a cloak designed with transformation principles would be invisible to EIT. This prediction was confirmed in practice by constructing a cloak from a 2D resistor network, which directly implements the required anisotropic conductivity profile (Yang *et al* 2012).

In parallel to electrostatics, we can also consider the magnetostatic limit. In the magnetostatics without free currents ($\nabla \times \mathbf{H} = 0$), we introduce the scalar magnetic potential $V_{\rm m}$ via $\mathbf{H} = -\nabla V_{\rm m}$. With $\mathbf{B} = \overline{\overline{\mu}} \cdot \mathbf{H}$ and $\nabla \cdot \mathbf{B} = 0$, the governing equation becomes

$$\nabla \cdot (\overline{\overline{\mu}} \cdot \nabla V_{\rm m}) = 0,$$

which is the same Laplace form as in electrostatics. Consequently, the transformation rule is identical to equation (3.41) with $\overline{\overline{\epsilon}}$ replaced by $\overline{\overline{\mu}}$. The anisotropic permeability (Magnus *et al* 2007) prescribed by the coordinate map bends static magnetic field lines around the cloaked region, providing the recipe for a dc magnetic cloak (Gömöry *et al* 2012).

3.6 Approaching geometrical optics

The material parameters derived from transformation optics also work in the geometrical-optics (GO) limit, for the same reason they work at zero frequency. When the wavelength is very short, material parameters vary only slowly over a wavelength and rays determine the behavior. Our goal is to describe how a light ray travels in such a transformation medium. As noted earlier, a ray entering a cylindrical cloak is squeezed sideways around the hidden region and then rejoins its original path. Below we give a ray-tracing recipe; the transformation will act on an effective metric.

Hamiltonian flow of scalar waves. First, a brief review of ray optics for a scalar wave. Consider a scalar wave that locally looks like a quasi-plane wave, $\exp(i\Phi(\mathbf{r}))$, with local wavevector $\mathbf{k}(\mathbf{r}) = \nabla\,\Phi(\mathbf{r})$. In a slowly varying medium at fixed frequency ω, the field obeys a *local* dispersion relation at every point:

$$\mathcal{H}(\mathbf{k},\, \mathbf{r},\, \omega) = 0, \tag{3.43}$$

obtained by localizing $\nabla \to i\mathbf{k}$ on its scalar wave equation. Equation (3.43) constrains $(\mathbf{k},\, \omega)$ to the allowed wave surface, exactly as a Hamiltonian constrains mechanical trajectories to the constant-energy surface in classical mechanics. In this analogy, the phase Φ plays the role of an action and the local wavevector $\mathbf{k} = \nabla\,\Phi$ plays the role of momentum; rays are the 'classical' trajectories of waves.

Comparing two neighboring points separated by $\Delta\mathbf{r}$ for the same solution at fixed ω,

$$\Delta\mathcal{H} = \Delta\mathbf{r}\cdot\left(\nabla\mathbf{k}\cdot\frac{\partial\mathcal{H}}{\partial\mathbf{k}} + \frac{\partial\mathcal{H}}{\partial\mathbf{r}}\right) = \Delta\mathbf{r}\cdot\left(\frac{\partial\mathcal{H}}{\partial\mathbf{k}}\cdot\nabla\,\mathbf{k} + \frac{\partial\mathcal{H}}{\partial\mathbf{r}}\right),$$

where the last equality uses $\nabla\times\mathbf{k} = 0 \Rightarrow \nabla\,\mathbf{k} = (\nabla\mathbf{k})^T$. Because (3.43) holds at every point, $\Delta\mathbf{r}$ is arbitrary and

$$\frac{\partial\mathcal{H}}{\partial\mathbf{k}}\cdot\nabla\,\mathbf{k} + \frac{\partial\mathcal{H}}{\partial\mathbf{r}} = 0.$$

Now consider a wave packet moving along a ray. The group velocity is

$$\frac{\mathrm{d}\mathbf{r}}{\mathrm{d}t} = \frac{\partial\omega(\mathbf{k},\,\mathbf{r})}{\partial\mathbf{k}} = -\frac{\partial\mathcal{H}}{\partial\mathbf{k}}\Big/\frac{\partial\mathcal{H}}{\partial\omega},$$

and, using the relation above,

$$\frac{\mathrm{d}\mathbf{k}}{\mathrm{d}t} = \frac{\mathrm{d}\mathbf{r}}{\mathrm{d}t}\cdot\nabla\,\mathbf{k} = \frac{\partial\mathcal{H}}{\partial\mathbf{r}}\Big/\frac{\partial\mathcal{H}}{\partial\omega}.$$

Introducing a ray parameter τ puts the equations in Hamiltonian form:

$$\begin{aligned}\frac{\mathrm{d}\mathbf{r}}{\mathrm{d}\tau} &= \frac{\partial\mathcal{H}}{\partial\mathbf{k}},\\ \frac{\mathrm{d}\mathbf{k}}{\mathrm{d}\tau} &= -\frac{\partial\mathcal{H}}{\partial\mathbf{r}},\\ \frac{\mathrm{d}t}{\mathrm{d}\tau} &= -\frac{\partial\mathcal{H}}{\partial\omega}.\end{aligned} \tag{3.44}$$

In short, a scalar wave with local dispersion relationship $\mathcal{H}(\mathbf{k},\,\mathbf{r},\,\omega) = 0$ induces the Hamiltonian dynamics in equation (3.44). The scaling degree of freedom on defining the Hamiltonian $\mathcal{H}$ rescales the ray parameter τ from the physical time t.

Dielectric media. As a concrete anchor, take a scalar wave in a frequency-non-dispersive isotropic refractive index profile $n(\mathbf{r})$, e.g., a dielectric medium profile, satisfying $(\nabla^2 + k_0^2 n^2)E = 0$ with $k_0 = \omega/c$. Localizing $\nabla \to i\mathbf{k}$ suggests

$$\mathcal{H} = \frac{1}{2k_0 n^2}\left(|\mathbf{k}|^2 - k_0^2 n^2\right). \tag{3.45}$$

Then, applying (3.44) gives

$$\frac{\mathrm{d}\mathbf{r}}{\mathrm{d}\tau} = \frac{\mathbf{k}}{k_0 n^2}, \qquad \frac{\mathrm{d}\mathbf{k}}{\mathrm{d}\tau} = \frac{k_0}{n} \nabla n, \qquad \frac{\mathrm{d}t}{\mathrm{d}\tau} = \frac{1}{c}.$$

Here, the scaling factor $1/(2k_0 n^2)$ is chosen so that $\tau = ct$, also means the optical path length ($|n\mathrm{d}\mathbf{r}/\mathrm{d}\tau| = 1$). Eliminating $\mathbf{k}$ gives

$$n \frac{\mathrm{d}}{\mathrm{d}\tau}\left(n^2 \frac{\mathrm{d}\mathbf{r}}{\mathrm{d}\tau}\right) = \nabla n.$$

If we reparametrize by arc length $\mathrm{d}s$ so that $|\frac{\mathrm{d}\mathbf{r}}{\mathrm{d}s}| = 1$, then $\mathrm{d}\tau = n\ \mathrm{d}s$ and we arrive the conventional eikonal equation:

$$\frac{\mathrm{d}}{\mathrm{d}s}\left(n \frac{\mathrm{d}\mathbf{r}}{\mathrm{d}s}\right) = \nabla n. \tag{3.46}$$

The Lagrangian follows from the Legendre transform (eliminating $\mathbf{k}$ in favor of $\mathrm{d}\mathbf{r}/\mathrm{d}\tau$).

$$\mathcal{S} = \int \mathcal{L}\left(\mathbf{r}, \frac{\mathrm{d}\mathbf{r}}{\mathrm{d}\tau}\right) \mathrm{d}\tau = \int (\mathbf{k} \cdot \mathrm{d}\mathbf{r} - \mathcal{H}\ \mathrm{d}\tau) = k_0 \int \mathrm{d}\tau = k_0 \int n\ \mathrm{d}s.$$

In obtaining the above, we have used $\mathcal{H} = 0$ and the Hamiltonian dynamics equations. This recovers Fermat's principle: the optical path $\int \mathrm{d}\tau = \int n \mathrm{d}s$ is stationary. This justifies the choice of $\mathcal{H}$.

Transformation media. After this review of GO in an ordinary dielectric, we move to a transformation medium which has $\overline{\overline{\epsilon}}(\mathbf{r}) = \overline{\overline{\mu}}(\mathbf{r})$ with its matrix form being symmetric in Cartesian components. We now go through the same process from dispersion to Hamiltonian to Lagrangian. Such a medium often arises from transforming optics (e.g., a cloak) where ray tracing can still be governed by a scalar-wave Hamiltonian (Schurig *et al* 2006). In this case, the Maxwell's equations decouple for the two fields $\mathbf{E} \pm \mathrm{i}\mathbf{H}$:

$$k_0 \overline{\overline{\epsilon}}(\mathbf{r}) \cdot (\mathbf{E} \pm \mathrm{i}\mathbf{H}) = \pm \nabla \times (\mathbf{E} \pm \mathrm{i}\mathbf{H}).$$

Localizing $\nabla \to \mathrm{i}\mathbf{k}$ in the equation gives rise to a common dispersion relationship for both fields as $\mathcal{H} = 0$, where the Hamiltonian $\mathcal{H}$ (with tensor $\overline{\overline{g}}$) can be chosen as

$$\mathcal{H} = \frac{1}{2k_0}\left(\mathbf{k} \cdot \overline{\overline{g}}^{-1} \cdot \mathbf{k} - k_0^2\right), \qquad \overline{\overline{g}} = (\det \epsilon)\ \overline{\overline{\epsilon}}^{-1}. \tag{3.47}$$

Here, $\det \epsilon$ is denotes the determinant of the Cartesian-component matrix of $\overline{\overline{\epsilon}}$.

In general coordinates one can write

$$\det \epsilon = \Omega^2 \det[\epsilon^{ij}],$$

where Ω is the Jacobian determinant of the Cartesian coordinates with respect to the general coordinates. Thus det ϵ is a scalar invariant under coordinate transformations, and therefore $\overline{\overline{g}}$ is a second-rank tensor. We note that if the medium has an additional spatial mirror symmetry and propagation is confined to the transverse plane, another choice of decoupling is the TE and TM modes; however, the Hamiltonian retains the same form as equation (3.47). Assuming $\overline{\overline{\epsilon}}$ is symmetric, the Hamiltonian dynamics by applying equation (3.44) gives

$$\begin{aligned}
\frac{\mathrm{d}\mathbf{r}}{\mathrm{d}\tau} &= \frac{1}{k_0}\overline{\overline{g}}^{-1}\cdot\mathbf{k}, \\
\frac{\mathrm{d}k_i}{\mathrm{d}\tau} &= -\frac{1}{2k_0}\mathbf{k}\cdot\frac{\partial\overline{\overline{g}}^{-1}}{\partial x^i}\cdot\mathbf{k} = \frac{k_0}{2}\frac{\mathrm{d}\mathbf{r}}{\mathrm{d}\tau}\cdot\frac{\partial\overline{\overline{g}}}{\partial x^i}\cdot\frac{\mathrm{d}\mathbf{r}}{\mathrm{d}\tau}, \\
\frac{\mathrm{d}(ct)}{\mathrm{d}\tau} &= 1 - \frac{1}{2k_0}\mathbf{k}\cdot\frac{\partial\overline{\overline{g}}^{-1}}{\partial k_0}\cdot\mathbf{k} = 1 + \frac{k_0}{2}\frac{\mathrm{d}\mathbf{r}}{\mathrm{d}\tau}\cdot\frac{\partial\overline{\overline{g}}}{\partial k_0}\cdot\frac{\mathrm{d}\mathbf{r}}{\mathrm{d}\tau}.
\end{aligned} \tag{3.48}$$

τ, the ray parameter, only has the same meaning of the actual time in frequency-non-dispersive media in which $\mathrm{d}(ct)/\mathrm{d}\tau = 1$ in the last line (used $\mathcal{H} = 0$ in derivation). Generally, τ is a fictitious time whose meaning will be settled soon. In this case, the action can be written as

$$\begin{aligned}
\mathcal{S} &= \int(\mathbf{k}\cdot\mathrm{d}\mathbf{r} - \mathcal{H}\,\mathrm{d}\tau) = \frac{k_0}{2}\int\left(\frac{\mathrm{d}\mathbf{r}}{\mathrm{d}\tau}\cdot\overline{\overline{g}}\cdot\frac{\mathrm{d}\mathbf{r}}{\mathrm{d}\tau} + 1\right)\mathrm{d}\tau \\
&= \frac{1}{2k_0}\int\left(\mathbf{k}\cdot\overline{\overline{g}}^{-1}\cdot\mathbf{k} + k_0^2\right)\mathrm{d}\tau = k_0\int\mathrm{d}\tau.
\end{aligned}$$

We have used equation (3.48) and $\mathcal{H} = 0$ in obtaining the second line. $\int\mathrm{d}\tau$ is still stationary, suggesting that τ is related to the optical path length or distance in some way. In fact, on the constraint $\mathcal{H} = 0$ we have $\mathbf{k}\cdot\overline{\overline{g}}^{-1}\cdot\mathbf{k} = k_0^2$, which implies

$$(\mathrm{d}\tau)^2 = \mathrm{d}\mathbf{r}\cdot\overline{\overline{g}}\cdot\mathrm{d}\mathbf{r}. \tag{3.49}$$

Thus, $\overline{\overline{g}}$ works like a metric and $\mathrm{d}\tau$ works like a distance in a curved space with metric $\overline{\overline{g}}$ effectively set by the medium. We call $\overline{\overline{g}}$ the medium metric. It should not be confused with the original unit metric used when Maxwell's equations are written. One may either describe the situation as Euclidean space (unit metric) filled with a medium, or, equivalently, as a curved free space (no medium) endowed with metric $\overline{\overline{g}}$, which already captures the medium's effect. Although the concept of the medium metric is established within the ray optics limit, it is not limited to that. In appendix A.5, we show the usage of medium metric in the level of Maxwell's equations.

By eliminating $\mathbf{k}$ from equation (3.48) (after some algebra) yields the ray equation

$$\frac{1}{2}\frac{\mathrm{d}x^k}{\mathrm{d}\tau}\frac{\partial g_{kl}}{\partial x^i}\frac{\mathrm{d}x^l}{\mathrm{d}\tau} - \frac{\mathrm{d}}{\mathrm{d}\tau}\left(g_{ij}\frac{\mathrm{d}x^j}{\mathrm{d}\tau}\right) = 0, \tag{3.50}$$

equivalently,

$$\frac{\mathrm{d}^2x^i}{\mathrm{d}\tau^2} + \Gamma^i{}_{\alpha\beta}\frac{\mathrm{d}x^\alpha}{\mathrm{d}\tau}\frac{\mathrm{d}x^\beta}{\mathrm{d}\tau} = 0, \qquad \Gamma^i{}_{\alpha\beta} = \frac{1}{2}g^{ij}\left(\frac{\partial g_{j\beta}}{\partial x^\alpha} + \frac{\partial g_{j\alpha}}{\partial x^\beta} - \frac{\partial g_{\alpha\beta}}{\partial x^j}\right). \tag{3.51}$$

Equation (3.51) has exactly the same form of geodesic equation for a given metric $\overline{\overline{g}}$.

The $\overline{\overline{g}}$ defined from the material tensors is not merely a convenient shorthand: because the geodesic law follows, $\overline{\overline{g}}$ behaves as a genuine metric tensor in the geometric sense. Such a picture is valid for symmetric impedance-matched media with $\overline{\overline{\epsilon}}(\mathbf{r}) = \overline{\overline{\mu}}(\mathbf{r})$. This is why transformation optics is often compared with general relativity. In general relativity, gravity is described by a curved spacetime metric that bends light rays. In transformation optics, a transformation medium can now be described by a metric $\overline{\overline{g}} = (\det \epsilon)\overline{\overline{\epsilon}}^{-1}$ that bends light rays as well. Both can capture the light-bending effect using a metric but there is also a subtle difference. In general relativity, a metric has intrinsic curvature (in spacetime). The metric interpretation of a medium in the ray optics limit is very useful in analog gravity: to use light to mimic certain aspects of general relativity, such as black holes in attracting light to be captured by the center of a radial index profile (Leonhardt and Philbin 2006; Genov *et al* 2009). On the other hand, we have to note that, in transformation optics, if you are transforming from a flat space (vacuum, without intrinsic curvature) to formulate a cloak, the resulting metric has no intrinsic curvature since it is just a relabeling of flat space.

Tensor transformation rule for the medium metric. Let $[\epsilon]$ denote the matrix with entries ϵ^{ij} and $[g]$ the matrix with entries g_{ij}. Let the row matrix $[k]$ hold k_i and the column matrix $[dr]$ hold dx^i. Then

$$\mathcal{H} = \frac{1}{2k_0}\left([k]\,[g]^{-1}[k]^T - k_0^2\right),$$

where

$$[g]^{-1} = \frac{1}{\det[\epsilon]}\,[\epsilon], \quad \left([g]^{-1}\right)^{ij} = \frac{\partial x^i}{\partial \mathbf{r}}\cdot\overline{\overline{g}}^{-1}\cdot\frac{\partial x^j}{\partial \mathbf{r}} = \frac{1}{\det \epsilon}\,\epsilon^{ij}.$$

Before transformation, assume Euclidean space ($\Omega = 1$) with Cartesian coordinates x^i (unprimed). We transform to new coordinates $x'^{i'}$ (primed) with Jacobian

$$\Lambda^{i'}{}_i = \frac{\partial x'^{i'}}{\partial x^i}.$$

Let $[\Lambda]$ be the matrix with row index i' and column index i; let $[dr']$ be the column matrix of $dx'^{i'}$ and $[k']$ the row matrix of $k'_{i'}$. Then

$$[k] = [k']\,[\Lambda], \qquad [dr'] = [\Lambda]\,[dr]. \tag{3.52}$$

In the transformed space, by the transformation-optics rule, the permittivity matrix obeys

$$[\epsilon'] = \frac{[\Lambda]\,[\epsilon]\,[\Lambda]^T}{\det[\Lambda]},$$

where the factor $1/\det[\Lambda]$ comes from reinterpreting the transformed space as Euclidean (form invariance). We have assumed the coordinate transformation is orientation preserving (positive $\det[\Lambda]$). Consequently,

$$\det[\epsilon'] = \frac{\det[\epsilon]}{\det[\Lambda]},$$

and similarly $[\mu'] = [\epsilon']$. Defining the corresponding medium metric in the transformed (reinterpreted Euclidean) space by

$$[g']^{-1} = \frac{1}{\det[\epsilon']}\,[\epsilon'], \tag{3.53}$$

we obtain

$$[g']^{-1} = \frac{[\Lambda]\,[\epsilon]\,[\Lambda]^T}{\det[\epsilon]} = [\Lambda]\,[g]^{-1}[\Lambda]^T, \tag{3.54}$$

i.e. the metric transforms with the conventional tensor rule. Thus $[g']$ is simply the matrix representation of the same tensor $\overline{\overline{g}}$ in primed coordinates. We have independently induced the medium-metric matrices from the material parameters in both spaces, and they are just different matrix representations of the same tensor $\overline{\overline{g}}$.

From the transformation rules for $[k]$, $[dr]$, and $[g]$ it follows directly that

$$[k]\,[dr] = [k']\,[dr'], \qquad [k]\,[g]\,[k]^T = [k']\,[g']\,[k']^T. \tag{3.55}$$

Invariance of ray properties. Consider a path (a segment of a ray trajectory) in the original, unprimed space and its pointwise-mapped counterpart in the primed space. The transformed space is reinterpreted as Euclidean, meaning $[dr']$ forms the vector $d\mathbf{r}'$ and $[k']$ forms the vector $\mathbf{k}'$. The matrix $[\epsilon']$, which already contains the factor $1/\det[\Lambda]$ due to this reinterpretation, represents the tensor $\overline{\overline{\epsilon}}'$, and $[g']$ represents $\overline{\overline{g}}'$.

An obvious invariant is the Hamiltonian $\mathcal{H}$ (a scalar):

$$\mathcal{H} = \frac{1}{2k_0}\left(\mathbf{k}\cdot\overline{\overline{g}}^{-1}\cdot\mathbf{k} - k_0^2\right) = \frac{1}{2k_0}\left(\mathbf{k}'\cdot\overline{\overline{g}}'^{-1}\cdot\mathbf{k}' - k_0^2\right), \tag{3.56}$$

and along any ray it evaluates to zero in both spaces.

Using the ray-tracing (Hamiltonian) equations separately in both spaces, together with the transformation rules for $[k]$, $[dr]$, and $[g]$, one finds (after some algebra)

$$\frac{dt}{d\tau} = \frac{dt}{d\tau'}, \qquad \frac{d[k']}{d\tau'}[\Lambda] = \frac{d[k]}{d\tau} - [k']\,\frac{d[\Lambda]}{d\tau}, \qquad [\Lambda]\frac{d[r]}{d\tau} = \frac{d[r']}{d\tau'}.$$

Therefore τ and τ' along the mapped paths differ at most by a global additive constant; without loss of generality we take the same parametrization:

$$\tau = \tau', \qquad [k] = [k']\,[\Lambda], \qquad [\Lambda]\frac{d[r]}{d\tau} = \frac{d[r']}{d\tau}. \tag{3.57}$$

As expected, the phase accumulated along the path is invariant:

$$\int \mathbf{k} \cdot d\mathbf{r} = \int \mathbf{k}' \cdot d\mathbf{r}'. \tag{3.58}$$

Since $\tau = \tau'$, the dispersion $\mathcal{H} = 0$ also implies the proper-length invariance

$$(d\tau)^2 = d\mathbf{r} \cdot \bar{\bar{g}} \cdot d\mathbf{r} = d\mathbf{r}' \cdot \bar{\bar{g}}' \cdot d\mathbf{r}'. \tag{3.59}$$

In other words, the 'proper distance' along a ray is the same in both spaces and may be evaluated using the metric of either space.

3.7 Problems

Problem 3.1 In cylindrical coordinates, is $\partial\mathbf{r}/\partial\theta$ a unit vector? If not, what is its length?

Problem 3.2 Check that in cylindrical coordinates the area element in the r–θ plane is r dr dθ and the volume element is r dr dθ dz, using only the scale factors.

Problem 3.3 Prove the index raising and lowering rule equation (3.12).

Problem 3.4 Verify the component forms of dot and cross products:

$$\mathbf{u} \cdot \mathbf{v} = u^i v_i = u_i v^i, \quad (\mathbf{u} \times \mathbf{v})^i = \varepsilon^{ijk} u_j v_k, \quad (\mathbf{u} \times \mathbf{v})_i = \varepsilon_{ijk} u^j v^k.$$

Problem 3.5 Show that

$$\frac{\partial q^i}{\partial \mathbf{r}} \times \frac{\partial q^j}{\partial \mathbf{r}} = \varepsilon^{ijk}\frac{\partial \mathbf{r}}{\partial q^k}, \qquad \frac{\partial \mathbf{r}}{\partial q^i} \times \frac{\partial \mathbf{r}}{\partial q^j} = \varepsilon_{ijk}\frac{\partial q^k}{\partial \mathbf{r}}, \qquad \text{and } \frac{\partial}{\partial q^i}\left(\frac{\partial q^i}{\partial \mathbf{r}}\Omega\right) = 0.$$

Problem 3.6 Using equation (3.13), write $\mathbf{B} = \bar{\bar{\mu}} \cdot \mathbf{H}$ in component form and show that it reduces to $B^i = \mu^{ij} H_j$.

Problem 3.7 Derive the line, area, and volume elements for cylindrical coordinates (r, θ, z):

(a) Line elements: $|\mathrm{d}\mathbf{r}| = \mathrm{d}r$, $|\mathrm{d}\mathbf{r}| = r\ \mathrm{d}\theta$, $|\mathrm{d}\mathbf{r}| = \mathrm{d}z$.

(b) Area elements: $\mathrm{d}A = r\ \mathrm{d}\theta\ \mathrm{d}z$, $\mathrm{d}A = \mathrm{d}r\ \mathrm{d}z$, $\mathrm{d}A = r\ \mathrm{d}r\ \mathrm{d}\theta$.
(c) Volume element: $\mathrm{d}V = r\ \mathrm{d}r\ \mathrm{d}\theta\ \mathrm{d}z$.
Then repeat for spherical coordinates (r, θ, ϕ):
(d) Line elements: $|\mathrm{d}\mathbf{r}| = \mathrm{d}r$, $|\mathrm{d}\mathbf{r}| = r\ \mathrm{d}\theta$, $|\mathrm{d}\mathbf{r}| = r \sin\theta\ \mathrm{d}\phi$.
(e) Area elements: $\mathrm{d}A = r^2 \sin\theta\ \mathrm{d}\theta\ d\phi$, $\mathrm{d}A = r \sin\theta\ \mathrm{d}r\ \mathrm{d}\phi$, $\mathrm{d}A = r\ \mathrm{d}r\ \mathrm{d}\theta$.
(f) Volume element: $\mathrm{d}V = r^2 \sin\theta\ \mathrm{d}r\ \mathrm{d}\theta\ \mathrm{d}\phi$.

Problem 3.8 Challenge (optional). Repeat the previous exercise for elliptical coordinates (u, v):

$$x = a \cosh u \cos v, \qquad y = a \sinh u \sin v,$$

with $u \geqslant 0$, $v \in [0, 2\pi]$, and $a \geqslant 0$.
Answer:

$$\mathrm{d}l = \frac{1}{\sqrt{2}} a \sqrt{\cosh(2u) - \cos(2v)}\ \mathrm{d}u, \quad \mathrm{d}l = \frac{1}{\sqrt{2}} a \sqrt{\cosh(2u) - \cos(2v)}\ \mathrm{d}v,$$

$$\mathrm{d}A = \frac{1}{2} a^2 (\cosh(2u) - \cos(2v))\ \mathrm{d}u\ \mathrm{d}v.$$

Problem 3.9 Poisson's equation in vacuum is $\nabla^2 V = -\rho$ with $\mathbf{E} = -\nabla V$. Derive its component form in general coordinates:

$$\frac{1}{\Omega} \frac{\partial}{\partial q^i} \left(\Omega\, g^{ij} \frac{\partial V}{\partial q^j} \right) = -\rho, \qquad E_i = -\frac{\partial V}{\partial q^i}.$$

Problem 3.10 Prove the component forms of the charge and energy continuity equations are

$$\partial_t \rho + \frac{1}{\Omega} \frac{\partial(\Omega J^i)}{\partial q^i} = 0,$$

and

$$\frac{1}{\Omega} \frac{\partial}{\partial q^i} (\Omega S^i) + \partial_t u = -J^i E_i,$$

with Poynting vector $S^i = c\ \varepsilon^{ijk} E_j H_k$ and energy density $u = (B^i H_i + D^i E_i)/2$. Also prove the Lorentz force density

$$f_i = \rho E_i + \frac{1}{c} \varepsilon_{ijk} J^j B^k.$$

Problem 3.11 Simple compression.

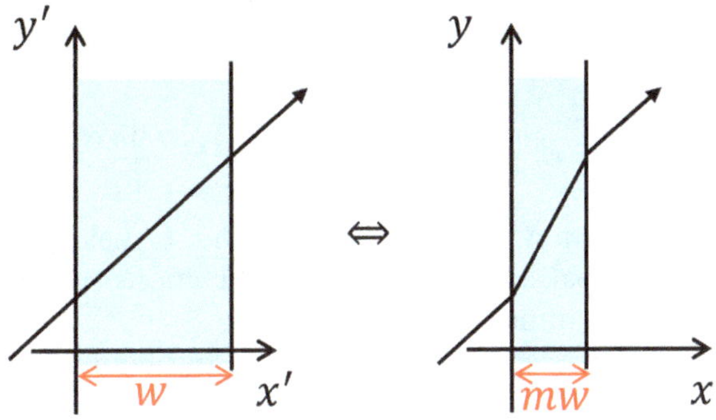

Suppose we compress a slab region of vacuum by a factor m (<1 for compression) in the x-direction. The transformed medium in the physical space has the only nonzero components $\mu_{xx} = \epsilon_{xx}$ and $\mu_{yy} = \mu_{zz} = \epsilon_{yy} = \epsilon_{zz}$.

(a) Derive the transformed material parameters in the physical space using equation (3.27).

(b) Using the area conservation law for **D** and the line conservation law for **E**, re-derive the same results in part (a), assuming a principal axes along x, y, z.

(c) Verify that the optical path is conserved for a light beam traveling separately along the x-, y-, and z-directions.

Problem 3.12 For a coordinate rotation by angle θ_0 about the z-axis,

$$x' = x\cos\theta_0 + y\sin\theta_0, \qquad y' = -x\sin\theta_0 + y\cos\theta_0, \qquad z' = z.$$

Show that a biaxial medium with

$$[\epsilon] = \begin{bmatrix} \epsilon_{xx} & 0 & 0 \\ 0 & \epsilon_{yy} & 0 \\ 0 & 0 & \epsilon_{zz} \end{bmatrix}$$

transforms to

$$[\epsilon'] = \begin{bmatrix} \epsilon_{xx}\cos^2\theta_0 + \epsilon_{yy}\sin^2\theta_0 & (\epsilon_{yy} - \epsilon_{xx})\sin\theta_0\cos\theta_0 & 0 \\ (\epsilon_{yy} - \epsilon_{xx})\sin\theta_0\cos\theta_0 & \epsilon_{yy}\cos^2\theta_0 + \epsilon_{xx}\sin^2\theta_0 & 0 \\ 0 & 0 & \epsilon_{zz} \end{bmatrix},$$

and note that the same transformation applies to $\overline{\overline{\mu}}$.

Problem 3.13 Bending transformation.

The bending transformation is specified by

$$x' = x'(r), \quad y' = b\theta, \quad z' = z,$$

which maps a rectangle in the primed coordinates to a circular arc in the unprimed coordinates.

(a) Construct the Jacobian $[\Lambda]$ using equation (3.40), with scale factors $h_1 = 1, h_2 = r, h_3 = 1$ (cylindrical physical space) and $h'_i = 1$ (Cartesian virtual space).

(b) Derive the permittivity through equation (3.39) by assuming a vacuum virtual space, and show

$$\overline{\overline{\epsilon}} = \hat{r}\hat{r}\,\frac{b}{r}\frac{\mathrm{d}r}{\mathrm{d}x'} + \hat{\theta}\hat{\theta}\,\frac{r}{b}\frac{\mathrm{d}x'}{\mathrm{d}r} + \hat{z}\hat{z}\frac{b}{r}\frac{\mathrm{d}x'}{\mathrm{d}r}.$$

(c) By using conservation laws for the **D** and **E** fields, and assuming the permittivity $\overline{\overline{\epsilon}}$ has principal axes in the r, θ, z-directions, prove the result in (b) without directly using the transformation rule.

Hint: you can first prove from the line integrals to obtain

$$E'_{x'}\mathrm{d}x' = E_r\mathrm{d}r, \quad bE'_{y'} = rE_\theta, \quad E'_{z'} = E_z,$$

and from the area integrals (about fluxes) to obtain

$$bD'^{x'} = rD^r, \quad D'^{y'}\mathrm{d}x' = D^\theta\mathrm{d}r, \quad D'^{z'}b\mathrm{d}x' = D^z r\mathrm{d}r.$$

References

Chen H and Chan C T 2007 Transformation media that rotate electromagnetic fields *Appl. Phys. Lett.* **90** 241105

Chen H, Hou B, Chen S, Ao X, Wen W and Chan C 2009 Design and experimental realization of a broadband transformation media field rotator at microwave frequencies *Phys. Rev. Lett.* **102** 183903

Genov D, Zhang S and Zhang X 2009 Mimicking celestial mechanics in metamaterials *Nat. Phys.* **5** 687–92

Greenleaf A, Lassas M and Uhlmann G 2003 Anisotropic conductivities that cannot be detected by EIT *Physiol. Meas.* **24** 413–9

Gömöry F, Solovyov M, Šouc J, Navau C, Prat-Camps J and Sanchez A 2012 Experimental realization of a magnetic cloak *Science* **335** 1466–8

Leonhardt U and Philbin T 2006 General relativity in electrical engineering *New J. Phys.* **8** 247

Magnus F, Wood B, Moore J, Morrison K, Perkins G, Fyson J, Wiltshire M, Caplin D, Cohen L and Pendry J 2007 A d.c. magnetic metamaterial *Nat. Mater.* **7** 295–7

Milton G, Briane M and Willis J 2006 On cloaking for elasticity and physical equations with a transformation invariant form *New J. Phys.* **8** 248

Pendry J B, Schurig D and Smith D R 2006 Controlling electromagnetic fields *Science* **312** 1780–2

Post E J 1962 *Formal Structure of Electromagnetics: General Covariance and Electromagnetics* (North-Holland)

Schurig D, Pendry J and Smith D R 2006 Calculation of material properties and ray tracing in transformation media *Opt. Express* **14** 9794–804

Ward A and Pendry J 1996 Refraction and geometry in Maxwellas equations *J. Mod. Opt.* **43** 773–93

Yang F, Mei Z, Jin T and Cui T 2012 Dc electric invisibility cloak *Phys. Rev. Lett.* **109** 053902

IOP Publishing

Essential Concepts in Metamaterials

Jensen Li

Chapter 4

Invisibility cloaks and other devices

In this chapter, we will use the transformation optics recipe developed in the previous chapter to design a variety of invisibility cloaks. We will also introduce several special maps and also techniques, such as the reduced-parameter approximation, that allow us to obtain materials with useful properties more easily, from natural media to simple metamaterials made of basic unit cells. Although invisibility cloaks serve as our main guide to exploring different coordinate maps, the same techniques can also be applied to design a wide range of optical elements, which we will present mainly as examples and problems in this chapter. These include light-harvesting devices, collimators, waveguide benders, and beam shifters.

4.1 Scaling map and invisibility cloak

There are three different approaches to hiding an object using coordinate transformations: compressing it into a tiny point, a thin wire, or a thin sheet, as illustrated in figure 4.1. In the transformed (virtual) space, the object is squashed to be infinitely small in three, two, or one dimension(s), respectively. In this limit, the scattering from the object becomes negligible, rendering it invisible to propagating electromagnetic waves. A variety of coordinate maps can be considered for these three approaches, ranging from simple scaling and affine maps to conformal and quasiconformal maps.

4.1.1 Scaling maps in orthogonal coordinate systems

To design the first two types of cloaks shown in figure 4.1, we can work in spherical or cylindrical coordinates and compress the object along the radial direction while leaving the other coordinates unchanged. This operation can be achieved by applying a simple scaling map to the radial coordinate.

For general purposes, we consider materials induced from a scaling map that transforms orthogonal coordinates q'^i in the virtual space into orthogonal coordinates q^i in the physical space. The scaling map takes the form

doi:10.1088/978-0-7503-1131-1ch4

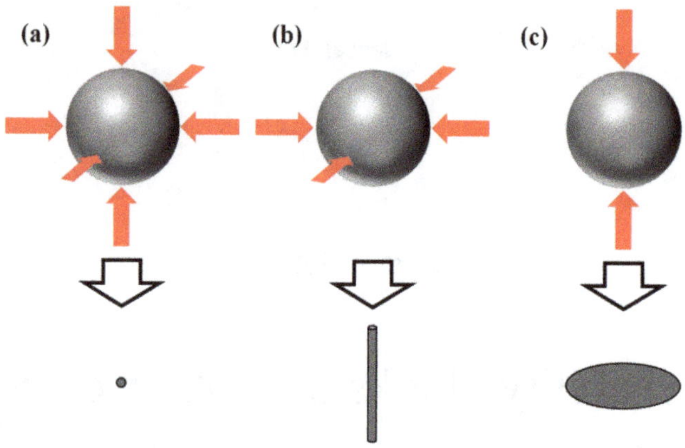

Figure 4.1. Three ways to make an object invisible: (a) squash an object into an infinitesimally small point, (b) squash it into an infinitesimally thin wire, and (c) squash it into an infinitesimally thin plate, corresponding to (a) a spherical cloak, (b) a cylindrical cloak, and (c) a carpet cloak.

$$q^1 = q^1(q'^1), \quad q^2 = q^2(q'^2), \quad q^3 = q^3(q'^3), \tag{4.1}$$

which scales each coordinate direction independently. If the principal axes of the medium in the virtual space are aligned with the coordinate lines, the permittivity tensor can be expressed as

$$[\epsilon'] = \epsilon'^{11}\hat{q}'_1\hat{q}'_1 + \epsilon'^{22}\hat{q}'_2\hat{q}'_2 + \epsilon'^{33}\hat{q}'_3\hat{q}'_3,$$

and the permeability tensor is expressed in the same form. Using the general transformation rule for constitutive tensors in equation (3.31), the induced permittivity in the physical space becomes

$$[\epsilon] = \frac{\Lambda^1{}_1}{\Lambda^2{}_2\Lambda^3{}_3}\epsilon'^{11}\hat{q}_1\hat{q}_1 + \frac{\Lambda^2{}_2}{\Lambda^1{}_1\Lambda^3{}_3}\epsilon'^{22}\hat{q}_2\hat{q}_2 + \frac{\Lambda^3{}_3}{\Lambda^1{}_1\Lambda^2{}_2}\epsilon'^{33}\hat{q}_3\hat{q}_3, \tag{4.2}$$

where the diagonal Jacobian elements are defined by

$$\Lambda^i_{i'} = \delta^i_{i'}\frac{h_i}{h'_i}\frac{\partial q^i}{\partial q'^i}, \tag{4.3}$$

with h_i and h'_i denoting the metric coefficients in the physical and virtual spaces, respectively. The tensors are expressed using their principal values in the matrix form (with the tilde sign omitted here for brevity). By substituting equation (4.3), the material parameters in the physical space can be directly expressed in terms of the coordinates. In particular, if the virtual space is vacuum ($\epsilon' = \mu' = 1$), the scaling map only introduces direction-dependent scaling of the material parameters.

As an important example, consider a radial scaling in cylindrical coordinates:

$$r = r(r'), \quad \theta = \theta', \quad z = z', \tag{4.4}$$

where only the radial coordinate is transformed while the angular and axial coordinates remain unchanged. Assuming vacuum in the virtual space and using the metric factors $h_r = 1$, $h_\theta = r$, $h_z = 1$, equation (4.3) gives

$$\Lambda^{r}_{r'} = \frac{\partial r}{\partial r'}, \quad \Lambda^{\theta}_{\theta'} = \frac{r}{r'}, \quad \Lambda^{z}_{z'} = 1.$$

Substituting these into equation (4.2), the tensor components of the induced medium are

$$\epsilon^{rr} = \mu^{rr} = \frac{r'}{r}\frac{\partial r}{\partial r'}, \quad \epsilon^{\theta\theta} = \mu^{\theta\theta} = \frac{r}{r'}\frac{\partial r'}{\partial r}, \quad \epsilon^{zz} = \mu^{zz} = \frac{r'}{r}\frac{\partial r'}{\partial r}. \tag{4.5}$$

4.1.2 Invisibility cloak from a scaling map

A two-dimensional (2D) cylindrical cloak can be designed by a scaling map that compresses a circular region into a single point in cylindrical coordinates. In this way, any object placed inside the circular region is effectively shrunk to zero size and produces no scattering. However, the medium outside the region is also modified by the map, which itself may introduce scattering. To achieve true invisibility, we want the induced medium outside the cloak to remain vacuum, so that electromagnetic waves pass through undisturbed.

This goal can be understood more clearly in reverse. Imagine starting with vacuum in the virtual space. We remove a point at the origin and expand it into a finite-sized hole in physical space, where an object can be placed. Since a point in virtual space has no scattering, transformation optics guarantees that the enlarged hole and any object hidden inside it will also produce no scattering. Thus, the object becomes invisible to an external observer. To make the device realizable, the mapping must be confined to a finite region, bounded by an outer radius. This motivates the design of a 2D cylindrical cloak using the following radial scaling map:

$$r = r(r') = \begin{cases} \frac{b-a}{b}r' + a & r' \leqslant b \\ r' & r' > b \end{cases}, \quad \theta = \theta', \quad z = z', \tag{4.6}$$

which transforms the disk region $r' \leqslant b$ in virtual space into a shell region $a \leqslant r \leqslant b$, while keeping the exterior $r' > b$ unchanged, as shown in figure 4.2. Any object placed inside the core $r \leqslant a$ will then be cloaked from external detection.

The material parameters induced by this mapping are obtained from equation (4.5). Inside the shell region $a \leqslant r \leqslant b$, we have $\partial r/\partial r' = (b-a)/b$. Substituting this into equation (4.5) gives

$$\epsilon^{rr} = \mu^{rr} = \frac{r'}{r}\frac{b-a}{b}, \quad \epsilon^{\theta\theta} = \mu^{\theta\theta} = \frac{r}{r'}\frac{b}{b-a}, \quad \epsilon^{zz} = \mu^{zz} = \frac{r'}{r}\frac{b}{b-a},$$

where the parameters are expressed in terms of both virtual and physical coordinates. Since the cloak is realized in physical space, we can eliminate r' in favor of r. From equation (4.6), we find $r' = b(r-a)/(b-a)$. Substituting this relation yields the final expressions for the material tensors:

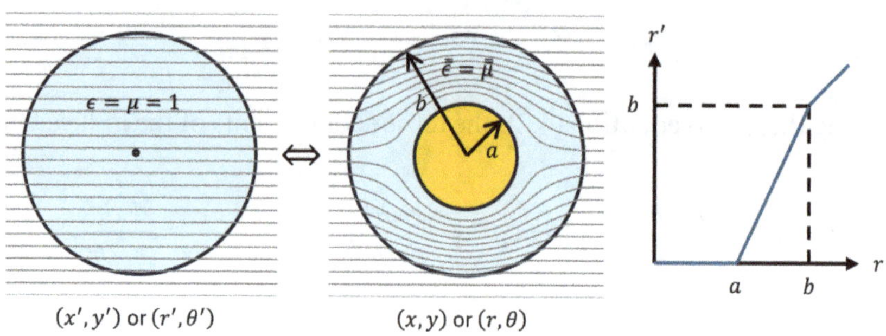

Figure 4.2. Radial map from virtual space (primed coordinates) to physical space (unprimed coordinates) used to construct a cylindrical cloak.

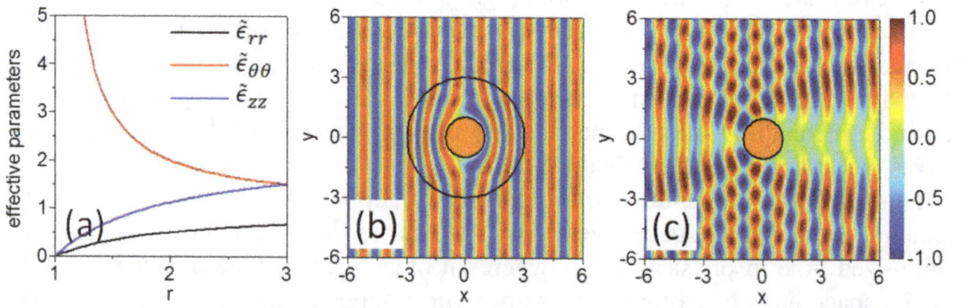

Figure 4.3. Ideal cylindrical cloak. (a) Induced effective material parameters. (b) Cloaking a PEC cylinder from a TE plane wave. (c) Control case: PEC cylinder without the cloak, showing a shadow.

$$\epsilon^{rr} = \mu^{rr} = \frac{r-a}{r}, \qquad \epsilon^{\theta\theta} = \mu^{\theta\theta} = \frac{r}{r-a}, \qquad \epsilon^{zz} = \mu^{zz} = \left(\frac{b}{b-a}\right)^2 \frac{r-a}{r}. \tag{4.7}$$

We have now derived the material parameters required for a 2D cylindrical invisibility cloak.

As an illustration, consider the case $a = 1$, $b = 3$. Figure 4.3(a) plots the induced effective medium parameters (only permittivity is shown; permeability has identical values). At the inner boundary $r = a$, the r- and z-components vanish, while the θ-component diverges. This singular behavior arises from compressing a finite-sized region into a line, but it is precisely what enables the cloak to work.

The performance of the cloak can be tested by placing a perfectly conducting (PEC) cylinder inside the cloaked region. For a TE-polarized plane wave with wavelength $\lambda = 1.2$, figure 4.3(b) shows that the fields outside the cloak are indistinguishable from those in free space, as if the PEC object were absent. In contrast, figure 4.3(c) shows the shadowing effect of the same PEC cylinder when the cloak is removed. Because the design is based on a coordinate transformation, the cloak functions for arbitrary excitations, such as point sources, not just plane waves. The cylindrical cloak has been first realized in the microwave regime and later in the optical regime (Cummer *et al* 2006; Schurig *et al* 2006; Cai *et al* 2006; Smolyaninov *et al* 2007).

4.1.3 Reduced parameters

We note that a general transformation medium obtained from a coordinate transformation requires both permittivity and permeability to be anisotropic and spatially varying. Taking the 2D cylindrical cloak as an example, the medium is impedance-matched at the outer boundary for both TE and TM modes, that is,

$$\sqrt{\frac{\tilde{\mu}_{\theta\theta}}{\tilde{\epsilon}_{zz}}}\Bigg|_{r=b} = \sqrt{\frac{\tilde{\mu}_{zz}}{\tilde{\epsilon}_{\theta\theta}}}\Bigg|_{r=b} = 1,$$

so no scattering is expected at the boundary.

In practice, however, realizing all the required tensor components is extremely challenging because they vary continuously throughout the device. To make the cloak more feasible, we can restrict ourselves to one polarization and apply a *reduced material parameter approximation*. For instance, under TE polarization (electric field along the z-direction), only the components $\tilde{\epsilon}_{zz}$, $\tilde{\mu}_{rr}$, $\tilde{\mu}_{\theta\theta}$ are relevant. We can then choose a reduced set of parameters by fixing $\tilde{\mu}_{\theta\theta} = 1$, giving

$$\tilde{\epsilon}_{zz} = \left(\frac{b}{b-a}\right)^2, \quad \tilde{\mu}_{rr} = \left(\frac{r-a}{r}\right)^2, \quad \tilde{\mu}_{\theta\theta} = 1.$$

This approximation preserves the effective indices $\sqrt{\tilde{\epsilon}_{zz}\tilde{\mu}_{rr}}$ and $\sqrt{\tilde{\epsilon}_{zz}\tilde{\mu}_{\theta\theta}}$, so the wave dispersion surfaces remain unchanged. An additional advantage is that both $\tilde{\epsilon}_{zz}$ and $\tilde{\mu}_{\theta\theta}$ are constants, and only the radial component of the permeability varies with position.

Such anisotropic permeability can be realized using metamaterials, particularly the split-ring resonator (SRR) structure. Figure 4.4 shows an implementation of a

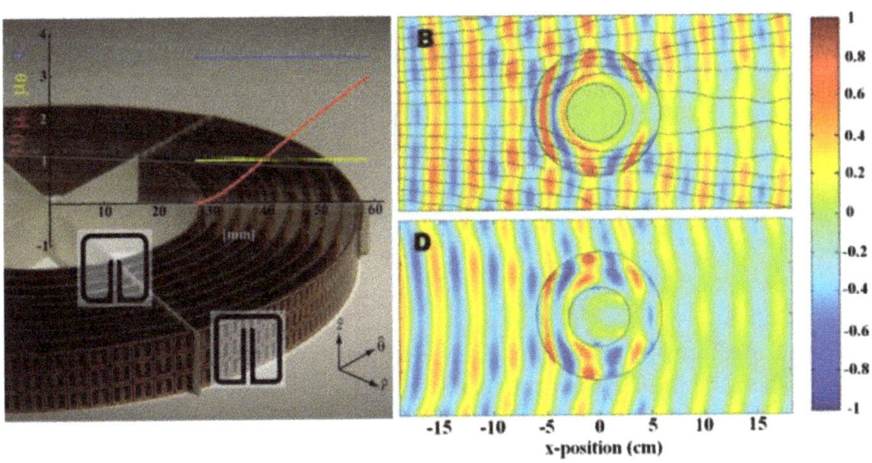

Figure 4.4. Cylindrical cloak based on reduced material parameters for TE polarization in the microwave regime. (a) SRR structure used to realize the reduced cloak. (b) Simulation results. (c) Experimental verification. Image reprinted from Schurig *et al* (2006). Reprinted with permission from AAAS.

reduced-parameter cloak in the microwave regime, including the SRR design, numerical simulation, and experimental measurement.

This reduced-parameter strategy illustrates how transformation optics can be adapted for practical realizations of cloaks, making experimental demonstrations possible even with limited metamaterial building blocks.

4.2 Conformal map with isotropic materials

Conformal maps have the advantage that the induced material parameters on the x–y plane remain isotropic and these maps are important as both an analytic tool and design tool in transformation optics (Leonhardt 2006; Leonhardt and Tyc 2009; Aubry *et al* 2010). In other words, if the medium in virtual space is vacuum, the induced material in physical space will also be isotropic, which can make realization easier.

For a 2D transformation between (x', y') and (x, y), with $z = z'$, we can define complex variables $w' = x' + iy'$ and $w = x + iy$ to describe the coordinates in virtual and physical spaces. The transformation can then be written as a conformal mapping,

$$w = f(w'), \tag{4.8}$$

provided that it satisfies the Cauchy–Riemann equations

$$\frac{\partial x}{\partial x'} = \frac{\partial y}{\partial y'}, \qquad \frac{\partial y}{\partial x'} = -\frac{\partial x}{\partial y'}.$$

A conformal map preserves the angle between two base vectors. Therefore, if the original coordinates are orthogonal, the transformed coordinates will also be orthogonal. Under such a map, the Jacobian matrix is

$$[\Lambda] = \begin{pmatrix} \frac{\partial x}{\partial x'} & \frac{\partial x}{\partial y'} & 0 \\ \frac{\partial y}{\partial x'} & \frac{\partial y}{\partial y'} & 0 \\ 0 & 0 & 1 \end{pmatrix} = \begin{pmatrix} \frac{\partial x}{\partial x'} & \frac{\partial x}{\partial y'} & 0 \\ -\frac{\partial x}{\partial y'} & \frac{\partial x}{\partial x'} & 0 \\ 0 & 0 & 1 \end{pmatrix}.$$

As a result, the determinant of the Jacobian is

$$\det[\Lambda] = \left(\frac{\partial x}{\partial x'}\right)^2 + \left(\frac{\partial x}{\partial y'}\right)^2 = \left|\frac{dw}{dw'}\right|^2.$$

Suppose the medium in virtual space is isotropic, with permittivity ϵ' and permeability μ'. Then the induced parameters in physical space are

$$\frac{\overline{\overline{\epsilon}}}{\epsilon'} = \frac{\overline{\overline{\mu}}}{\mu'} = \frac{1}{\det[\Lambda]}[\Lambda][\Lambda]^T = \begin{pmatrix} 1 & 0 & 0 \\ 0 & 1 & 0 \\ 0 & 0 & \frac{1}{\det[\Lambda]} \end{pmatrix}.$$

We see that, for the components on the x–y plane, the induced material parameters remain unchanged, while only the out-of-plane components ϵ_{zz} and μ_{zz} vary. Furthermore, for in-plane wave propagation the TE (E_z) and TM (H_z) polarizations are decoupled. For the TE mode, only a dielectric response is required. In particular, by choosing $\mu' = 1$ (non-magnetic), the physical medium needs

$$\epsilon = \frac{\epsilon'}{\det[\Lambda]}.$$

It is also convenient to describe the transformation in terms of the refractive index. If the refractive index in virtual space is $n' = \sqrt{\epsilon'}$ with unit permeability in virtual space, then the induced refractive index in physical space is

$$n = n'\left|\frac{\mathrm{d}w'}{\mathrm{d}w}\right| = \frac{n'}{\sqrt{\det[\Lambda]}}. \tag{4.9}$$

4.2.1 Example: collimator from conformal map

A useful demonstration of a conformal map is the design of a 2D directional radiator from a line source (Yao and Jiang 2011; Xu and Chen 2015). Normally, a line source produces cylindrical wavefronts that spread out in all directions. By introducing a suitable mapping, we can reshape these wavefronts into a nearly parallel beam. This can be done with a bilinear, or Möbius, transformation

$$w = \mathrm{i}\,\frac{\mathrm{i} + w'}{\mathrm{i} - w'},$$

which maps the lower semicircle region ($y' < 0, |w'| \leqslant 1$) in virtual space into the upper semicircle region in physical space. Rays that are radial in the virtual coordinates are converted into parallel rays in the transformed space, as sketched in figure 4.5(a).

Using equation (4.9), the required index distribution is

$$n = \left|\frac{\mathrm{d}w'}{\mathrm{d}w}\right| = \left|-\frac{2}{(\mathrm{i} + w)^2}\right| = \frac{2}{x^2 + (y + 1)^2},$$

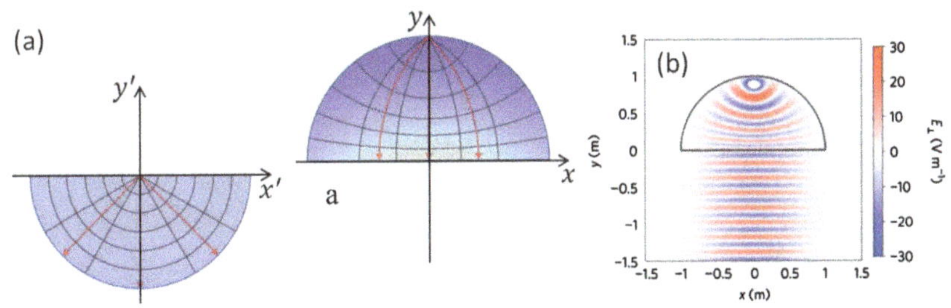

Figure 4.5. (a) Möbius transformation used to construct a directional radiator. (b) Performance of the directional radiator. A line source is placed at the top of the semicircle. Image reproduced from Xu and Chen (2015), with permission from Springer Nature.

assuming $n' = 1$ in virtual space (vacuum). This profile varies smoothly without singularities. Figure 4.5(a) uses color shading to indicate regions of higher and lower index.

The performance is illustrated in figure 4.5(b), where a line source placed at (0, 1) is transformed into a well-collimated beam propagating downward.

For comparison, classical directive devices such as parabolic reflectors or dielectric lenses also convert diverging waves into a beam, but they do so by shaping surfaces or bulk materials. Here, the same functionality is achieved by remapping the coordinates and realizing the corresponding index distribution. This example shows how conformal maps provide a systematic design rule for compact, flat optical elements.

4.2.2 Example: localize SPP from a conformal map

Conformal maps can also be used to design light-harvesting devices that remain effective over a broad frequency range (Pendry *et al* 2012). In general, broadband collection of energy is easier in systems of very large or infinite extent, which is impractical for compact devices. A conformal map allows us to fold an infinite structure into a finite one, while keeping the broadband response inherited from the virtual system.

As an illustration, consider the transformation

$$w = \frac{g^2}{w'^*},$$

where g is a constant. This map converts an infinitely extended thin metal sheet in virtual space into a finite crescent-shaped structure in physical space, as shown in figure 4.6(a). In virtual space, surface plasmons travel indefinitely along the metal sheet in the y'-direction. After mapping, these waves are bent and concentrated toward the origin in physical space. Since the infinite system supports plasmons across a wide frequency band, the crescent inherits this broadband feature.

In this case, we consider the TM mode for plasmonics with E-field on-plane and magnetic field out-of-plane. The transformed medium has the in-plane permittivity values preserved, i.e. still the same metal permittivity within crescent. The out-of-

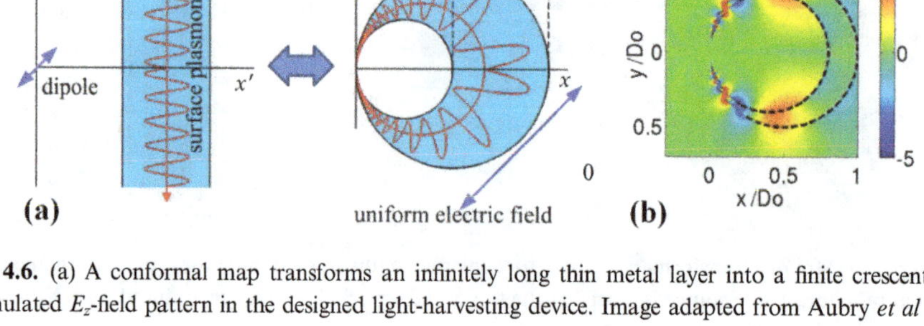

Figure 4.6. (a) A conformal map transforms an infinitely long thin metal layer into a finite crescent shape. (b) Simulated E_z-field pattern in the designed light-harvesting device. Image adapted from Aubry *et al* (2010).

plane permeability (from value one before transformation) becomes $\mu_{zz} = 1/\det[\Lambda]$. However, in practice we do not need to implement such permeability in the near-field approximation as objects are very small in the plasmonics regime, comparing to freespace wavelength. Therefore, we simply need to have the crescent shape realized using the same metal (before transformation). From a physical point of view, the crescent acts as a collector that channels wave energy into its narrow tip. Instead of spreading out along an infinite surface, the energy is naturally directed into a finite focal region, leading to strong localization of fields that is highly desirable for harvesting.

4.3 Quasiconformal map and carpet cloak

Quasiconformal maps provide an alternative way to design transformation optical devices when finding a suitable conformal map is challenging, especially for boundaries with prescribed shapes. They allow the anisotropy of the material to be kept small, nearly constant, and close to unity within the device. Moreover, by avoiding sharp corners in the boundary design, the required range of refractive indices can also be reduced. This makes quasiconformal mapping particularly useful for practical devices such as waveguide benders, couplers, and surface-conformal antennas, where low or negligible anisotropy in the index profile is desired.

4.3.1 Generating a quasiconformal map

Quantifying anisotropy

Here, we consider waves propagating in the x–y plane, while both the fields and the medium remain uniform along the z direction. We therefore restrict to a 2D coordinate transformation between two coordinate systems (x', y') (virtual space) and (x, y) (physical space), with $z' = z$. The transformation is represented by the Jacobian matrix, and a corresponding metric,

$$\Lambda = \begin{pmatrix} \partial_{x'}x & \partial_{y'}x \\ \partial_{x'}y & \partial_{y'}y \end{pmatrix}, \quad g = \Lambda^T\Lambda, \tag{4.10}$$

which describes how a small displacement in the virtual space is mapped to the physical space.

For definiteness, we take the E_z-polarization (electric field along z). Transformation optics gives the constitutive parameters in the physical space as

$$\begin{aligned} \epsilon &= \frac{\epsilon'}{\det\Lambda}, \\ \mu &= \frac{\Lambda\,\mu'\,\Lambda^T}{\det\Lambda}, \end{aligned} \tag{4.11}$$

where ϵ is the out-of-plane permittivity and μ is the in-plane 2×2 permeability tensor. For the complementary H_z-polarization, the roles of ϵ and μ are interchanged.

To quantify anisotropy, we denote the eigenvalues of μ as μ_T and μ_L (the two principal values). Let us assume isotropic and positive μ'. A convenient measure is

$$\frac{\mathrm{Tr}(\mu)}{\sqrt{\det \mu}} = \sqrt{\frac{\mu_T}{\mu_L}} + \sqrt{\frac{\mu_L}{\mu_T}} = \frac{\mathrm{Tr}(g)}{\sqrt{\det g}},$$

which equals two when μ is isotropic and grows larger as anisotropy increases. The second equality shows that this measure depends only on the metric g of the coordinate transformation if μ' is isotropic.

Minimizing anisotropy

A *conformal module M* of a quadrilateral is the aspect ratio of a rectangle to which it can be conformally mapped. Two quadrilaterals can be conformally mapped to each other if and only if they share the same conformal module. If their modules differ, a conformal map does not exist, but a *quasiconformal map* of constant dilation can still be constructed. In quasiconformal maps we use here, infinitesimal squares are mapped to rectangles of constant aspect ratio, i.e., with constant dilation.

To make this concrete, we introduce isothermal coordinates (u, v) for the quadrilateral in (x, y) and (u', v') for that in (x', y'). These coordinates satify the Cauchy–Riemann relations for conformal maps:

$$\begin{aligned} \nabla u &= -\hat{z} \times \nabla v, \\ \nabla' u' &= -\hat{z} \times \nabla' v'. \end{aligned} \tag{4.12}$$

We have chosen to write the relations in vector form which will be useful later. Such reparametrizations for the two domains are shown in figure 4.7, where u' is chosen to range from 0 on the left boundary to M' on the right boundary, and v' ranges from 0 on the bottom boundary to 1 on the top boundary. Similarly, u ranges from 0 to M and v from 0 to 1. M' and M are the conformal modules of the quadrilaterals.

With the auxiliary coordinates (u, v) and (u', v'), The transformation between the two quadrilaterals are cascaded by Λ_1: $(x', y') \to (u', v')$, Λ_2: $(u', v') \to (u, v)$ and Λ_3: $(u, v) \to (x, y)$. The first and last parts of the transformation are conformal maps while the middle part $(u', v') \to (u, v)$ is the transformation we would like to optimize to minimize anisotropy. Its Jacobian and metric is defined by

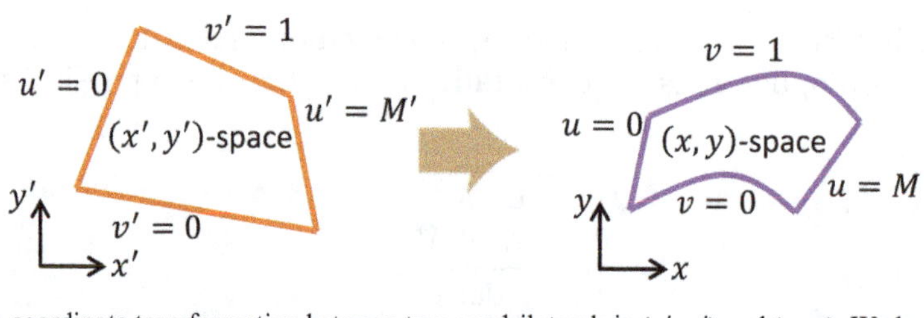

Figure 4.7. A coordinate transformation between two quadrilaterals in (x', y') and (x, y). We have introduced isothermal coordinates (u, v) and (u', v') for the two quadrilaterals for the ease to describe the transformation that maps corners to corners and edges to edges

$$\Lambda_2 = \begin{pmatrix} \partial_{u'}u & \partial_{v'}u \\ \partial_{u'}v & \partial_{v'}v \end{pmatrix}, \quad g_2 = \Lambda_2^T \Lambda_2. \tag{4.13}$$

In this case, due to the conformal map nature for the first and last parts, the anisotropy is only induced from the middle part of transformation. We can prove

$$\frac{\mathrm{Tr}(g)}{\sqrt{\det g}} = \frac{\mathrm{Tr}(g_2)}{\sqrt{\det g_2}},$$

which is what we would like to minimize. We start with the inequality in giving the lower bound to be written in terms of the integrated value of $\mathrm{Tr}(g_2)$ in the (u', v') space:

$$\max\left(\frac{\mathrm{Tr}(g_2)}{\sqrt{\det g_2}}\right) \geqslant \frac{1}{M}\int \frac{\mathrm{Tr}(g_2)}{\sqrt{\det g_2}}\, du\, dv = \frac{1}{M}\int \mathrm{Tr}(g_2)\, du'\, dv'.$$

Next, by applying the generalized mean inequalities

$$\sqrt{\frac{1}{M'}\int\left(\frac{\partial u}{\partial u'}\right)^2 du'} \geqslant \frac{1}{M'}\int \frac{\partial u}{\partial u'}\, du' = \frac{M}{M'},$$

and

$$\sqrt{\int\left(\frac{\partial v}{\partial v'}\right)^2 dv'} \geqslant \int \frac{\partial v}{\partial v'}\, dv' = 1,$$

for a sense-preserving map (i.e., a one-to one mapping that preserves orientation), we have the following inequality for $\frac{1}{M}\int \mathrm{Tr}(g_2)\, du'\, dv'$:

$$\frac{1}{M}\int\left[\left(\frac{\partial u}{\partial u'}\right)^2 + \left(\frac{\partial v}{\partial v'}\right)^2\right] du'\, dv' \geqslant \frac{1}{M}\left[M'\left(\frac{M}{M'}\right)^2 + M'\right] = \frac{M}{M'} + \frac{M'}{M}.$$

Combining the above inequalities, we arrive

$$\max\left(\frac{\mathrm{Tr}(\mu)}{\sqrt{\det \mu}}\right) \geqslant \frac{M}{M'} + \frac{M'}{M}. \tag{4.14}$$

Equality acutally holds for the optimal map

$$\frac{u}{M} = \frac{u'}{M'}, \qquad v = v', \tag{4.15}$$

which minimizes the maximum anisotropy. The optimal map in minimizing the maximum anisotropy, consists of three parts: a conformal map from (x', y') to (u', v'), a scaling map from (u', v') to (u, v), and another conformal map from (u, v) to (x, y). The scaling map is the only part that induces anisotropy, which is constant across the device. The anisotropy measure is given by the right-hand side of equation (4.14).

Such a map is called a specific kind of *quasiconformal map* that has a constant dilation, which sends tiny squares in one domain to tiny rectangles in another domain with constant aspect ratio.

Generating quasiconformal maps with Laplace's equation

To generate a quasiconformal map with constant dilation (figure 4.8), we introduce a single set of auxiliary coordinates (U, V) and relate them to the normalized coordinates in the physical and virtual domains by

$$\frac{u}{M} = \frac{u'}{M'} = \frac{U - U_1}{U_2 - U_1}, \qquad v = v' = \frac{V - V_1}{V_2 - V_1}.$$

In this way, the coordinate pair (cU, V) is conformally mapped to (x, y), and $(c'U, V)$ is conformally mapped to (x', y'), where the scaling constants

$$c = \frac{M(V_2 - V_1)}{U_2 - U_1}, \qquad c' = \frac{M'(V_2 - V_1)}{U_2 - U_1}$$

ensure the correct aspect ratios in each domain. The corresponding Cauchy–Riemann equations take the form

$$\nabla(cU) = -\hat{z} \times \nabla V, \qquad \nabla'(c'U) = -\hat{z} \times \nabla' V. \tag{4.16}$$

From equation (4.16), we have

$$\nabla^2 U = \nabla'^2 U = \nabla^2 V = \nabla'^2 V = 0, \tag{4.17}$$

so V and U satisfy Laplace's equation with boundary values prescribed on opposite edges. This shows that a quasiconformal map can be obtained by solving a Laplace equation for (U, V) with appropriate boundary conditions, either numerically or analytically.

By integrating equation (4.16) along a path AB across the domain, we obtain the ratio of conformal modules (a constant) through c'/c, defined as γ:

$$\gamma = \frac{M'}{M} = \frac{\int_A^B \mathrm{d}\mathbf{l} \cdot (\hat{z} \times \nabla U)}{\int_A^B \mathrm{d}\mathbf{l}' \cdot (\hat{z} \times \nabla' U)} = \frac{\int_A^B \mathrm{d}\mathbf{l}' \cdot (\hat{z} \times \nabla' V)}{\int_A^B \mathrm{d}\mathbf{l} \cdot (\hat{z} \times \nabla V)}. \tag{4.18}$$

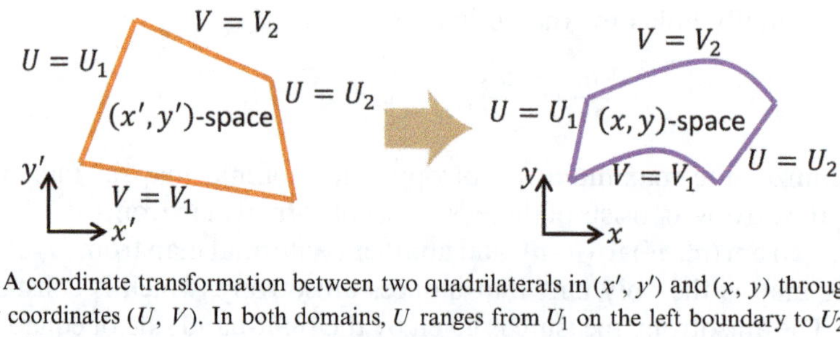

Figure 4.8. A coordinate transformation between two quadrilaterals in (x', y') and (x, y) through a single set of auxiliary coordinates (U, V). In both domains, U ranges from U_1 on the left boundary to U_2 on the right boundary, and V ranges from V_1 on the bottom boundary to V_2 on the top boundary.

By applying equation (4.11), we can obtain the constitutive parameters for E_z-polarization as

$$\epsilon = \epsilon' \frac{1}{\gamma} \frac{|\nabla U|^2}{|\nabla' U|^2} = \epsilon' \gamma \frac{|\nabla V|^2}{|\nabla' V|^2}, \quad \mu = \mu' \left(\frac{1}{\gamma} \hat{U}\hat{U} + \gamma \hat{V}\hat{V} \right), \tag{4.19}$$

where $\hat{U} = \nabla U / |\nabla U|$ and $\hat{V} = \nabla V / |\nabla V|$ are orthogonal unit vectors. The constant γ means the dilation factor of the quasiconformal map. It also gives a constant anisotropy of the permeability in the (x, y) domain.

In fact, we do not need to generate both U and V. By solving Laplace's equation for either U or V with appropriate boundary conditions, we can directly obtain the conformal module ratio γ from equation (4.18), and then calculate the constitutive parameters from equation (4.19). Through this approach, we can generate a quasiconformal map with constant dilation and obtain the corresponding material parameters for transformation optics, minimizing the induced anisotropy from the coordinate transformation.

4.3.2 Quasiconformal map devices

Example: carpet cloak from a quasiconformal map

A carpet cloak is designed to hide an object under a protrusion (or bump) on a reflective surface, making the reflection appear as if it came from a perfectly flat mirror (Li and Pendry 2008), in which the transformation medium can be obtained by the quasiconformal map using grid generation and more conveniently by solving an auxillary Laplace's equation (Chang *et al* 2009). This can be achieved using the quasiconformal mapping framework developed in the previous section.

Step 1: Define the transformation. We begin with a rectangular region of dimensions $w \times h$ (here $w = 3$, $h = 2$) in the virtual space and map it to the physical region shown in figure 4.9(a), where the bottom boundary is curved upward to cover the

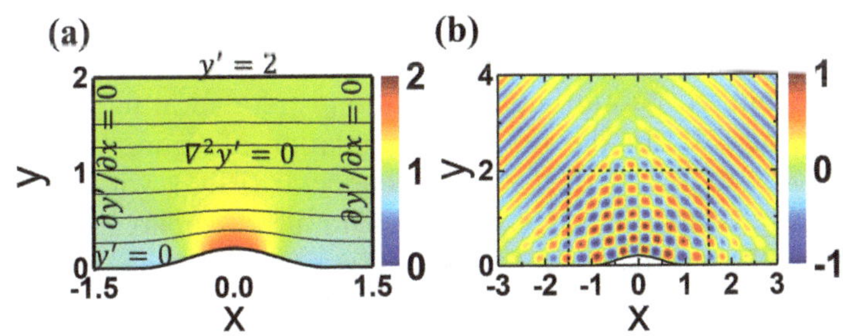

Figure 4.9. A quasiconformal carpet cloak. (a) Induced permittivity profile of the cloak. (b) Full-wave simulation of the E_z-field for an incident beam at 45° to the curved PEC surface with the cloak placed on top at a free-space wavelength of 0.4. (c) Scanning electron microscope image of the first infrared carpet cloak, fabricated from a hole lattice of varying density on a silicon-on-insulator wafer. (d) Experimental result: an infrared Gaussian beam incident on the carpet cloak produces a reflected beam profile almost identical to that from a flat surface.

hidden object. A direct analytic conformal mapping is not possible here. Instead, we use a quasiconformal map with constant dilation, obtained by solving Laplace's equation for auxiliary coordinates as described previously.

Step 2: Solve Laplace's equation. Using the intermediate coordinates $(U, V) = (x', y')$, we set $y' = 0$ on the bottom boundary and $y' = 2$ on the top boundary, while imposing Neumann boundary conditions on the left and right edges. In this electrostatic analogy, y' plays the role of a potential satisfying

$$\nabla^2 y' = 0.$$

The resulting equipotential lines of constant y' are shown in figure 4.9(a). Since we assume a quasiconformal map, solving for y' alone (or equivalently x' alone) is sufficient to determine the material profile.

We note that the original development for a quasiconformal map to minimize anisotropy has more relaxed constraint on the boundary conditions that only require matching edges to edges and corners to corners. However, for practical applications, we often want to impose simple boundary conditions, e.g., adopting Neumann boundary conditions on the left and right sides here. In many cases, this gives a good approximation to a map with very small anisotropy. The advantage of this approach is that we can hand over to numerical procedure to solve Laplace's equations in getting the map which can avoid requesting materials of large anisotropy.

Step 3: Calculate the dilation factor. The constant dilation factor γ (see equation (4.18)) can be extracted by integrating along the top boundary of the cloak:

$$\frac{1}{\gamma} = \frac{1}{w} \int \frac{\partial y'}{\partial y} \, \mathrm{d}x. \tag{4.20}$$

Step 4: Obtain material parameters. From equation (4.19), the two principal values of the permeability are $1/\gamma$ and γ, aligned with the coordinate lines. The permittivity is scalar and given by

$$\epsilon = \gamma |\nabla y'|^2.$$

Results. The induced permittivity distribution is plotted in figure 4.9(a), varying between 0.65 and 1.91. In the full-wave simulation of figure 4.9(b), the bottom boundary is modeled as a perfect electric conductor (PEC), with the cloak placed above it. For this example, $\gamma = 0.96$, very close to the isotropic case $\gamma = 1$. Thus, the anisotropy is weak, and the cloak can be well approximated by taking an isotropic permeability $\mu = 1$. An incident beam at 45° reflects as if the PEC surface were flat, canceling the scattering from the bump. A small residual aberration (phase distortion or lateral shift (Zhang *et al* 2010a)) may occur due to the isotropic approximation, but this vanishes as the cloak dimensions w, h increase.

The carpet cloak has been demonstrated in the microwave regime using metamaterials (Liu *et al* 2009) and in the optical regime using dielectric platform (Valentine *et al* 2009; Gabrielli *et al* 2009; Gharghi *et al* 2011). By revolving the 2D carpet cloak around the vertical axis, a 3D carpet cloak can be obtained (Ma and Cui 2010; Ergin *et al* 2010). It can also be extended to THz regime (Zhou *et al* 2011), and other domains such as acoustics (Popa *et al* 2011; Zigoneanu *et al* 2014).

Example: Beam collimator from quasiconformal map
In the previous section, we generated quasiconformal maps by solving Laplace's equation (equation (4.17)) for auxiliary coordinates (U, V) with appropriate boundary conditions. This corresponds to an electrostatic analogy with fixed potentials on opposite boundaries. An alternative approach is to use a *Poisson equation* with a point source. This method is particularly useful when we want to design devices where the phase fronts originate from a single point, such as collimators.

Consider transforming a circular disk of radius 1 in the virtual (primed) space into a regular pentagon in the physical (unprimed) space. The goal is that, when a point source is placed at the center, the resulting wavefronts reach the outer boundary with equal phase, so that each flat edge of the pentagon collimates the wave into a separate beam.

To construct this mapping, we solve a Poisson-like equation for the auxiliary coordinate V in both domains:

$$\begin{aligned} \nabla'^2 V &= 2\pi q' \, \delta(\mathbf{r}\,'), \\ \nabla^2 V &= 2\pi q \, \delta(\mathbf{r}), \end{aligned} \tag{4.21}$$

with the boundary condition $V = 0$ at the outer boundary. Here, δ is the Dirac delta function and q' and q are the effective source strength in the virtual and physical space.

It can be shown that in the circular virtual space, the solution is $V = \ln(r')$. Using the closed-path integration in equation (4.18), together with the material relation in equation (4.19), the resulting medium parameters are

$$\epsilon = \gamma |\nabla r'|^2, \qquad \gamma = \frac{q'}{q}. \tag{4.22}$$

The current method can be easily adapted for other shapes for collimation (Zentgraf *et al* 2010) or other applications such as focusing, radiation control Mei *et al* 2011; Yang *et al* 2011; García-Meca *et al* 2011) or on-chip wave guiding (Gabrielli *et al* 2012; Liu *et al* 2013; Wu and Werner D 2012).

The constant dilation of the quasiconformal map is determined by γ. By varying the effective charge q (by setting $q' = 1$), we can tune the anisotropy: smaller q produces stronger anisotropy, while $q = 1$ recovers the strictly conformal case with $\gamma = 1$.

Figure 4.10(a) shows the constant-r' coordinate lines obtained from numerically solving equation (4.21) for the circular-to-pentagon mapping. The resulting material profile ensures that when a point source is placed at the origin, the wavefront reaches each side of the pentagon simultaneously, just as it would in the circular virtual space. Each flat edge then collimates the beam in one direction, producing five beams in total, as confirmed by the full-wave simulation in figure 4.10(b).

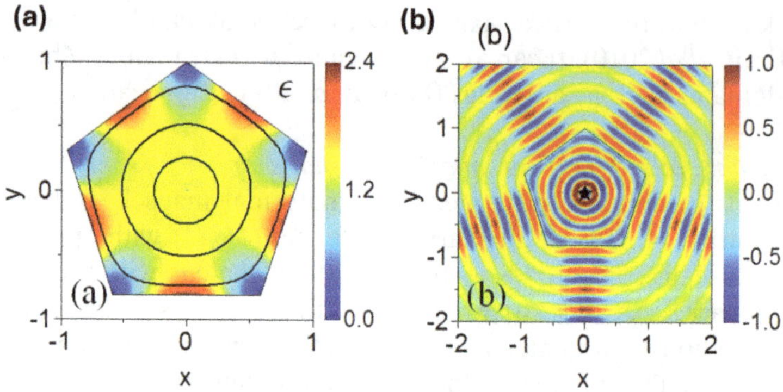

Figure 4.10. Quasiconformal TO medium obtained from solving a Poisson equation with a point source. (a) Induced permittivity profile of the cloak (color map) in the physical space. Solid lines show constant $r' = 0.3$, 0.6, and 0.9. (b) A 5-way collimator designed with this TO medium, excited by a point source at the origin at a wavelength of 0.3.

4.4 Affine map and carpet cloak

Linear maps can produce homogeneous but anisotropic media, in contrast to conformal and quasiconformal maps that generate inhomogeneous but effectively isotropic dielectric materials.

An affine transformation in the x–y plane is the simplest type of coordinate map used in transformation optics. It is described by a constant Jacobian matrix, so the resulting medium is homogeneous, though generally anisotropic. For 2D wave problems, the transformation can be expressed as

$$\begin{pmatrix} x \\ y \\ z \end{pmatrix} = \begin{pmatrix} s_{xx} & s_{xy} & 0 \\ s_{yx} & s_{yy} & 0 \\ 0 & 0 & 1 \end{pmatrix} \begin{pmatrix} x' \\ y' \\ z' \end{pmatrix} + \begin{pmatrix} c_x \\ c_y \\ 0 \end{pmatrix}, \tag{4.23}$$

where the constants s_{ij} specify stretching/tilting in the plane and c_x, c_y are shifts. Define

$$\Delta_s \equiv s_{xx}s_{yy} - s_{xy}s_{yx}.$$

With vacuum in the virtual space, the induced tensors are

$$\overline{\overline{\epsilon}} = \overline{\overline{\mu}} = \frac{1}{\Delta_s} \begin{pmatrix} s_{xx}^2 + s_{xy}^2 & s_{xx}s_{yx} + s_{xy}s_{yy} & 0 \\ s_{xx}s_{yx} + s_{xy}s_{yy} & s_{yx}^2 + s_{yy}^2 & 0 \\ 0 & 0 & 1 \end{pmatrix}.$$

For TM (H_z) polarization, this reduces to

$$\epsilon = \frac{1}{\Delta_s} \begin{pmatrix} s_{xx}^2 + s_{xy}^2 & s_{xx}s_{yx} + s_{xy}s_{yy} \\ s_{xx}s_{yx} + s_{xy}s_{yy} & s_{yx}^2 + s_{yy}^2 \end{pmatrix}, \qquad \mu = \frac{1}{\Delta_s}, \tag{4.24}$$

while for TE (E_z) polarization the roles of ϵ and μ are interchanged.

In practice, realizing magnetic response can be challenging. The reduced-parameter approximation avoids this by setting $\mu = 1$ and rescaling ϵ:

$$\epsilon = \frac{1}{\Delta_s^2}\begin{pmatrix} s_{xx}^2 + s_{xy}^2 & s_{xx}s_{yx} + s_{xy}s_{yy} \\ s_{xx}s_{yx} + s_{xy}s_{yy} & s_{yx}^2 + s_{yy}^2 \end{pmatrix}, \qquad \mu = 1. \tag{4.25}$$

This makes implementation feasible using only anisotropic dielectrics.

A carpet cloak can be designed using a linear map with only two domains of homogeneous media, which is different from the one created using a quasiconformal map. As shown in figures 4.11(a) and (b), a bilinear transformation is applied to compress the triangular region along the y'-axis into a polygonal region.

$$x = x', \quad y = \frac{h_2 - h_1}{h_2}y' + \frac{w - |x'|}{w}h_1.$$

By using equation (4.25), we obtain the permittivity and permeability profiles for the H_z-polarization with reduce parameter approximation as

$$\epsilon = \begin{pmatrix} \dfrac{h_2^2}{(h_2 - h_1)^2} & -\dfrac{h_1h_2^2/w}{(h_2 - h_1)^2}\mathrm{sgn}(x) \\ -\dfrac{h_1h_2^2/w}{(h_2 - h_1)^2}\mathrm{sgn}(x) & 1 + \dfrac{h_1^2h_2^2/w^2}{(h_2 - h_1)^2} \end{pmatrix}, \qquad \mu = 1, \tag{4.26}$$

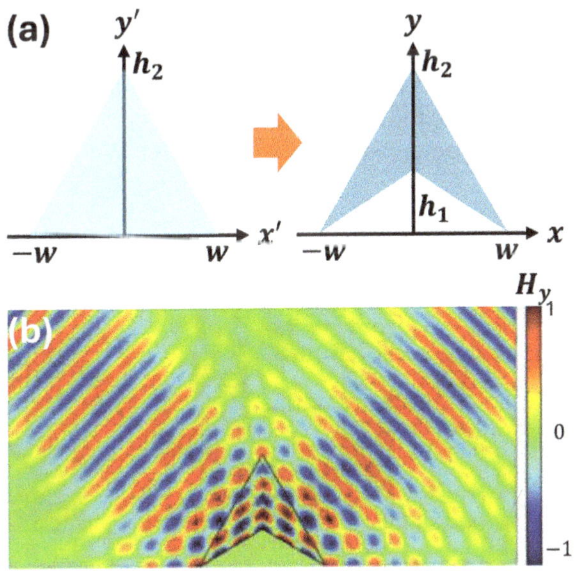

Figure 4.11. (a) The linear transformation for the design of a carpet cloak using an affine map. (b) Full-wave simulation of H_z-profiles with an incident beam at 45° to the homogeneous carpet cloak.

to realize the carpet cloak. If we change the background medium to a dielectric medium in an actual experiment, the permittivity tensor above is scaled by the background permittivity. Additionally, the reduced parameter approximation may cause a small impedance mismatch or reflection on the interface between the cloak and the background medium, but this is usually negligible. To illustrate, let's consider a transverse-magnetic (TM) polarized Gaussian beam incident obliquely on a carpet cloak with anisotropic dielectric material, placed on top of a flat surface. The cloak guides the beam around the bump, resulting in an output beam that propagates in the same way as if it were reflected from a flat surface, as shown in figures 4.11(b).

It is important to note that the permittivity tensor described in equation (4.26) can be diagonalized by rotating the optical axis. As a result, it is possible to use natural birefringent crystals or metamaterials with effective uniaxial material profiles, such as dielectric grating structures, to realize these types of cloaks (Chen *et al* 2010; Zhang *et al* 2010b, 2011a; Liang *et al* 2011; Chen and Zheng 2012). The use of dielectric grating structures offers more freedom in terms of engineering the geometrical parameters, such as the filling factor, to achieve desired anisotropy levels not typically found in natural materials (Zhang *et al* 2011a; Xu *et al* 2011). The same configuration with mirrored copy below the flat surface can be used as a unidirectional cloak for wave coming from the left (Xi *et al* 2009; Landy and Smith 2013). Similar linear maps can also be applied to control guided waves in a waveguide (Zhang *et al* 2010c; Xu *et al* 2012; Zhang *et al* 2011b) and imaging (Wee *et al* 2016).

4.4.1 Exercise: square cloak

A square cloak (He *et al* 2011) is shown in figure 4.12. It is sufficient to derive the mapping and material parameters for the right-hand slab; the other three follow by 90°, 180°, and 270° rotations about the cloak center. Let a and b denote the half-side lengths of the inner and outer squares, respectively. We assume vacuum in the virtual space.

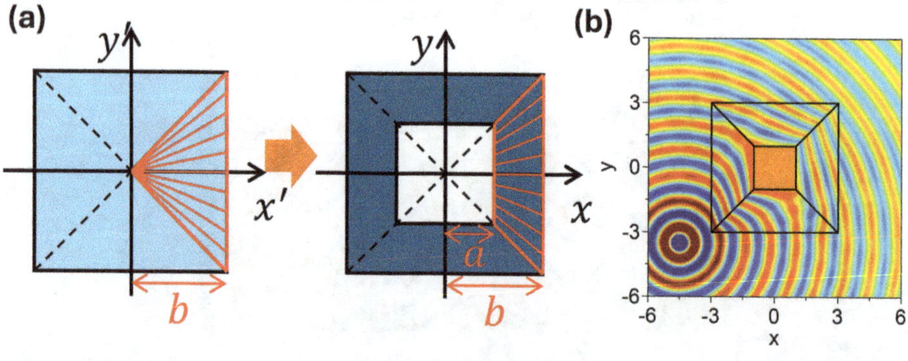

Figure 4.12. (a) The linear transformation for the design of a square cloak into four pieces using affine maps. (b) Full-wave simulation of H_z-profiles with a point source outside the square cloak.

Let $\alpha = (b - a)/b$. The map that sends the virtual triangular sector to the physical trapezoid is

$$x = a + \alpha\, x', \qquad y = \frac{x}{x'}\, y', \qquad z = z'. \tag{4.27}$$

This mapping reduces to the identity on the outer boundary ($x' = b \Rightarrow x = b,\ y = y'$), ensuring reflectionless matching there, while it expands the inner point to the square boundary ($x = a$).

The Jacobian $\Lambda = \partial(x, y, z)/\partial(x', y', z')$ expressed in physical coordinates (x, y) is

$$\Lambda = \begin{pmatrix} \alpha & 0 & 0 \\ -\dfrac{a\,\alpha}{x(x-a)}\,y & \dfrac{\alpha\, x}{x-a} & 0 \\ 0 & 0 & 1 \end{pmatrix}, \qquad \det\Lambda = \frac{\alpha^2\, x}{x-a}. \tag{4.28}$$

As the map is not exactly an affine map, only the x–x component of the Jacobian matrix is constant. With vacuum in the virtual space, the induced (non-reduced) tensors are

$$\overline{\overline{\epsilon}} = \overline{\overline{\mu}} = \frac{1}{\det\Lambda}\,\Lambda\,\Lambda^{\mathrm{T}} = \begin{pmatrix} 1 - \dfrac{a}{x} & -\dfrac{a\,y}{x^2} & 0 \\ -\dfrac{a\,y}{x^2} & \dfrac{x^4 + a^2y^2}{x^3(x-a)} & 0 \\ 0 & 0 & \left(\dfrac{b}{b-a}\right)^2 \dfrac{x-a}{x} \end{pmatrix}. \tag{4.29}$$

For practical implementation, one often uses the reduced-parameter approximation for TM (H_z), setting $\mu = 1$ and rescaling ϵ (cf equation (4.25)). Using $\epsilon = (\Lambda\Lambda^{\mathrm{T}})/(\det\Lambda)^2$, the in-plane permittivity becomes

$$\epsilon_{\mathrm{TM}} = \left(\frac{b}{b-a}\right)^2 \begin{pmatrix} \left(\dfrac{x-a}{x}\right)^2 & -\dfrac{a\,y\,(x-a)}{x^3} \\ -\dfrac{a\,y\,(x-a)}{x^3} & 1 + \left(\dfrac{a\,y}{x^2}\right)^2 \end{pmatrix}, \qquad \mu = 1. \tag{4.30}$$

(For 2D TM (H_z), only the in-plane ϵ is required; the out-of-plane ϵ_{zz} is not needed.)

The media for the top, left, and bottom slabs are obtained by rotating the right-hand result by 90°, 180°, and 270°, respectively, and expressing the tensors in the global (x, y) coordinates. A point source outside the cloak (e.g., at the lower-left region) excites a cylindrical wave that, upon transmission through the cloak, recovers the free-space cylindrical wavefront.

4.5 Internal-symmetry transformation

4.5.1 Identifying gauge fields

In transformation optics (TO), constant rotations and translations are global symmetries of Maxwell's equations in vacuum. Promoting them to local symmetries (with the Jacobian matrix varying with position) generates new terms in the constitutive tensors, leading to anisotropy with $\overline{\overline{\epsilon}} = \overline{\overline{\mu}}$. The family of anisotropic $\overline{\overline{\epsilon}} = \overline{\overline{\mu}}$ media remains in the same form under further coordinate transformations. This closure property identifies a prevalent form of TO media that is both suitable for ray tracing and wave manipulation and still manageable to analyze. In section 3.6 we saw how TO media allow ray tracing via Fermat's principle, with decoupled wave equations for the two polarizations $\mathbf{E} \pm \mathrm{i}\mathbf{H}$. Each polarization thus behaves like a scalar wave, making the analysis much simpler than in general anisotropic media. The two polarizations also share the same dispersion relation, so they follow the same ray trajectories. They also allow a geometric interpretation in terms of geodesics in a curved space equipped with a metric.

Now, for gauge transformations (internal symmetry transformations of the fields), we follow a similar logic. First, we identify global symmetry transformations that mix **E** and **H** while leaving Maxwell's equations in vacuum invariant (see section A.3). Promoting these to local symmetries (spatially varying transformations) induces new terms in the constitutive tensors, which are identified as gauge fields. Substituting these back, we find that the medium with the gauge-field terms retains its form under further application of the same transformation. Thus, just as in TO, we can then identify a closed family of media as *gauge-field media*: a natural platform for exploring new physics in wave manipulation and ray tracing, while still being manageable to analyze.

We now illustrate the above logic with two examples: the 3D duality transformation and the 2D polarization rotation.

3D duality transformation. Section A.3 shows that the only global symmetry transformation mixing **E** and **H** for waves propagating in all three dimensions is the duality transformation, which we now write as

$$\begin{bmatrix} \mathbf{E} \\ \mathbf{H} \end{bmatrix} = \begin{bmatrix} \cos(k_0\phi) & \sin(k_0\phi) \\ -\sin(k_0\phi) & \cos(k_0\phi) \end{bmatrix} \begin{bmatrix} \mathbf{E}' \\ \mathbf{H}' \end{bmatrix},$$

where ϕ is an abstract rotation angle. By promoting ϕ to a spatially varying function $\phi(\mathbf{r})$, new terms appear in the permittivity and permeability tensors. These terms can be collected into the form

$$\overline{\overline{\epsilon}} = \overline{\overline{\mu}} = \begin{pmatrix} n_1 & -\mathrm{i}A_z & \mathrm{i}A_y \\ \mathrm{i}A_z & n_2 & -\mathrm{i}A_x \\ -\mathrm{i}A_y & \mathrm{i}A_x & n_3 \end{pmatrix} = \overline{\overline{n}} + \mathrm{i}\,\mathbf{A} \times \overline{\overline{I}}.$$

When the duality transformation is applied to the macroscopic Maxwell's equations with the medium above, the medium still remains in the same form with the gauge field shifted as

$$\mathbf{A} \rightarrow \mathbf{A} + \nabla\,\phi,$$

while all the other material parameters remain the same. The diagonal elements n_1, n_2, n_3 are written down from the consideration of what kind of medium, instead of vacuum, can remain invariant under the global duality transformation. This is actually done before promoting ϕ to a spatially varying function and helps to make the medium the most general form with a set of gauge fields. The off-diagonal elements A_x, A_y, A_z are identified as gauge fields. By looking at the transformation matrix we know it is a $U(1)$ gauge transformation, characterized by a single scalar parameter ϕ.

In this case, the macroscopic Maxwell's equations decouple into two independent helicity channels:

$$(\nabla \pm \mathrm{i}k_0\mathbf{A}) \times (\mathbf{E} \pm \mathrm{i}\mathbf{H}) = \pm k_0\, \overline{\overline{n}} \cdot (\mathbf{E} \pm \mathrm{i}\mathbf{H}).$$

We have also identified the covariant derivative $\nabla \to \nabla \pm \mathrm{i}k_0\mathbf{A}$.

Although we have successfully identified one possible gauge-field medium, $\overline{\overline{\epsilon}} = \overline{\overline{\mu}}$ is not symmetric and requires non-reciprocal responses, which are difficult to achieve in practice. Therefore, in the following we will focus on the second example, which is more practical.

2D polarization rotation. If the waves travel in the x–y plane with z-invariance, section A.3 shows that the global symmetry transformation mixing **E** and **H** can be relaxed from the 3D case to

$$\begin{bmatrix} E_z \\ \mathrm{i}H_z \end{bmatrix} = u \begin{bmatrix} E_z' \\ \mathrm{i}H_z' \end{bmatrix}, \qquad \begin{bmatrix} \mathrm{i}H_T \\ E_T \end{bmatrix} = u \begin{bmatrix} \mathrm{i}H_{\mathrm{T}}' \\ E_{\mathrm{T}}' \end{bmatrix},$$

where u is any 2×2 unitary matrix. This corresponds to an $SU(2)$ gauge symmetry up to an exponential phase factor. Here E_{T} is a column vector holding E_x, E_y and H_{T} is a column vector holding H_x, H_y.

Without going into more details, promoting u to a spatially varying function generally leads to non-reciprocal responses. One simplification is to restrict u to be a real rotation matrix, effectively reducing the gauge symmetry from $SU(2)$ to $U(1)$. In this case, the medium induced from the gauge transformation remains reciprocal (Liu and Li 2015). The gauge transformation can be written as

$$\begin{bmatrix} E_z \\ \mathrm{i}H_z \end{bmatrix} = \begin{bmatrix} \cos(k_0\phi) & \sin(k_0\phi) \\ -\sin(k_0\phi) & \cos(k_0\phi) \end{bmatrix} \begin{bmatrix} E_z' \\ \mathrm{i}H_z' \end{bmatrix},$$

$$\begin{bmatrix} \mathrm{i}\mathbf{H}_{\mathrm{T}} \\ \mathbf{E}_{\mathrm{T}} \end{bmatrix} = \begin{bmatrix} \cos(k_0\phi) & \sin(k_0\phi) \\ -\sin(k_0\phi) & \cos(k_0\phi) \end{bmatrix} \begin{bmatrix} i\mathbf{H}_{\mathrm{T}}' \\ \mathbf{E}_{\mathrm{T}}' \end{bmatrix},$$

where ϕ is an abstract rotation angle.

By promoting ϕ to a spatially varying function $\phi(x, y)$, new terms appear in the permittivity and permeability tensors:

$$\overline{\overline{\epsilon}} = \begin{bmatrix} \epsilon_{xx} & \epsilon_{xy} & A_y \\ \epsilon_{xy} & \epsilon_{yy} & -A_x \\ A_y & -A_x & \epsilon_{zz} \end{bmatrix}, \qquad \overline{\overline{\mu}} = \begin{bmatrix} \epsilon_{xx} & \epsilon_{xy} & -A_y \\ \epsilon_{xy} & \epsilon_{yy} & A_x \\ -A_y & A_x & \epsilon_{zz} \end{bmatrix}. \tag{4.31}$$

An application of the transformation shifts $\mathbf{A} \to \mathbf{A} + \nabla\,\phi$, so this gauge-field medium is also closed under repeated transformations. We call this the *tilted-anisotropy medium*, since the principal axis of the structures producing these tensors is tilted with respect to the z-axis. The other permittivity and permeability elements are chosen by asking what kind of medium, instead of vacuum, can remain invariant under the global symmetry transformation. It turns out that we only need these values to be the same in $\overline{\overline{\epsilon}}$ and $\overline{\overline{\mu}}$, with symmetric in-plane tensors. The off-diagonal elements A_x, A_y are identified as gauge fields.

In the tilted medium, the natural decoupled variables are

$$\psi_{\pm} = E_z \pm H_z.$$

Eliminating the in-plane fields gives

$$(\nabla \pm \mathrm{i}k_0\mathbf{A}) \cdot m^{-1} \cdot (\nabla \pm \mathrm{i}k_0\mathbf{A})\,\psi_{\pm} + k_0^2\,\epsilon_{zz}\,\psi_{\pm} = 0, \tag{4.32}$$

with

$$m = \begin{pmatrix} \epsilon_{yy} & -\,\epsilon_{xy} \\ -\,\epsilon_{xy} & \epsilon_{xx} \end{pmatrix}.$$

Each pseudo-spin $\psi_{\pm}$ thus satisfies its own scalar wave equation with a covariant derivative $\nabla \to \nabla \pm \mathrm{i}k_0\mathbf{A}$.

Figure 4.13(a) shows a sample block with one slice of structures on the x–z plane tiled by 45 degrees from the principal axes, it is designed to have a dielectric tensor:

$$\overline{\overline{\epsilon}} = \begin{bmatrix} n_0^2 & 0 & 2n_0A_y \\ 0 & n_0^2 & 0 \\ 2n_0A_y & 0 & n_0^2 + 4A_y^2 \end{bmatrix}, \tag{4.33}$$

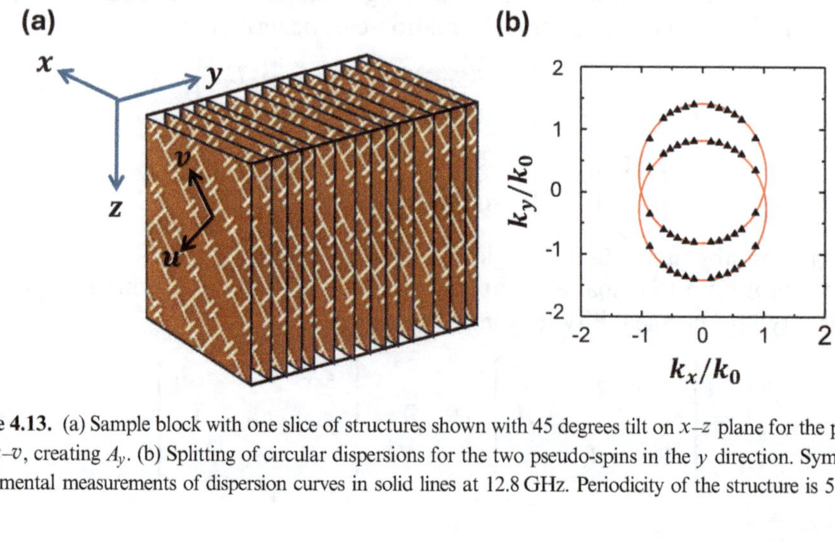

Figure 4.13. (a) Sample block with one slice of structures shown with 45 degrees tilt on x–z plane for the principal axes u–v, creating A_y. (b) Splitting of circular dispersions for the two pseudo-spins in the y direction. Symbols are experimental measurements of dispersion curves in solid lines at 12.8 GHz. Periodicity of the structure is 5 mm.

with $\mu = 1$. The medium represents a reduced parameter approximation for the gauge-field medium in equation (4.31), in the sense that it also creates the same gauge field A_y in splitting two circular dispersions in the y-direction (Liu *et al* 2019). Figure 4.13(b) shows the splitting of circular dispersions for the two pseudo-spins. The symbols are experimental measurements of dispersion curves in solid lines.

4.5.2 Pseudo-magnetic field for photon

We begin with a simple reciprocal gauge-field medium,

$$\bar{\bar{\epsilon}} = \begin{bmatrix} n & 0 & A_y \\ 0 & n & -A_x \\ A_y & -A_x & n \end{bmatrix}, \qquad \bar{\bar{\mu}} = \begin{bmatrix} n & 0 & -A_y \\ 0 & n & A_x \\ -A_y & A_x & n \end{bmatrix}. \tag{4.34}$$

For this isotropic choice ($\epsilon_{xx} = \epsilon_{yy} = \epsilon_{zz} = n$ and $\epsilon_{xy} = 0$), the 2×2 matrix in equation (4.32) reduces to $m = n\,\mathbb{I}_2$, and the decoupled pseudo-spins $\psi_\pm = E_z \pm H_z$ satisfy

$$\frac{1}{n}\,(\nabla \pm ik_0\mathbf{A}) \cdot (\nabla \pm ik_0\mathbf{A})\,\psi_\pm + k_0^2\, n\, \psi_\pm = 0. \tag{4.35}$$

From equation (4.35) we obtain, under the eikonal approximation, the dispersion relation

$$(\mathbf{k} \mp k_0\mathbf{A}(\mathbf{r}))^2 = (k_0 n)^2, \tag{4.36}$$

which is simply the minimal substitution $\mathbf{k} \to \mathbf{k} \mp k_0\mathbf{A}$. Applying Hamilton's equations (3.43) gives the ray law

$$\frac{\mathrm{d}}{\mathrm{d}s}\left(n\,\frac{\mathrm{d}\mathbf{r}}{\mathrm{d}s}\right) = \nabla n \mp \hat{s} \times \mathbf{B}_{\text{eff}}, \qquad \mathbf{B}_{\text{eff}} = \nabla \times \mathbf{A}, \tag{4.37}$$

where $\hat{s}$ is the unit tangent to the ray. Compared with the TO case, the new term is exactly a pseudo-Lorentz force acting on the two pseudo-spins with opposite signs.

In a uniform medium ($n =$ const.) and a uniform pseudo-magnetic field $\mathbf{B}_{\text{eff}} = B_0\hat{z}$, equation (4.37) integrates to circular orbits of radius

$$R = \frac{n}{B_0}, \tag{4.38}$$

with opposite senses of rotation for ψ_+ and ψ_-.

As a specific realization, we choose a Coulomb-gauge vector potential corresponding to a uniform field inside a disk:

$$\mathbf{A} = \begin{cases} \frac{1}{2}B_0\, r\, \hat{\phi}, & r \leqslant r_0, \\ \frac{1}{2}B_0\, \frac{r_0^2}{r}\, \hat{\phi}, & r > r_0, \end{cases}$$

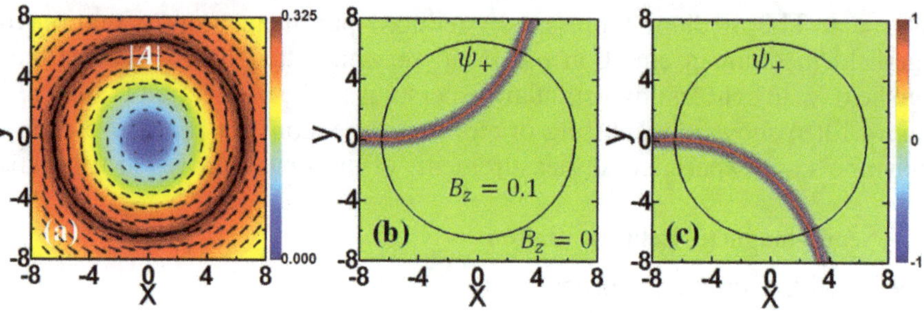

Figure 4.14. (a) Profile of the vector potential $|\mathbf{A}|$ and direction of $\mathbf{A}$. (b,c) Cyclotron bending of $\psi_\pm$ at a shorter wavelength $\lambda_0 = 0.15$, confirming the frequency independence of the trajectory. Reprinted figure with permission from Liu and Li (2014), Copyright (2014) by the American Physical Society.

where $\hat{\phi}$ is the azimuthal unit vector. Full-wave simulations (COMSOL) of a Gaussian beam confirm the predicted behavior: ψ_+ bends anti-clockwise, ψ_- bends clockwise, both following circular arcs of radius $R = n/B_0$ (figure 4.14). Upon exiting the disk, the beams resume straight-line propagation.

The bending radius $R = n/B_0$ is frequency independent as long as the medium parameters are frequency non-dispersive. In our example the maximum $|\mathbf{A}|$ is only 0.325, ensuring broadband validity. The same cyclotron trajectories are observed at $\lambda_0 = 0.15$ (figure 4.14).

This construction directly associates the vector potential $\mathbf{A}$ with reciprocal anisotropic parameters, giving a strong and geometrically transparent pseudo-magnetic effect. Alternative approaches, such as generating $\mathbf{A}$ from derivatives of bianisotropic or gyromagnetic parameters, typically rely on semiclassical corrections and produce weaker deflections.

4.6 Spacetime transformation

Transformation techniques can be extended beyond spatial transformations to include the temporal domain, providing the capability to manipulate temporal phenomena. For example, by transforming both space and time simultaneously, an *event cloak* can be designed to make a space–time event undetectable.

Transformation rules

To establish transformation optics under spacetime coordinate transformations, it is more convenient to use the 4D covariant formulation of Maxwell's equations. Within a source-free region, we have

$$\frac{\partial}{\partial x^\alpha}\left(\Omega \mathcal{D}^{\alpha\beta}\right) = 0, \qquad \frac{\partial}{\partial x^\alpha}\left([\alpha\beta\gamma\delta]\ \mathcal{F}_{\gamma\delta}\right) = 0, \tag{4.39}$$

with the field tensors in matrix form:

$$[\Omega\mathcal{D}^{\alpha\beta}] = \begin{pmatrix} 0 & -\Omega D^1 & -\Omega D^2 & -\Omega D^3 \\ \Omega D^1 & 0 & -H_3 & H_2 \\ \Omega D^2 & H_3 & 0 & -H_1 \\ \Omega D^3 & -H_2 & H_1 & 0 \end{pmatrix},$$
$$\left[\mathcal{F}_{\alpha\beta}\right] = \begin{pmatrix} 0 & E_1 & E_2 & E_3 \\ -E_1 & 0 & -\Omega B^3 & \Omega B^2 \\ -E_2 & \Omega B^3 & 0 & -\Omega B^1 \\ -E_3 & -\Omega B^2 & \Omega B^1 & 0 \end{pmatrix}. \tag{4.40}$$

The constitutive relation can be formally written as

$$\mathcal{D}^{\alpha\beta} = \mathcal{K}^{\alpha\beta\gamma\delta}\mathcal{F}_{\gamma\delta}. \tag{4.41}$$

Here $[\alpha\beta\gamma\delta]$ denotes the 4D Levi–Civita symbol, which evaluates to the sign of a valid permutation and 0 otherwise. The coordinates (x^0, x^1, x^2, x^3) correspond to (ct, x, y, z). We have adopted Heaviside–Lorentz units. Greek indices label 4D quantities, and Latin indices label 3D quantities. In the above form, we have a metric-induced parameter (Jacobian determinant)

$$\Omega = \sqrt{-\det[g_{\alpha\beta}]}$$

appearing in various expressions. In Minkowski spacetime, where $\left[g_{\alpha\beta}\right] = \mathrm{diag}(1, -1, -1, -1)$, we have $\Omega = 1$, so such a multiplicative factor is only needed when we deviate from Minkowski coordinates. More details are given in appendix A.5 for the 4D covariant formulation of Maxwell's equations, definitions of the Levi-Civita symbol/tensor and the relationship between the constitutive tensor in equation (4.41) to the more conventional usage of permittivity, permeability and bianisotropy tensors..

A direct consequence of using the covariant formulation is that we can transform the various tensors to a new set of general coordinates (ct', x', y', z') using the Jacobian of the transformation (target coordinates with respect to original coordinates):

$$\Lambda_{\alpha}^{\alpha'} = \frac{\partial x^{\alpha'}}{\partial x^{\alpha}}.$$

The metric transforms as

$$g^{\alpha'\beta'} = \Lambda^{\alpha'}{}_{\alpha}\,\Lambda^{\beta'}{}_{\beta}\,g^{\alpha\beta}. \tag{4.42}$$

The field tensors transform as

$$\begin{aligned} \Omega'\mathcal{D}^{\alpha'\beta'} &= \frac{1}{\det[\Lambda]}\Lambda^{\alpha'}{}_{\alpha}\,\Lambda^{\beta'}{}_{\beta}\,\Omega\mathcal{D}^{\alpha\beta}, \\ \mathcal{F}_{\alpha\beta} &= \mathcal{F}_{\alpha'\beta'}\,\Lambda^{\alpha'}{}_{\alpha}\,\Lambda^{\beta'}{}_{\beta}, \end{aligned} \tag{4.43}$$

and the constitutive tensor transforms as

$$\Omega'\mathcal{K}^{\alpha'\beta'\gamma'\delta'} = \frac{1}{\det[\Lambda]} \Lambda^{\alpha'}{}_{\alpha} \Lambda^{\beta'}{}_{\beta} \Lambda^{\gamma'}{}_{\gamma} \Lambda^{\delta'}{}_{\delta} \Omega \mathcal{K}^{\alpha\beta\gamma\delta}, \tag{4.44}$$

where matrix $[\Lambda]$ stores the components $\Lambda^{\alpha'}{}_{\alpha}$, and we have used

$$\frac{\Omega'}{\Omega} = \frac{\sqrt{-\det[g^{\alpha\beta}]}}{\sqrt{-\det[g^{\alpha'\beta'}]}} = \frac{1}{\det[\Lambda]}.$$

We choose to include the multiplicative factor Ω on $\mathcal{D}^{\alpha\beta}$, D^i, and B^i, and $\Omega' = \sqrt{-\det[g_{\alpha'\beta'}]}$ on $\mathcal{D}^{\alpha'\beta'}$, $D^{i'}$, and $B^{i'}$ in writing the transformation rules. Defining

$$\mathcal{D}'^{\alpha'\beta'} = \Omega'\mathcal{D}^{\alpha'\beta'}, \quad \mathcal{F}'_{\alpha'\beta'} = \mathcal{F}_{\alpha'\beta'}, \quad \mathcal{K}'^{\alpha'\beta'\gamma'\delta'} = \Omega'\mathcal{K}^{\alpha'\beta'\gamma'\delta'},$$

and

$$D'^{i'} = \Omega' D^{i'}, \quad B'^{i'} = \Omega' B^{i'}, \quad E'_{i'} = E_{i'}, \quad H'_{i'} = H_{i'},$$

we finally obtain

$$\begin{gathered} \frac{\partial}{\partial x^{\alpha'}}\mathcal{D}'^{\alpha'\beta'} = 0, \qquad \frac{\partial}{\partial x^{\alpha'}}([\alpha'\beta'\gamma'\delta']\mathcal{F}'_{\gamma'\delta'}) = 0, \\ \mathcal{D}'^{\alpha'\beta'} = \mathcal{K}'^{\alpha'\beta'\gamma'\delta'}\mathcal{F}'_{\gamma'\delta'}. \end{gathered} \tag{4.45}$$

These equations take the same form as Maxwell's equations in Minkowski spacetime. Hence, transforming from Minkowski spacetime ($\Omega = 1$) to a general coordinate system leads to an equivalent description where $\mathcal{D}'^{\alpha'\beta'}$, $\mathcal{F}'_{\alpha'\beta'}$, and $\mathcal{K}'^{\alpha'\beta'\gamma'\delta'}$ are interpreted as the displacement field tensor, electromagnetic field tensor, and constitutive tensor in the transformed coordinates. This constitutes the rule of transformation optics under spacetime coordinate transformations. Although its four-dimensional form is compact, it is not convenient for direct operational comparison. Nevertheless, it provides the essential foundation for deriving simplified transformation rules.

The 3+1 form

To connect back to our normal usage of the different fields and tensors in space and time coordinates, the components of the Jacobian matrix $[\Lambda]$ (storing $\Lambda^{\alpha'}{}_{\alpha}$, transformed coordinates with respect to the original coordinates) can be written as

$$[\Lambda^{\alpha'}{}_{\alpha}] = \begin{pmatrix} \gamma & -\gamma[w]^{\mathrm{T}} \\ -\gamma[v] & \mathrm{Sp}[\Lambda] \end{pmatrix}, \tag{4.46}$$

where $\mathrm{Sp}[\Lambda]$ extracts the spatial block from a 4×4 matrix $[\Lambda]$ to a 3×3 matrix by omitting the elements with index 0. We also note that the γ here is overloaded to represent the Lorentz factor (a scalar) instead of a Greek index in 4D. w and v are

3-by-1 column matrices. The overbar on $\overline{\text{Sp}}$ indicates a complementary operation defined by

$$\overline{\text{Sp}}[\Lambda] = \left(\text{Sp}([\Lambda]^{-1})\right)^{-1}. \tag{4.47}$$

With the definitions of various components of the Jacobian matrix in equation (4.46), equation (4.43) can be rewritten. The electromagnetic and auxiliary fields can be proved to transform according to

$$\begin{aligned} \begin{pmatrix}[E]\\ [\Omega B]\end{pmatrix} &= [L]^{\text{T}}\begin{pmatrix}[E']\\ [\Omega' B']\end{pmatrix}, \\ \begin{pmatrix}-[\Omega' D']\\ [H']\end{pmatrix} &= \frac{1}{\det[\Lambda]}[L]\begin{pmatrix}-[\Omega D]\\ [H]\end{pmatrix}, \end{aligned} \tag{4.48}$$

where $[L]$ is a 6×6 matrix given by

$$[L] = \begin{pmatrix}\gamma\ \overline{\text{Sp}}[\Lambda] & -\gamma\ \text{Sp}[\Lambda]\ [w\times]\\ \gamma[v\times]\ \overline{\text{Sp}}[\Lambda] & \det(\text{Sp}[\Lambda])\ \overline{\text{Sp}}[\Lambda]^{\text{T}}\end{pmatrix}, \tag{4.49}$$

and $[w\times]$ abbreviates the cross-product operation $(\hat{x}w_1 + \hat{y}w_2 + \hat{z}w_3)\times$ as

$$[w\times] = \begin{pmatrix}0 & -w_3 & w_2\\ w_3 & 0 & -w_1\\ -w_2 & w_1 & 0\end{pmatrix}.$$

On the other hand, the constitutive relation can be written using a 6×6 matrix $[K]$ (instead of the fourth-rank tensor $K^{\alpha\beta\gamma\delta}$) as

$$\begin{pmatrix}-[\Omega D]\\ [H]\end{pmatrix} = [K]\begin{pmatrix}[E]\\ [\Omega B]\end{pmatrix}, \quad \begin{pmatrix}-[\Omega' D']\\ [H']\end{pmatrix} = [K']\begin{pmatrix}[E']\\ [\Omega' B']\end{pmatrix}. \tag{4.50}$$

According to the transformation rules for the electromagnetic and auxiliary fields, the constitutive matrix transforms as

$$[K'] = \frac{1}{\det[\Lambda]}[L][K][L]^{\text{T}}. \tag{4.51}$$

This form is convenient for practical use. However, later we will seek further simplification of this transformation rule for specific types of media.

Lorentz transformation. Let us now consider the Lorentz transformation—a common spacetime transformation—with velocity $\mathbf{v} = \hat{x}v_1 + \hat{y}v_2 + \hat{z}v_3$ (velocity normalized to the speed of light c):

$$\begin{aligned} &\gamma = \frac{1}{\sqrt{1 - v^2}}, \qquad [w] = [v], \\ &\text{Sp}[\Lambda] = [I] + \frac{\gamma^2}{1+\gamma}[v][v]^{\text{T}}, \qquad \overline{\text{Sp}}[\Lambda] = [I] - \frac{\gamma}{1+\gamma}[v][v]^{\text{T}}. \end{aligned} \tag{4.52}$$

In this case,

$$\det[\Lambda] = 1, \qquad \det(\mathrm{Sp}[\Lambda]) = \gamma.$$

The matrix $[L]$ becomes

$$[L] = \begin{pmatrix} \gamma\left([I] - \dfrac{\gamma}{1+\gamma}[v][v]^{\mathrm{T}}\right) & -\gamma[v\times] \\ \gamma[v\times] & \gamma\left([I] - \dfrac{\gamma}{1+\gamma}[v][v]^{\mathrm{T}}\right) \end{pmatrix}. \tag{4.53}$$

The metric is invariant under this transformation, i.e.

$$[g_{\alpha\beta}] = [g_{\alpha'\beta'}] = \mathrm{diag}(1, -1, -1, -1), \qquad \Omega = \Omega' = 1.$$

The field transformation rules simplify to

$$\begin{aligned} [D'] &= \gamma\left([I] - \frac{\gamma}{1+\gamma}[v][v]^{\mathrm{T}}\right)[D] + \gamma[v\times][H], \\ [H'] &= \gamma\left([I] - \frac{\gamma}{1+\gamma}[v][v]^{\mathrm{T}}\right)[H] - \gamma[v\times][D], \\ [E] &= \gamma\left([I] - \frac{\gamma}{1+\gamma}[v][v]^{\mathrm{T}}\right)[E'] - \gamma[v\times][B'], \\ [B] &= \gamma\left([I] - \frac{\gamma}{1+\gamma}[v][v]^{\mathrm{T}}\right)[B'] + \gamma[v\times][E']. \end{aligned} \tag{4.54}$$

These are the expected results from special relativity: the longitudinal field components (along $\mathbf{v}$) remain invariant, while the transverse components are mixed through cross-products between velocity and the counterpart field, with an overall scaling by γ. From the field transformation (or equivalently from the transformation rule for $[K]$), it is straightforward to see that the vacuum is invariant under Lorentz transformation.

4D Transformation optical medium

The transformation rule of constitutive matrices has been established for a general medium, which we now specify through the following constitutive relation:

$$\begin{pmatrix} [D] \\ [B] \end{pmatrix} = \begin{pmatrix} [\varepsilon] & [\xi] \\ [\zeta] & [\mu] \end{pmatrix} \begin{pmatrix} [E] \\ [H] \end{pmatrix}, \tag{4.55}$$

where all matrices are expressed in Cartesian components. In the same spirit as the setup of transformation media for purely spatial transformations, we do not need to consider the most general case. Instead, we ask: what is the simplest type of medium that can be regarded as being transformed from free space? We define such media as transformation media.

In the present case of spacetime transformation, free space can be regarded as a medium invariant under a Lorentz transformation. Such a medium, although not the most general form, can be chosen as

$$[\varepsilon]\eta = [\mu]/\eta = [I], \qquad [\xi] = [\zeta] = [0]. \tag{4.56}$$

It still has a unit refractive index but we have relaxed slightly from the vacuum condition. Here, η represents the intrinsic impedance of free space, which need not be unity as in vacuum. Applying the transformation rule equation (4.51) to this free-space medium, it is straightforward to show that the transformed medium satisfies

$$[\varepsilon]\eta = [\mu]/\eta, \qquad [\varepsilon] = [\varepsilon]^{\mathrm{T}}, \qquad [\mu] = [\mu]^{\mathrm{T}}, \tag{4.57}$$

and

$$[\xi] = -[\zeta], \qquad [\xi] = -[\xi]^{\mathrm{T}}, \qquad [\zeta] = -[\zeta]^{\mathrm{T}}. \tag{4.58}$$

We define such a medium as a *4D transformation optical (TO) medium*. Compared with a 3D TO medium, it includes an additional antisymmetric bianisotropy tensor satisfying $[\xi] = -[\zeta]$. Therefore, a spacetime transformation preserves the symmetry and antisymmetry properties of the four constitutive matrices, and it can be shown that the intrinsic impedance η is likewise conserved, so that

$$[\varepsilon'] = [\mu'] \quad \text{if} \quad [\varepsilon] = [\mu],$$

a result consistent with what we already know from 3D TO media.

Equivalent curved free space of a 4D TO medium

The antisymmetric nature of the bianisotropic tensor of a TO medium allows us to define a vector

$$\boldsymbol{\alpha} = \hat{x}\,\alpha_1 + \hat{y}\,\alpha_2 + \hat{z}\,\alpha_3,$$

or equivalently a column matrix $[\alpha]$ storing the components, such that

$$[\xi] = -[\zeta] = [\alpha\times], \tag{4.59}$$

where $[\alpha\times]$ represents the 3×3 antisymmetric matrix corresponding to the cross product with $\boldsymbol{\alpha}$. A very useful representation of the same medium is an equivalent curved free-space metric $[\mathcal{G}]$ (the matrix form storing the covariant components $\mathcal{G}_{\alpha\beta}$), chosen as

$$[\mathcal{G}] = \frac{1}{\eta^2\sqrt{\det[\varepsilon]}}\begin{pmatrix} 1 & [\alpha]^{\mathrm{T}} \\ [\alpha] & -\eta^2(\det[\varepsilon])\,[\varepsilon]^{-1} - [\alpha][\alpha]^{\mathrm{T}} \end{pmatrix}. \tag{4.60}$$

A curved free space with such a metric, together with an intrinsic impedance η (a scalar), satisfies

$$K^{\alpha\beta\gamma\delta} = \frac{1}{\eta}\,\mathcal{G}^{\alpha\beta}\mathcal{G}^{\gamma\delta}. \tag{4.61}$$

We use $\mathcal{G}$ (instead of g) here to stress the constitutive $\mathcal{K}$ tensor can be written in the form of equation (4.61). By converting this into the 3+1 form, we can obtain the 6×6 constitutive matrix $[K]$ that relates $(-[\Omega D], [H])$ to $([E], [\Omega B])$, and subsequently recover the constitutive matrices $[\varepsilon]$, $[\mu]$, $[\xi]$, and $[\zeta]$. These matrices coincide with those originally defined for the 4D TO medium, thereby proving their equivalence. In fact, the equivalent curved free-space metric $[\mathcal{G}]$ can be defined up to an arbitrary scalar multiplication factor without changing the corresponding constitutive matrices or the tensor form of Maxwell's equations (relevant for field computation or ray tracing). Without loss of generality, we adopt the convention

$$\det[\mathcal{G}] = -\frac{1}{\eta^2}, \tag{4.62}$$

which fixes the overall normalization of $[\mathcal{G}]$. In this way, we may treat $[\mathcal{G}]$ as an equivalent representation of the 4D TO medium. The TO medium possesses six degrees of freedom (DOFs) from the symmetric permittivity or permeability tensors, one DOF from the intrinsic impedance, and three DOFs from the antisymmetric bianisotropy tensor, matching the ten DOFs of a symmetric 4×4 metric $[\mathcal{G}]$.

Conversely, given a metric matrix $[\mathcal{G}]$, the constitutive matrices can be retrieved as

$$\begin{aligned} &[\varepsilon]\eta = [\mu]/\eta = -\frac{\sqrt{-\det[\mathcal{G}]}\ \mathrm{Sp}\big([\mathcal{G}]^{-1}\big)}{\mathcal{G}_{00}}, \\ &[\xi] = -[\zeta] = [\alpha\times], \qquad \alpha_i = \frac{\mathcal{G}_{i0}}{\mathcal{G}_{00}}, \end{aligned} \tag{4.63}$$

where $1/\eta = \sqrt{-\det[\mathcal{G}]}$. The right-hand sides of the formulas are expressed such that they remain invariant under any overall multiplicative scaling of $[\mathcal{G}]$. When $\eta = 1$, this reduces to the well-known formula for free curved space with impedance matched to vacuum (Leonhardt and Philbin, 2006). Here, we have generalized the expression to include an arbitrary intrinsic impedance η, thereby defining such a 'generalized free space' using equation (4.61). Similar to the case of 3D TO media, this allows ε and μ to differ by a scaling factor, which is equivalent to changing the intrinsic impedance η here.

Suppose now that $[\alpha]$ vanishes. Then the equivalent metric $[\mathcal{G}]$ has only nonzero components $\mathcal{G}_{00}$ and $\mathrm{Sp}[\mathcal{G}]$, while all off-diagonal elements are zero. Dividing $\mathcal{G}_{00}$ from $\mathrm{Sp}[\mathcal{G}]$ yields

$$-\frac{\mathrm{Sp}[\mathcal{G}]}{\mathcal{G}_{00}} = (\det[\eta\varepsilon])\,[\eta\varepsilon]^{-1}. \tag{4.64}$$

which reduces to the equivalent metric of a 3D transformation-optics medium, equation (3.53), when η approaches one to have equal permittivity and permeability.

4D transformation rule in terms of the equivalent curved-space metric

The advantage of using the $[\mathcal{G}]$ representation for a 4D TO medium is that it satisfies a tensor transformation rule for a given coordinate transformation with Jacobian matrix $[\Lambda]$ (storing the components $\Lambda^{\alpha'}{}_{\alpha}$). However, since $[\mathcal{G}]$ is unique only up to a multiplicative scalar factor, it is unavoidable that $[\mathcal{G}]$ and $[\mathcal{G}']$ (the equivalent metric of the transformed medium) can only be related by the conventional tensor transformation rule up to a multiplicative factor.

For our chosen convention in defining $[\mathcal{G}]$, we have

$$[\mathcal{G}']^{-1} = \frac{1}{\sqrt{\det[\Lambda]}} [\Lambda] [\mathcal{G}]^{-1} [\Lambda]^{\mathrm{T}}, \tag{4.65}$$

so that the metric determinant, or equivalently the intrinsic impedance, is conserved upon transformation. In deriving this rule, we have assumed $\mathcal{G}_{00} > 0$ and $\det\left(\mathrm{Sp}([\mathcal{G}]^{-1})\right) < 0$, with similar assumptions for $[\mathcal{G}']$. Otherwise, an additional minus sign must be included. From the resulting $[\mathcal{G}']$, one can directly obtain the corresponding constitutive matrices $[\varepsilon']$, $[\mu']$, $[\xi']$, and $[\zeta']$. Finally, by considering the case of vanishing $[\alpha]$ in $[\mathcal{G}]$ together with a space-only transformation, it is straightforward to verify that the above rule reduces to the familiar transformation rules for 3D transformation-optics media.

4.6.1 Event cloak

We now apply the established transformation rules, equations (4.43) and (4.45), to design an *event cloak*, or *space–time cloak*, which hides an event by engineering a spatiotemporal profile of material parameters. Figure 4.15 illustrates such a space-time transformation, mapping the virtual space (t', x') domain to the physical space (t, x) domain.

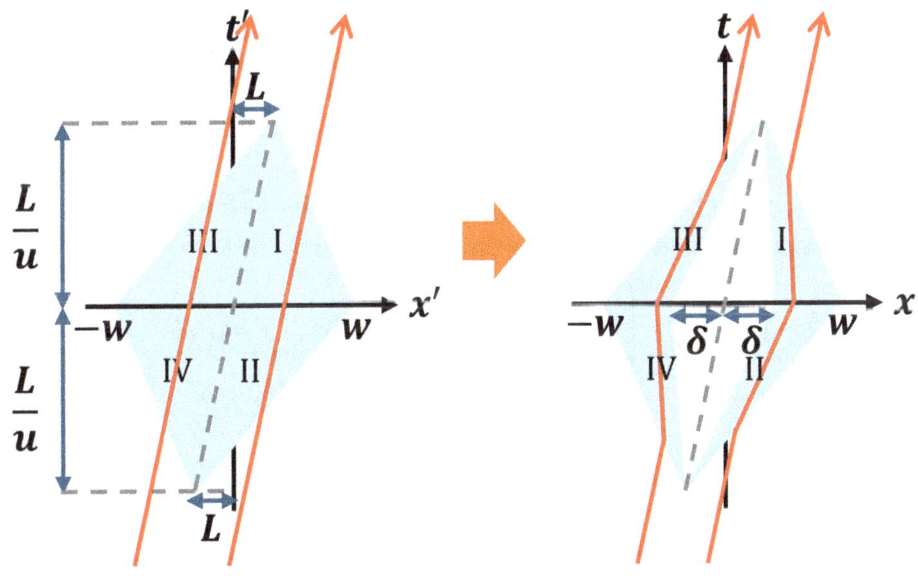

Figure 4.15. Event cloak from affine maps. The left panel shows the virtual space (t', x') and the right panel the physical space (t, x). A blue rhombic region of height $2L/u$ and width $2w$ is divided into four triangles (I–IV), each undergoing an affine transformation that keeps the outer boundaries fixed while shifting one vertex of each region by a distance δ, creating a spacetime void. Light rays (at constant velocity u) are bent accordingly but emerge at the same space–time coordinates, leaving the central event hidden.

Suppose the entire virtual space is filled with a homogeneous dielectric having a constant light speed u, smaller than the vacuum speed of light c. The blue-colored region in the figure represents the portion of the virtual space that is transformed into the corresponding blue region—the cloak—in the physical space. In this case, the transformation opens a spacetime void around the origin in the physical domain. The four outer boundaries of the cloak are designed to remain invariant under the transformation, similar to the design principle of a spatial cloak. Light traveling at a constant velocity u in the virtual domain, along trajectories such as $x' = ut' + \text{const.}$ (e.g., the two red paths in the figure), is transformed into the physical space such that the light still propagates entirely within the blue regions. It enters and exits the cloak at the same spatial and temporal coordinates as before, thus following its original trajectory outside the cloak. The spacetime void created in this manner is therefore never intersected by any light rays, and any event occurring within this void remains undetectable to an external observer. The concept of the event cloak was first proposed by McCall (McCall *et al* 2011) and later experimentally demonstrated in an optical-fiber-based system that cloaked an event in time (Fridman *et al* 2011).

For simplicity, we divide the blue rhombic area into four triangular regions, labeled I to IV, and apply four different affine mappings to each of them, as shown in figure 4.15. These mappings are defined in the (t', x') domain (the virtual space), with the corresponding transformed triangular regions shown in blue and labeled identically in the (t, x) domain (the physical space). Since the mappings are affine in nature, the transformed material parameters in the four regions are expected to be constant. To derive these parameters from transformation optics, let $2w$ and $2L/u$ denote the width (in x') and height (in t'), respectively, of the entire rhombic region. The center of the rhombus is set at the origin, while its top vertex is located at $(t', x') = (L/u, L)$. In the physical domain, the spacetime void created by the transformation has a width of 2δ (with $\delta < w$) and the same height $2L/u$.

Taking region I as an example, we seek an affine mapping $(t', x') \to (t, x)$ satisfying

$$(0, 0) \mapsto (0, \delta), \qquad (0, w) \mapsto (0, w), \qquad (L/u, L) \mapsto (L/u, L).$$

This mapping, by requiring it as a linear map, is found to be

$$\begin{pmatrix} ct \\ x - \delta \end{pmatrix} = [\Lambda]\begin{pmatrix} ct' \\ x' \end{pmatrix} = \begin{pmatrix} 1 & 0 \\ \dfrac{(L-w)\delta}{Lwn} & \dfrac{w-\delta}{w} \end{pmatrix}\begin{pmatrix} ct' \\ x' \end{pmatrix}, \qquad y = y', \qquad z = z', \tag{4.66}$$

with

$$\det[\Lambda] = \frac{w-\delta}{w}, \qquad n = \frac{c}{u}, \qquad \eta = \frac{1}{n}.$$

This transformation preserves the boundaries interfacing the cloak and the background, while the lower-left corner of the region is shifted in the positive x direction by δ. Note that in this section we have swapped the role of the primed and unprimed coordinates compared to the previous section, so that the virtual space is now represented by the primed coordinates.

The virtual space (t', x', y', z') has

$$[\varepsilon'] = n^2[\mu'] = n^2[I], \qquad [\mathcal{G}']^{-1} = \begin{pmatrix} n & 0 \\ 0 & -n^{-1}[I] \end{pmatrix}. \tag{4.67}$$

The scaling of metric is chosen to satisfy $\det[\mathcal{G}'] = -1/\eta^2 = -n^2$. Applying the transformation rule for the equivalent metric, equation (4.65) with swapped primed and unprimed quantities, i.e.,

$$[\mathcal{G}]^{-1} = \frac{1}{\sqrt{\det[\Lambda]}}[\Lambda]\,[\mathcal{G}']^{-1}\,[\Lambda]^{\mathrm{T}},$$

we have the equivalent metric in the physical space (t, x, y, z) satisfying

$$\begin{aligned}
\mathrm{Sp}([\mathcal{G}]^{-1}) &= -\frac{\sqrt{w}}{n\sqrt{w-\delta}}\,\mathrm{diag}\\
&\quad \left(\frac{L-\delta}{L}\,\frac{L(w-\delta)+(w-L)\delta}{Lw},\ 1,\ 1\right),\\
\mathcal{G}_{00} &= -\frac{\det\left(\mathrm{Sp}([\mathcal{G}]^{-1})\right)}{\eta^2} = \frac{\sqrt{w}}{n\sqrt{w-\delta}}\,\frac{L-\delta}{L}\left(1+\frac{w-L}{w-\delta}\frac{\delta}{L}\right),\\
\mathcal{G}_{00}[\alpha] &= -\frac{\sqrt{w}}{\sqrt{w-\delta}}\,\frac{w-L}{w-\delta}\,\frac{\delta}{L}\begin{pmatrix}1\\0\\0\end{pmatrix}.
\end{aligned} \tag{4.68}$$

Therefore, the transformed constitutive parameters in the physical domain are

$$\begin{aligned}
\eta^2[\varepsilon] &= [\mu] = \frac{w-\delta}{w}\,\mathrm{diag}\left(1,\ \frac{L^2 w}{(L-\delta)\,[L(w-2\delta)+w\delta]},\right.\\
&\qquad \left.\frac{L^2 w}{(L-\delta)\,[L(w-2\delta)+w\delta]}\right),\\
[\xi] &= -[\zeta] = [\alpha\times].
\end{aligned} \tag{4.69}$$

The other three constant media corresponding to regions II to IV can be derived in a similar manner.

4.6.2 Frequency shifter

A spacetime transformation can be used to design time-varying media with novel functionalities. An example is a frequency shifter, a device that shifts the frequency of an electromagnetic wave as it propagates through the medium (Cummer and Thompson 2011). Since frequency is tied to the temporal coordinate, compressing or stretching time leads to an increase or decrease of frequency, respectively.

For simplicity, consider electromagnetic waves propagating in the x-direction, with the virtual space being vacuum. We aim to vary the frequency linearly within $0 \leqslant x < d$. This is achieved by applying the following space–time transformation:

$$
\begin{aligned}
ct &= \begin{cases} ct', & x' < 0, \\ ct' + \dfrac{x'}{d}(s-1)ct', & 0 \leqslant x' < d, \\ sct', & x' \geqslant d, \end{cases} \\
x &= x', \qquad y = y', \qquad z = z',
\end{aligned}
\tag{4.70}
$$

where the constant s denotes the final time compression/stretch ratio. As we shall see, s also determines the rate of frequency change.

Within the region $0 \leqslant x < d$, the Jacobian matrix $[\Lambda]$ (from (ct', x', y', z') to (ct, x, y, z)) is

$$
[\Lambda] = \begin{pmatrix} \Lambda_{00} & \Lambda_{01} & 0 & 0 \\ 0 & 1 & 0 & 0 \\ 0 & 0 & 1 & 0 \\ 0 & 0 & 0 & 1 \end{pmatrix}, \quad \Lambda_{00} = 1 + \frac{x(s-1)}{d}, \quad \Lambda_{01} = \frac{(s-1)\,ct}{d + (s-1)x}.
$$

Using equation (4.65) (swapping primed and unprimed coordinates), the equivalent metric in the physical space is

$$
\begin{aligned}
[\mathcal{G}]^{-1} &= \frac{1}{\sqrt{\Lambda_{00}}} \begin{pmatrix} \Lambda_{00}^2 - \Lambda_{01}^2 & -\Lambda_{01} & 0 & 0 \\ -\Lambda_{01} & -1 & 0 & 0 \\ 0 & 0 & -1 & 0 \\ 0 & 0 & 0 & -1 \end{pmatrix}, \\
[\mathcal{G}] &= \sqrt{\Lambda_{00}} \begin{pmatrix} \Lambda_{00}^{-2} & -\Lambda_{01}\Lambda_{00}^{-2} & 0 & 0 \\ -\Lambda_{01}\Lambda_{00}^{-2} & -1 + \Lambda_{01}^2\Lambda_{00}^{-2} & 0 & 0 \\ 0 & 0 & -1 & 0 \\ 0 & 0 & 0 & -1 \end{pmatrix},
\end{aligned}
\tag{4.71}
$$

where

$$
\mathcal{G}_{00} = \sqrt{\Lambda_{00}}\,\Lambda_{00}^{-2}, \qquad -\det[\mathcal{G}] = \frac{1}{\eta^2} = 1.
$$

From the equivalent metric, the transformed constitutive parameters are obtained as

$$
\begin{aligned}
[\varepsilon] &= [\mu] = -\frac{\mathrm{Sp}\left([\mathcal{G}]^{-1}\right)}{\mathcal{G}_{00}\eta} = \Lambda_{00}[I], \\
[\xi] &= -[\zeta] = [\alpha\times], \qquad [\alpha] = -\,\Lambda_{01}\begin{pmatrix} 1 \\ 0 \\ 0 \end{pmatrix}.
\end{aligned}
\tag{4.72}
$$

The permittivity and permeability tensors are isotropic and time-independent but vary linearly with position, while the electromagnetic coupling term is both space- and time-dependent. For $x' \geqslant d$, the induced parameters are $[\varepsilon] = [\mu] = s[I]$, with no bianisotropy.

In this medium, the frequency varies along x within $0 \leqslant x < d$. Substituting the material parameters in equation (4.72) into Maxwell's equations for a plane wave (E_y, H_z polarization) with propagation factor $e^{i(kx-\omega t)}$ gives

$$\omega(x) = \frac{ck}{\Lambda_{00}}.$$

Thus, the frequency inside the medium is inversely proportional to the local time compression/stretch. At $x = 0$, $\omega = ck$, while at $x = d$, $\omega = ck/s$. Hence, the factor s directly specifies the frequency shift: a time compression ($s < 1$) increases frequency, whereas a stretch ($s > 1$) decreases it. Since the induced materials are frequency-independent—a general property of transformation optics—the device performs broadband frequency shifting with the same factor across all frequencies.

4.7 Problems

Problem 4.1 A biaxial medium in virtual space has permittivity

$$[\epsilon'] = \text{diag}\left\{\epsilon'_x, \epsilon'_y, \epsilon'_z\right\}$$

in its principal axes (Cartesian coordinates). Prove that after applying a scaling map

$$x = ax', \quad y = by', \quad z = cz',$$

the induced permittivity in the physical space is still biaxial with the same principal axes, but with rescaled values

$$\epsilon_x = \frac{a}{bc}\epsilon_x', \quad \epsilon_y = \frac{b}{ac}\epsilon_y', \quad \epsilon_z = \frac{c}{ab}\epsilon_z'.$$

Problem 4.2 Show that the identity map

$$q^1 = q'^1, \quad q^2 = q'^2, \quad q^3 = q'^3,$$

where q^i and q'^i refer to the same coordinate system, leaves the material parameters unchanged. That is, verify that

$$[\epsilon'] = \epsilon'^{11}\hat{q}'_1\hat{q}'_1 + \epsilon'^{22}\hat{q}'_2\hat{q}'_2 + \epsilon'^{33}\hat{q}'_3\hat{q}'_3$$

in the virtual space is identical to the permittivity in the physical space.

Problem 4.3 A 3D invisibility cloak is designed by compressing an object into a point along the r-direction using the map

$$r = r(r'), \quad \theta = \theta', \quad \phi = \phi'.$$

(a) Derive the permittivity assuming a vacuum virtual space:

$$\epsilon = \hat{r}\hat{r}\frac{r'^2}{r^2}\frac{dr}{dr'} + \hat{\theta}\hat{\theta}\frac{dr'}{dr} + \hat{z}\hat{z}\frac{dr'}{dr}$$

(b) Alternatively, use conservation of the D- and E-fields, assuming the permittivity tensor ϵ has principal axes in the radial, angular, and z-directions. Show that this gives the same result without explicitly using the transformation rule.

Problem 4.4 A 2D invisibility cloak can also be designed in elliptic cylindrical coordinates (ξ, η, z), related to Cartesian coordinates (x, y, z) by

$$x = p \cosh \xi \cos \eta, \quad y = p \sinh \xi \sin \eta,$$

where $(\pm p, 0)$ are the foci of the ellipse, as shown in the figure (Jiang *et al* 2008).

(a) Write down the coordinate map that linearly scales ξ while keeping η and z fixed, to construct a cloak in the region $\xi_1 \leqslant \xi \leqslant \xi_2$.

(b) Find the Jacobian matrix and the induced material parameters, assuming air in virtual space.

(c) Compare these material parameters with those of the cylindrical cloak. Discuss any advantages of using an elliptic map.

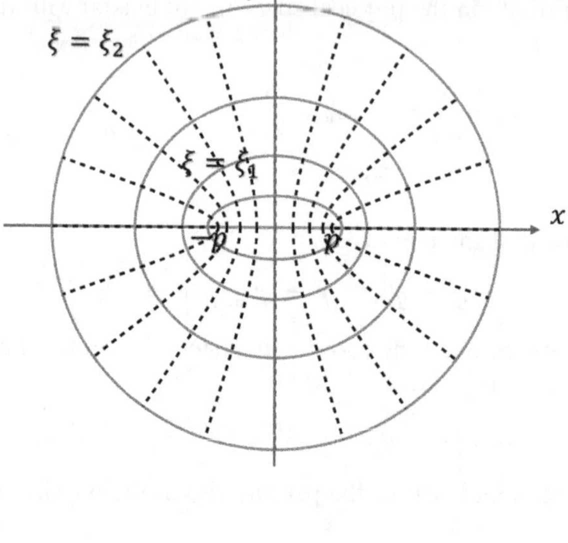

Problem 4.5 A scaling map can be used to design a field concentrator (Rahm *et al* 2008b). This device compresses the region $r' \leqslant a'$ in virtual space into a smaller region $r \leqslant a$ in physical space, thereby concentrating the fields inside $r \leqslant a$. The mapping is

$$r' = r'(r) = \begin{cases} a'r/a & r \leqslant a \\ a' + (b - a')\dfrac{r - a}{b - a} & a < r \leqslant b, \\ r & r > b \end{cases} \quad \theta' = \theta, \quad z' = z.$$

Here, the scaling differs in the core and shell regions.

(a) Derive the induced material parameters for both the core and the shell.

(b) Plot the material parameters as functions of r for the case $a = 1$, $a' = 2$, $b = 3$.

(c) This device enhances sensing by concentrating more energy in the core. For the parameters in (b), calculate how much more energy is confined in the core compared to the case without the concentrator.

Problem 4.6 A hyperlens can transmit near fields into the far field while magnifying subwavelength information (Fang *et al* 2005; Li *et al* 2009a; Kildishev and Shalaev 2007). It enables far-field imaging beyond the diffraction limit. A hyperlens can be designed using coordinate transformations (Kildishev and Narimanov 2007). In cylindrical coordinates, one example stretches a thin ring $a < r' < b'$ near the sources in virtual space into a much thicker ring $a < r < b$ in physical space, as illustrated in the figure. The fields at the boundary $r = b$ then correspond to the near-field distribution at $r' = b'$ in virtual space and can be used for far-field imaging. Write down the mapping and the corresponding induced material parameters.

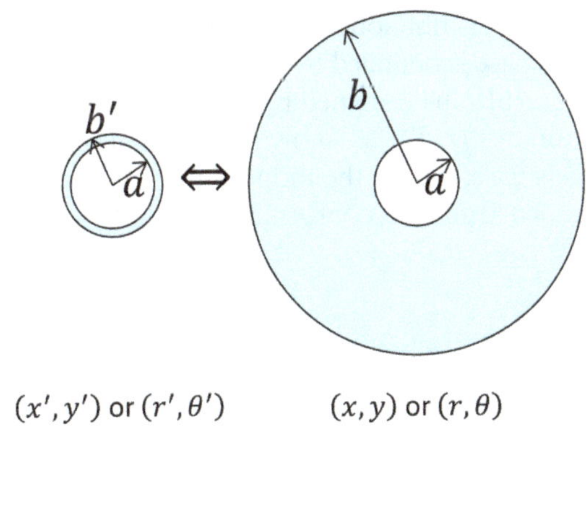

Problem 4.7 A 2D cylindrical cloak can also be designed using a quadratic map (Cai *et al* 2007)

$$r' = r'(r) = \frac{b}{2a}(2a - b + \sqrt{b^2 - 4a(b - r)}), \qquad 0 < r' \leqslant b.$$

(a) Derive the induced material parameters in physical space and plot them as functions of r.

(b) Apply the reduced material parameter approximation for the TM mode (where only anisotropic permittivity is required), and plot the reduced material parameters as functions of r.

Problem 4.8 Show that the following maps are conformal (i.e., they satisfy the Cauchy–Riemann equations). Then, assuming the refractive index in virtual space is n', find the induced refractive index n in physical space.

(a) $w = w'^3 - 1$;

(b) $w = w'^2 + w'$;

(c) $w = w' + \frac{g^2}{w'}$ (Zhukowski map), where g is a constant.

Problem 4.9 Prove that a conformal map preserves isotropy on the x–y plane.

Hint: Show that the principal values of the permittivity (or permeability) tensor on the x–y plane remain unchanged under a conformal transformation.

Problem 4.10 A waveguide bend can be designed using the conformal map (Yao and Jiang 2011; Landy and Padilla 2009)

$$w = e^{w'/A},$$

where the constant A determines the bending angle and size in physical space.

(a) Determine the transformed shape in physical space for a rectangular region in virtual space bounded by $0 \leqslant x' \leqslant 1$ and $0 \leqslant y' \leqslant 3$, using $A = 1$.

(b) What refractive index distribution is needed to realize this waveguide bend? Assume the waveguide in virtual space is bounded by two vertical PEC plates and filled with air.

(c) The same bend can also be designed by mapping Cartesian coordinates (x', y') into cylindrical coordinates (r, θ), as shown in the figure below. Write down the coordinate map for this case, derive the induced material parameters, and compare them with those obtained from the conformal map.

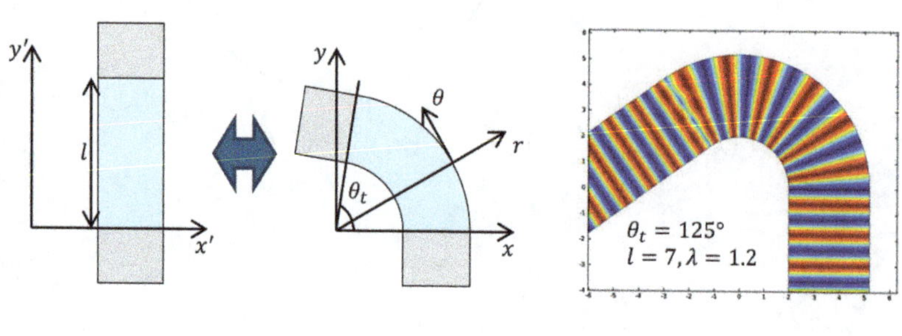

Problem 4.11 A unidirectional cloak can be designed using an analytic quasiconformal map

$$x' = \frac{b^2(r^2 + a^2)}{(b^2 + a^2)r^2}x, \quad y' = \frac{b^2(r^2 - a^2)}{(b^2 - a^2)r^2}y,$$

where $r^2 = x^2 + y^2$. It transforms a disk of radius b in the primed space to a circular ring of $a \leqslant r \leqslant b$, the cloak, in the unprimed space. An identity map is used outside the cloak for $r \geqslant b$.

(a) Calculate the constant dilation factor γ.

(b) By taking $(U, V) = (x', y')$, write down the induced permittivity.

Problem 4.12 A beam shifter

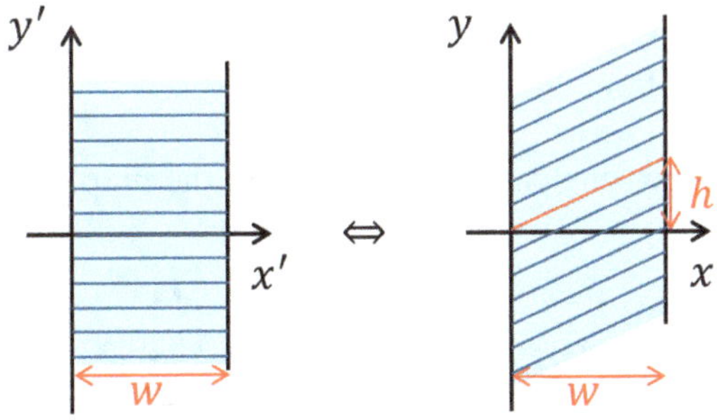

The linear map can be used to design a beam shifter (Rahm *et al* 2008a) using

$$x = x', \quad y = y' + x'h/w, \quad z = z',$$

where w and h are constants indicating the vertical shift, as shown in the figure. Prove that the transformed medium is an anisotropic dielectric with an in-plane tensor

$$\overline{\overline{\epsilon}} = \begin{pmatrix} 1 & h/w \\ h/w & 1 + (h/w)^2 \end{pmatrix}.$$

Draw the trajectory of a beam incident horizontally onto the device in the physical space (unprimed coordinates).

Problem 4.13 Design the beam shifter independently for TM and TE modes to obtain a TE/TM splitter. What are the material parameters for a splitter that deflects TE upward by 30° and TM downward by 45°? (Use the reduced forms.)

Problem 4.14 A planar waveguide adapter compresses/expands energy from a waveguide of width d_1 into another of width d_2. The design uses two linear maps (two colored regions).

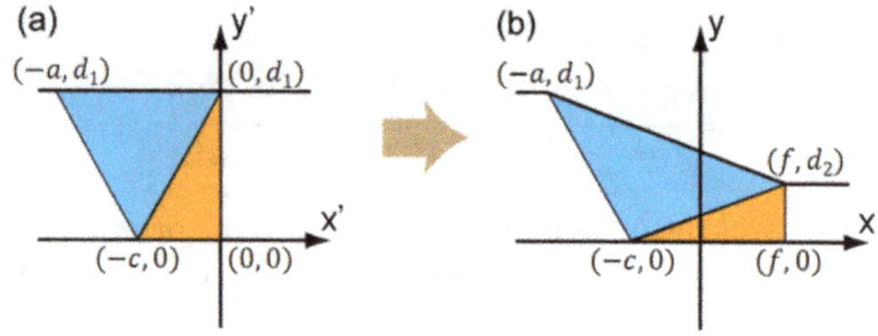

(a) Write down the linear maps for the upper (blue) and lower (orange) pieces.

(b) Derive the material parameters needed for the adapter.

(c) What are the permittivities along the principal axes if we adopt the reduced-parameter approximation?

(d) The non-magnetic condition can also be obtained directly from the mapping. What requirement on the mapping parameters c and f ensures $\mu = 1$ without using the reduced approximation?

Problem 4.15 Super-focusing can be designed via the map shown in the figure.

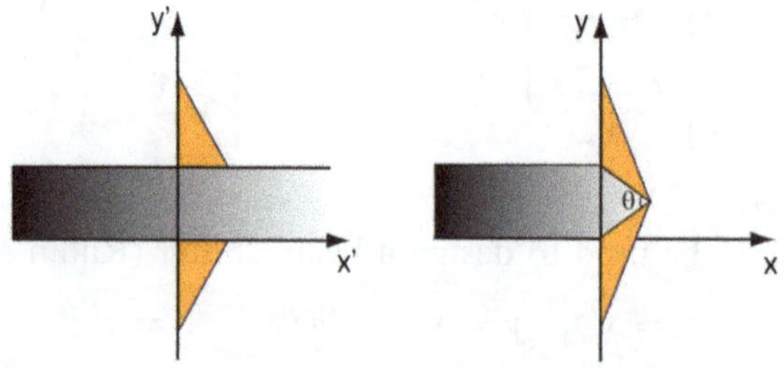

(a) Write down the linear maps for the orange regions.

(b) Derive the material parameters needed for this device.

Problem 4.16 A field rotator can be designed from a linear map, as shown in the figure below.

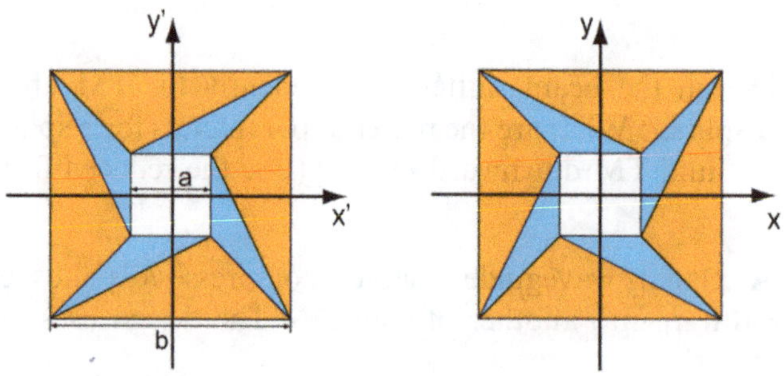

(a) Write down the linear maps for the blue, orange, and white (center) regions.

(b) Derive the material parameters for the three regions.

(c) Why is the center region still air?

(d) What do the scattered fields look like if there is a scatterer inside the white region?

Problem 4.17 Beam-width expander and/or shrinker (Liang *et al* 2012). A beam-width expander can be constructed by the map shown in the figure below, where a vertical slab undergoes an x-dependent expansion in y.

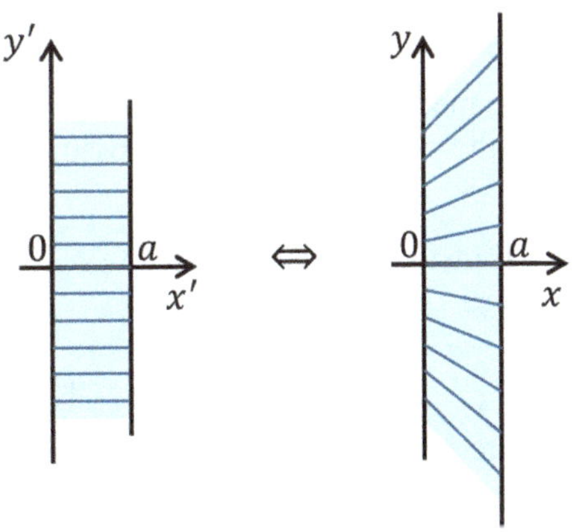

(a) Find the Jacobian matrix and the induced material parameters if the virtual space is air.

(b) This gives another waveguide adapter. Compare it with the linear-map adapter in terms of advantages and disadvantages.

Problem 4.18 The linear transformation can be further applied to design an omnidirectional cloak by expanding a line segment into a cloaked region through stretching or compressing the space in two orthogonal directions, as illustrated in the figure below (Li *et al* 2009b). The cloak shown is achieved by applying two different linear transformations. The gray regions are obtained by applying a space compression along the $+y'$- and $-y'$-directions, while the pink regions are more complicated: the spaces are expanded along the y'-axis and then compressed along the x'-axis.

(a) Write down the linear maps for the upper right pieces of both gray and pink color. Expand them for all the pieces.

(b) Derive the material parameters needed for the cloak.

(c) What is the permittivity in the principal directions if we adopt the reduced material parameter approximation?

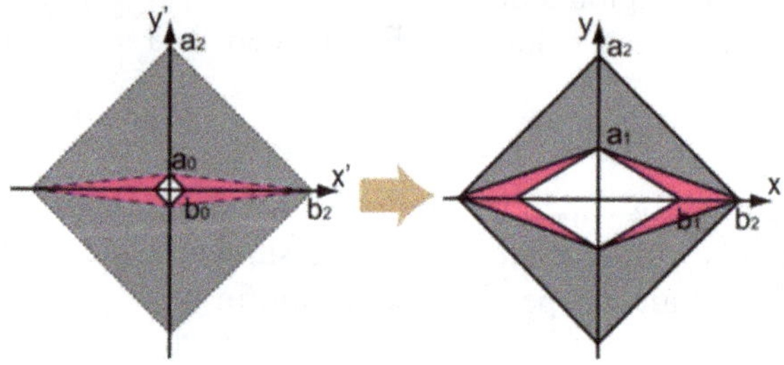

Problem 4.19 Show that the medium in equation (4.33) indeed creates a splitting for two circular dispersions in the y-direction with two pseduo spins decoupled.

Problem 4.20 Space-time transformation

(a) For a mixed space-time transformation

$$ct = s \quad ct'x', \qquad x = x', \qquad y = px' + qy', \qquad z = z',$$

prove that the induced material has permittivity and permeability tensors as

$$[\epsilon] = [\mu] = \frac{sx}{q}\begin{pmatrix} 1 & p & 0 \\ p & p^2 + q^2 & 0 \\ 0 & 0 & 1 \end{pmatrix},$$

and the electromagnetic coupling term as

$$[\alpha] = (-ct/x, 0, 0)^T.$$

The material parameters are space- and time-dependent because the time transformation is complicated.

(b) Now consider another mixed space-time transformation

$$ct = a \quad ct' + bx', \qquad x = px' + qy', \qquad y = y', \qquad z = z',$$

where only linear transformations are involved. Prove that the induced material has constant permittivity and permeability tensors as

$$[\epsilon] = [\mu] = \frac{a}{q}\begin{pmatrix} 1 & p & 0 \\ p & p^2 + q^2 & 0 \\ 0 & 0 & 1 \end{pmatrix},$$

and constant electromagnetic coupling term as

$$[\alpha] = (-b, 0, 0)^T.$$

References

Aubry A, Lei D, Fernández-Domínguez A, Sonnefraud Y, Maier S and Pendry J 2010 Plasmonic light-harvesting devices over the whole visible spectrum *Nano Lett.* **10** 2574–9

Cai W, Chettiar U, Kildishev A and Shalaev V 2006 Optical cloaking with metamaterials *Nat. Photon.* **1** 224–7

Cai W, Chettiar U, Kildishev A, Shalaev V and Milton G 2007 Nonmagnetic cloak with minimized scattering *Appl. Phys. Lett.* **91** 111105

Chang Z, Zhou X, Hu J and Hu G 2009 Design method for quasi-isotropic transformation materials based on inverse Laplace's equation with sliding boundaries *Opt. Express* **18** 6089–96

Chen H and Zheng B 2012 Broadband polygonal invisibility cloak for visible light *Sci. Rep.* **2** 255

Chen X, Luo Y, Zhang J, Jiang K, Pendry J and Zhang S 2010 Macroscopic invisibility cloaking of visible light *Nat. Commun.* **2** 176

Cummer S, Popa B, Schurig D, Smith D R and Pendry J 2006 Full-wave simulations of electromagnetic cloaking structures *Phys. Rev.* **74** 036621

Cummer S and Thompson R T 2011 Frequency conversion by exploiting time in transformation optics *J.* Optics **13** 024007

Ergin T, Stenger N, Brenner P, Pendry J and Wegener M 2010 Three-dimensional invisibility cloak at optical wavelengths *Science* **328** 337–9

Fang N, Lee H, Sun C and Zhang X 2005 Sub-diffraction-limited optical imaging with a silver superlens *Science* **308** 534–7

Fridman M, Farsi A, Okawachi Y and Gaeta A 2011 Demonstration of temporal cloaking *Nature* **481** 62–5

Gabrielli L, Cardenas J, Poitras C and Lipson M 2009 Silicon nanostructure cloak operating at optical frequencies *Nat. Photon.* **3** 461–3

Gabrielli L, Liu D, Johnson S G and Lipson M 2012 On-chip transformation optics for multimode waveguide bends *Nat. Commun.* **3** 1217

García-Meca C, Martínez A and Leonhardt U 2011 Engineering antenna radiation patterns via quasi-conformal mappings *Opt. Express* **19** 23743–50

Gharghi M, Gladden C, Zentgraf T, Liu Y, Yin X, Valentine J and Zhang X 2011 A carpet cloak for visible light *Nano Lett.* **11** 2825–8

He C, Zhang X, Feng L, Lu M-H and Chen Y-F 2011 One-way cloak based on nonreciprocal photonic crystal *Appl. Phys. Lett.* **99** 151112

Jiang W, Cui T, Yang X, Cheng Q, Liu R and Smith D R 2008 Invisibility cloak without singularity *Appl. Phys. Lett.* **93** 194102

Kildishev A and Narimanov E 2007 Impedance-matched hyper-lens *Opt. Lett.* **32** 3432–4

Kildishev A and Shalaev V 2007 Engineering space for light via transformation optics *Opt. Lett.* **33** 43–5

Landy N and Padilla W J 2009 Guiding light with conformal transformations *Opt. Express* **17** 14872–9

Landy N and Smith D R 2013 A full-parameter unidirectional metamaterial cloak for microwaves *Nat. Mater.* **12** 25–8

Leonhardt U 2006 Optical conformal mapping *Science* **312** 1777–80

Leonhardt U and Philbin T 2006 General relativity in electrical engineering *New J. Phys.* **8** 247

Leonhardt U and Tyc T 2009 Broadband invisibility by non-Euclidean cloaking *Science* **323** 110–2

Li J, Fok L, Yin X, Bartal G and Zhang X 2009a Experimental demonstration of an acoustic magnifying hyperlens *Nat. Mater.* **8** 931–4

Li J and Pendry J 2008 Hiding under the carpet: a new strategy for cloaking *Phys. Rev. Lett.* **101** 203901

Li W, Guan J, Sun Z, Wang W and Zhang Q 2009b A near-perfect invisibility cloak constructed with homogeneous materials *Opt. Express* **17** 23410–6

Liang D, Gu J, Han J, Yang Y, Zhang S and Zhang W 2011 Robust large dimension terahertz cloaking *Adv. Mater.* **24** 916–21

Liang Z, Jiang X, Miao F, Guenneau S and Li J 2012 Transformation media with variable optical axes *New J. Phys.* **14** 103042

Liu D, Gabrielli L, Lipson M and Johnson S G 2013 Transformation inverse design *Opt. Express* **21** 14223–43

Liu F and Li J 2015 Gauge field optics with anisotropic media *Phys. Rev. Lett.* **114** 103902

Liu F, Xu T, Wang S, Hang Z and Li J 2019 Polarization beam splitting with gauge field metamaterials Adv *Opt. Mater.* **7** 1801582

Liu R, Ji C, Mock J, Chin J Y, Cui T and Smith D R 2009 Broadband ground-plane cloak *Science* **323** 366–9

Ma H and Cui T 2010 Three-dimensional broadband ground-plane cloak made of metamaterials *Nat. Commun.* **1** 21

McCall M, Favaro A, Kinsler P and Boardman A 2011 A spacetime cloak, or a history editor *J.* Optics **13** 024003

Mei Z, Bai J and Cui T 2011 Experimental verification of a broadband planar focusing antenna based on transformation optics *New J. Phys.* **13** 063028

Pendry J, Aubry A, Smith D R and Maier S 2012 Transformation optics and subwavelength control of light *Science* **337** 549–52

Popa B, Zigoneanu L and Cummer S 2011 Experimental acoustic ground cloak in air *Phys. Rev. Lett.* **106** 253901

Rahm M, Cummer S, Schurig D, Pendry J and Smith D R 2008a Optical design of reflectionless complex media by finite embedded coordinate transformations *Phys. Rev. Lett.* **100** 063903

Rahm M, Schurig D, Roberts D A, Cummer S, Smith D R and Pendry J 2008b Design of electromagnetic cloaks and concentrators using form-invariant coordinate transformations of Maxwellas equations *Photon. Nanostruct: Fundam. Appl.* **6** 87–95

Schurig D, Mock J, Justice B, Cummer S, Pendry J, Starr A and Smith D R 2006 Metamaterial electromagnetic cloak at microwave frequencies *Science* **314** 977–80

Smolyaninov I, Hung Y and Davis C 2007 Two-dimensional metamaterial structure exhibiting reduced visibility at 500 nm *Opt. Lett.* **33** 1342–4

Valentine J, Li J, Zentgraf T, Bartal G and Zhang X 2009 An optical cloak made of dielectrics *Nat. Mater.* **8** 568–71

Wee W H, Ye Y and Luo Y 2016 Towards a practical compact magnifying superlens–a simple simplical design *J. Opt.* **18** 044011

Wu Q, Turpin J and Werner D 2012 Integrated photonic systems based on transformation optics enabled gradient index devices *Light: Sci. Appl.* **1** e38

Xi S, Chen H, Wu B-I and Kong J 2009 One-directional perfect cloak created with homogeneous material *IEEE Microw. Wirel. Compon. Lett.* **19** 131–3

Xu H, Zhang B, Yu T, Barbastathis G and Sun H 2012 Dielectric waveguide bending adapter with ideal transmission: practical design strategy of area-preserving affine transformation optics *J. Opt. Soc. Am.* B **29** 1287–90

Xu L and Chen H 2015 Conformal transformation optics *Nat. Photon.* **9** 15–23

Xu X, Feng Y, Xiong S, Fan J, Zhao J and Jiang T 2011 Broad band invisibility cloak made of normal dielectric multilayer *Appl. Phys. Lett.* **99** 154104

Yang R, Tang W and Hao Y 2011 A broadband zone plate lens from transformation optics *Opt. Express* **19** 12348–55

Yao K and Jiang X 2011 Designing feasible optical devices via conformal mapping *J. Opt. Soc. Am.* B **28** 1037–42

Zentgraf T, Valentine J, Tapia N, Li J and Zhang X 2010 An optical 'janus' device for integrated photonics *Adv. Mater.* **22** 2651–4

Zhang B, Chan T and Wu B-I 2010a Lateral shift makes a ground-plane cloak detectable *Phys. Rev. Lett.* **104** 233903

Zhang B, Luo Y, Liu X and Barbastathis G 2010b Macroscopic invisibility cloak for visible light *Phys. Rev. Lett.* **106** 033901

Zhang J, Liu L, Luo Y, Zhang S and Mortensen N 2011a Homogeneous optical cloak constructed with uniform layered structures *Opt. Express* **19** 8625–31

Zhang J, Luo Y and Mortensen N 2010c Transmission of electromagnetic waves through sub-wavelength channels *Opt. Express* **18** 3864–70

Zhang J, Xiao S, Wubs M and Mortensen N 2011b Surface plasmon wave adapter designed with transformation optics *ACS nano* **5** 4359–64

Zhou F, Bao Y, Cao W, Stuart C, Gu J, Zhang W and Sun C 2011 Hiding a realistic object using a broadband terahertz invisibility cloak *Sci. Rep.* **1** 78

Zigoneanu L, Popa B and Cummer S 2014 Three-dimensional broadband omnidirectional acoustic ground cloak *Nat. Mater.* **13** 352–5

IOP Publishing

Essential Concepts in Metamaterials

Jensen Li

Chapter 5

Metasurfaces

The concept of metamaterials is powerful because it frees us from the limited palette of material parameters. Once we assume that we have the ability to have any material parameters we wish, we can use macroscopic descriptions to design optical devices, and a generic framework like transformation optics emerges as a result. Recent advances have shown that even a single layer of metamaterial atoms, called metasurfaces, can play a similar role.

What makes metasurfaces distinct from bulk metamaterials is that the wavefront can be reshaped through abrupt phase changes or effective boundary conditions across a sheet much thinner than the wavelength. This allows us to replace a thick volume of material with a patterned interface, leading to ultrathin and lightweight devices. From microwave reflectarrays to optical holograms, metasurfaces have rapidly expanded the possibilities of wave manipulation.

However, to move beyond isolated demonstrations, we still need a coherent framework to describe and design metasurfaces. One such framework is the generalized laws of reflection and refraction, which show how a prescribed phase gradient across the surface modifies Snell's law. Another is the use of effective surface response matrices, where metasurfaces are treated as discontinuities relating incident and transmitted fields. In this chapter, we will explore how metasurfaces can be constructed from simple elements, how their local responses can be engineered, and how they can be described within such effective theories to systematically manipulate waves.

5.1 Geometric and propagation phases

Let us start with a metasurface consisting of unit cells such as those shown in figure 5.1. Examples include a metallic split-ring aperture on a conducting surface, or a dielectric rectangular pillar on a substrate. The transmission through such a metasurface can be described using the Jones matrix method. In both cases in panels (a) and (b), we define a rotation angle α for the unit cell, while the structure in panel (c),

doi:10.1088/978-0-7503-1131-1ch5

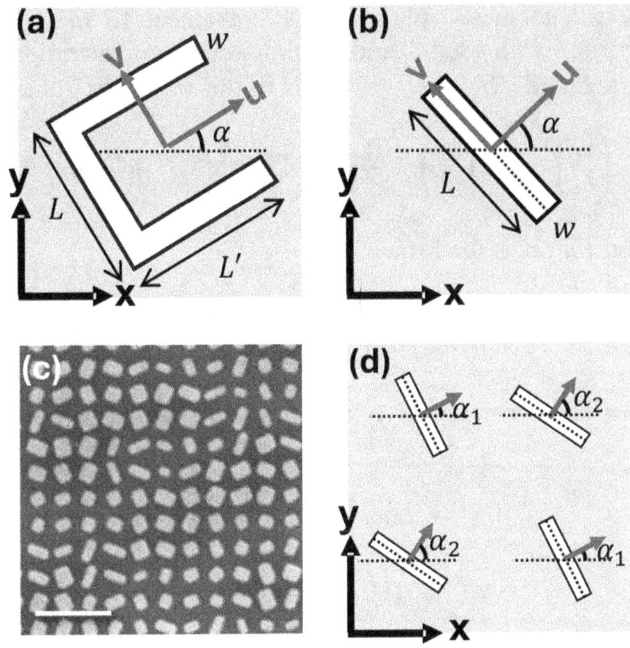

Figure 5.1. Examples of unit cells in a square lattice. (a) A metallic split ring aperture at an orientation angle α; (b) amorphous Si ($n_{\text{Si}} = 3.44$) nanopillar with width (W), length (L), and orientation angle (α) on a SiO_2 ($n_{\text{SiO2}} = 1.45$) substrate (gray background). (c) Zoomed-in SEM image of the sample in (b). (d) A unit cell with two pairs of slots of same dimensions but different orientation angles α_1 and α_2. The scale bar is 1 μm.

which involves two orientation angles, will be discussed later. Suppose the incident light propagates along the positive z-direction (normal incidence on the metasurface). Without rotation ($\alpha = 0$), the polarization response of each atom in the linear polarization basis $\{\hat{x}, \hat{y}\}$ is

$$\begin{pmatrix} b_x \\ b_y \end{pmatrix} = \begin{pmatrix} t_{xx}^{(0)} & t_{xy}^{(0)} \\ t_{yx}^{(0)} & t_{yy}^{(0)} \end{pmatrix} \begin{pmatrix} a_x \\ a_y \end{pmatrix},$$

where the input polarization is $a_x\hat{x} + a_y\hat{y}$ and the output polarization is $b_x\hat{x} + b_y\hat{y}$. When the structure is rotated by an angle α, the polarization response becomes

$$\begin{pmatrix} b_x \\ b_y \end{pmatrix} = J_{\text{LP}} \begin{pmatrix} a_x \\ a_y \end{pmatrix} = R(-\alpha) \begin{pmatrix} t_{xx}^{(0)} & t_{xy}^{(0)} \\ t_{yx}^{(0)} & t_{yy}^{(0)} \end{pmatrix} R(\alpha) \begin{pmatrix} a_x \\ a_y \end{pmatrix}, \tag{5.1}$$

where J_{LP} is the Jones matrix in the linear polarization basis and

$$R(\theta) = \begin{pmatrix} \cos\theta & \sin\theta \\ -\sin\theta & \cos\theta \end{pmatrix}.$$

To better reveal geometric phases, it is convenient to move to the circular polarization basis, with left- and right-circular polarizations defined as $|L\rangle = (\hat{x} + \mathrm{i}\hat{y})/\sqrt{2}$ and $|R\rangle = (\hat{x} - \mathrm{i}\hat{y})/\sqrt{2}$. In this basis, the Jones matrix is

$$\begin{pmatrix} b_L \\ b_R \end{pmatrix} = J_{\mathrm{CP}} \begin{pmatrix} a_L \\ a_R \end{pmatrix} = \frac{1}{2} \begin{pmatrix} 1 & -\mathrm{i} \\ 1 & \mathrm{i} \end{pmatrix} J_{\mathrm{LP}} \begin{pmatrix} 1 & 1 \\ \mathrm{i} & -\mathrm{i} \end{pmatrix} \begin{pmatrix} a_L \\ a_R \end{pmatrix}. \tag{5.2}$$

The resulting matrix takes the form

$$J_{\mathrm{CP}} = \begin{pmatrix} t_{LL} & t_{LR} \\ t_{RL} & t_{RR} \end{pmatrix} = \begin{pmatrix} |t|\mathrm{e}^{\mathrm{i}\varphi} & |t'|\mathrm{e}^{\mathrm{i}(\varphi'-2\alpha)} \\ |t'|\mathrm{e}^{\mathrm{i}(\varphi'+2\alpha)} & |t|\mathrm{e}^{\mathrm{i}\varphi} \end{pmatrix}. \tag{5.3}$$

If the x- and y-axes align with the principal axes of the structure ($t_{xy}^{(0)} = t_{yx}^{(0)} = 0$), then

$$\begin{aligned} t_{LL} &= t_{RR} = \frac{1}{2}\left(t_{xx}^{(0)} + t_{yy}^{(0)}\right) = |t|\mathrm{e}^{\mathrm{i}\varphi}, \\ \sqrt{t_{LR}t_{RL}} &= \frac{1}{2}\left(t_{xx}^{(0)} - t_{yy}^{(0)}\right) = |t'|\mathrm{e}^{\mathrm{i}\varphi'}. \end{aligned}$$

Mirror symmetry of the unit cell ensures $t_{LL} = t_{RR}$ and $|t_{LR}| = |t_{RL}|$. The additional phase factor $\pm 2\alpha$, arising from the orientation angle, is called the geometric phase (or Pancharatnam–Berry phase). It only appears for cross-polarized transmission and flips sign depending on whether the input is LCP ($\sigma = +1$) or RCP ($\sigma = -1$), summarized as $2\sigma\alpha$.

By contrast, the phases φ and φ' are propagation phases, which originate from the resonant response of the structure itself and can be tuned by adjusting the geometry (e.g., width W, length L). In metasurface design, both the geometric and propagation phases can be combined, providing flexible control over the transmitted wavefront. Table 5.1 compares the different aspects of geometric and propagation phases utilized with metasurfaces.

Therefore, if the orientation angle α varies spatially across the metasurface, the wavefront of the cross-polarized light can be modulated through a corresponding spatially varying geometric phase discontinuity. Importantly, this phase discontinuity does not arise from frequency dispersion, as in the case of sweeping through a series of resonant structures. Instead, it is purely geometric in nature, linked directly to the

Table 5.1. Comparison between geometric and propagation phases.

Geometric phase	Propagation phase
Arises from rotation of the unit cell orientation	Arises from the resonant response of the structure itself
Gives phase shift $2\sigma\alpha$ in cross-polarized transmission	Phase shift depends on structural dimensions, e.g., width W, length L, or height
Controlled by adjusting orientation angle α	Controlled by tuning structural dimensions

optical-axis rotation angle by a factor of two. This contribution is known as the Pancharatnam–Berry phase, or simply the geometric phase. A key advantage of this scheme is that the imposed phase profile is broadband and independent of frequency.

5.2 Beam steering and q-plate

5.2.1 Generalized Snell's law

Unlike propagation phases which are more sensitive to structural dimensions and material parameters due to the resonating nature of the constituting metamaterial structures, the geometric phase is more robust against fabrication errors due to its geometric nature merely coming from the orientation of the metamaterial atoms. With the same geometric phase elements but with different orientations at different positions, we can directly transform the wavefront at different positions in a parallel fashion.

Suppose the orientation profile of the geometric phase elements is a linear profile:

$$\alpha(x) = \frac{1}{2}(\Delta k)\, x + \alpha_0,$$

where Δk is interpreted as the momentum imparted by the inhomogeneous material. Figure 5.2 shows an example metasurface with such a one-dimensional orientation profile.

Generalized Snell's law as a unifying view. For a metasurface imposing a spatially varying total phase discontinuity $\Phi_{\text{tot}}(x)$, conservation of tangential momentum gives

$$n_t \sin\theta_t - n_i \sin\theta_i = \frac{\lambda}{2\pi}\frac{\mathrm{d}\Phi_{\text{tot}}}{\mathrm{d}x}, \tag{5.4}$$

where $\theta_i(\theta_t)$ are the incident (refracted) angles and $n_i(n_t)$ are the refractive indices of the two media on incident and transmitted side. This is called the *generalized Snell's law* (Yu *et al* 2011; Ni *et al* 2012). Figure 5.2 illustrates the beam steering effect based on geometric phase elements (Kang *et al* 2012; Huang *et al* 2012).

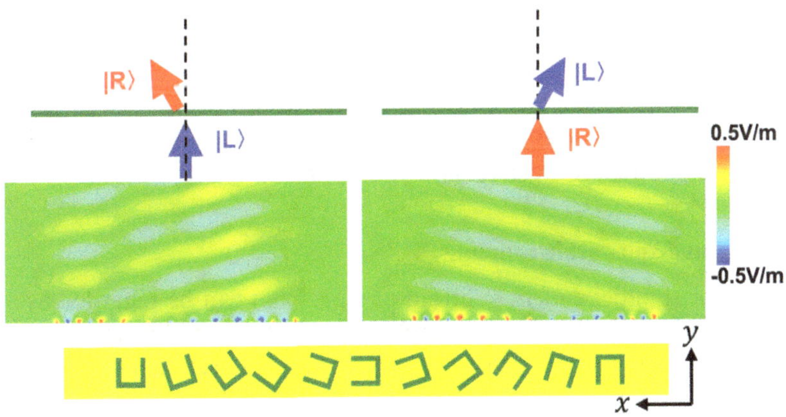

Figure 5.2. An example of a 1D metasurface array with a linear orientation profile for beam steering. Reprinted with permission from Kang *et al* (2012) © Optical Society of America.

In general, the total phase can be decomposed as

$$\Phi_{\text{tot}}(x) = \phi_{\text{prop}}(x) + 2\sigma\ \alpha(x), \qquad \sigma = \begin{cases} +1, & \text{LCP} \\ -1, & \text{RCP}, \end{cases}$$

where $\phi_{\text{prop}}(x)$ is the propagation phase from structural resonance and $2\sigma\alpha(x)$ is the geometric (Pancharatnam–Berry) phase (Berry 1987; Bomzon *et al* 2002). In this section, we set $\phi_{\text{prop}} = 0$, so that the phase gradient comes solely from the geometric phase.

Substituting the linear profile $\alpha(x)$ into equation (5.4) gives

$$k_{x,t} = k_{x,i} \pm \Delta k, \tag{5.5}$$

where $k_{x,t}$ ($k_{x,i}$) are the transverse wave numbers of the transmitted (incident) waves. The upper (lower) sign corresponds to LCP (RCP) incidence.

This spin-dependent steering is a manifestation of the optical spin Hall effect (OSHE) (Onoda *et al* 2004; Hosten and Kwiat 2008; Liberman and Zel'dovich 1992; Bliokh and Bliokh 2004; Tomita and Chiao 1986; Shitrit *et al* 2011; Gorodetski *et al* 2008, 2009), where opposite spins acquire opposite transverse momentum shifts. A metasurface can make the effect much more pronounced (Yin *et al* 2013; Li *et al* 2015; Luo *et al* 2015).

Since the geometric phase depends only on orientation, the imparted momentum Δk is wavelength-independent, although the resulting steering angle still varies with wavelength. Such devices are often called *geometric-phase metasurfaces*. Using typical split-ring dimensions in the microwave regime, full-wave simulations (figure 5.2) show the beam steering effect: the left panel for LCP incidence and the right panel for RCP incidence, in agreement with theory.

When the incident spin flips, the steering direction reverses. Moreover, in this case the co-polarized transmission can be designed to vanish, leaving only the cross-polarized transmitted wavefront. From equation (5.3), the cross-polarization conversion efficiency, defined as the ratio between the cross-polarization transmittance and the total transmittance, is given by

$$\eta = \frac{|t_{xx}^{(0)} - t_{yy}^{(0)}|^2}{|t_{xx}^{(0)} - t_{yy}^{(0)}|^2 + |t_{xx}^{(0)} + t_{yy}^{(0)}|^2}.$$

By designing the structure so that $t_{xx}^{(0)} + t_{yy}^{(0)} = 0$, effectively realizing a half-wave plate, the cross-polarization conversion can approach 100%.

5.2.2 Q-plate and spin–orbit coupling

Polarization- or spin-dependent splitting of wavefront can be regarded as a generic form of spin–orbit coupling. Apart from spin-splitting of linear momentum for beam-steering, a geometric phase in polarization space can also be used to manipulate a structured light beam.

After discussing linear orientation profiles for beam steering, we now turn to azimuthal orientation profiles. A linear gradient in $\alpha(x)$ imparts a constant momentum shift Δk, steering the beam in real space. If instead we choose an azimuthally

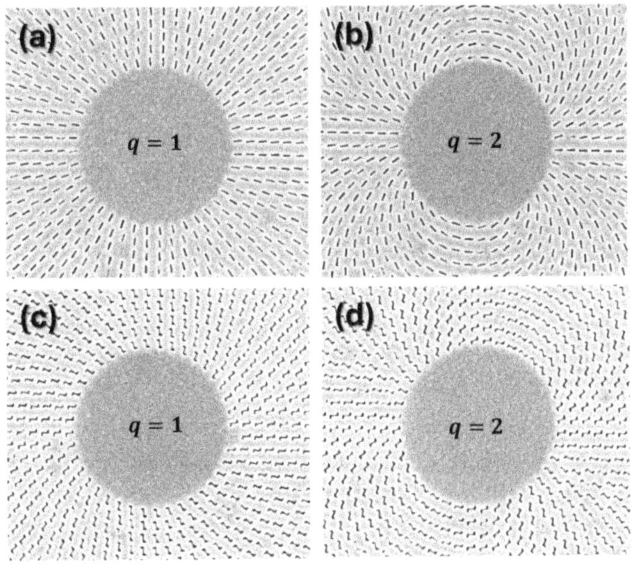

Figure 5.3. SEM images of metasurface q-plates made of rectangular slits (a) and (b) or Z-shape slits (c) and (d) on a metallic surface.

varying orientation $\alpha(\phi)$, the metasurface imparts an azimuthal phase twist, steering the beam in angular momentum space. This leads naturally to the concept of a q-plate, a geometric-phase element that couples the spin of light (circular polarization) to its orbital angular momentum (OAM) (Marrucci *et al* 2006).

A beam with OAM possesses a helical wavefront comprising an azimuthal profile $\propto \exp(\mathrm{i}l\phi)$ (Allen *et al* 1992, 1996; O'Neil *et al* 2002). Here, ϕ is the azimuthal angle and l is an integer, simply called the OAM of the beam. In the current case of a geometric-phase metasurface, the orientation profile of the metamaterial atoms is designed as a linear function of the azimuthal angle, termed as a q-plate:

$$\alpha = q\phi + \alpha_0. \tag{5.6}$$

q is called the topological charge of the metasurface and α_0 is the initial orientation at zero azimuthal angle. Figure 5.3 shows several examples of metasurface q-plates made of rectangular slits or Z-shape slits on a metallic surface.

Figure 5.3(c) shows an example constructed from Z-shape slits on a metallic surface with $q = 1$ and $\alpha_0 = 0$ (Karimi *et al* 2014; Li *et al* 2013). For a general q, including half-integer values, the orientation profile $\alpha(\phi)$ imparts opposite geometric phase profiles ($\pm 2\alpha$) for the two circular polarizations. Here the term 'topological charge' refers to the winding number of the metasurface orientation profile, which should not be confused with the topological charge (orbital angular momentum) carried by an incident light beam. The total phase on transmission becomes $l\phi \pm 2\alpha$, and thus the transmitted cross-polarized beam acquires a change of OAM:

$$l' = l \pm 2\partial_\phi \alpha = l \pm 2q. \tag{5.7}$$

For example, when light with zero OAM passes through a $q = 1$ plate in figure 5.4, the transmitted beam acquires an OAM of $l' = \pm 2$. To show the spin-splitting and reveal

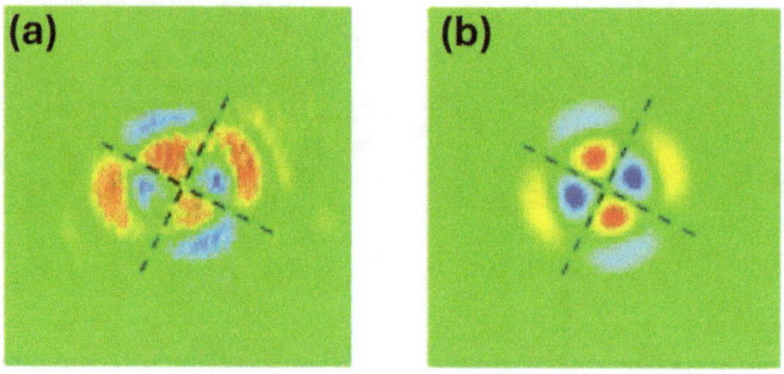

Figure 5.4. (a) Experimental results of spin-splitting of OAM using Stokes parameter S_3 profile in the far-field for the metasurface shown in figure 5.3(c): a topological charge $q = 1$ and an initial orientation $\alpha_0 = 0$. (b) Corresponding numerical results for the Stokes parameter. Reprinted (adapted) with permission from Li *et al* (2013). Copyright (2013) American Chemical Society.

the OSHE, one can use the residual beam (the co-polarization transmission without geometric phase, hence with zero OAM) as a reference. This reference beam interferes with the cross-polarized transmitted beam. For y-polarized incidence (containing both spins as input), the Stokes parameter $S_3 = I_L - I_R$, with I_L and I_R being the LCP and RCP intensities, can be measured in the far-field. The results (experiment and simulation) are shown in figures 5.4(a) and (b), displaying clear angular splitting as direct evidence of the OSHE. The two different colors in the figure represent the two spin states, and the spin angular momentum is transferred to the OAM of the transmitted beam.

Unlike the liquid-crystal or crystal-based q-plates that have been used traditionally, metasurface q-plates are built from nanostructures patterned directly on a surface. This makes them extremely thin—just a tiny fraction of the wavelength of light—yet they can still produce strong coupling between spin and orbital angular momentum. Because the phase shift they use is geometric rather than resonant, their operation does not rely heavily on material dispersion and can work across a broad range of wavelengths. The fine, subwavelength features of a metasurface also give much more freedom in design: the topological charge q can be chosen flexibly, even with fractional values, and complex wavefronts can be shaped with high precision. Another important advantage is that metasurface q-plates are flat and compatible with chip-scale optics, making them well suited for integration into compact photonic systems.

The mechanism of OAM change in equation (5.7) also indicates an interesting possibility when a metasurface of topological charge q interacts with an incident beam composed of only two OAM states:

$$E_0 = a_L|l_1L\rangle + a_R|l_2R\rangle, \tag{5.8}$$

with the requirement $q = (l_2 - l_1)/2$. In this case, the spin-flip in the cross-polarization simply swaps the two OAM values l_1 and l_2 without generating additional OAM states. The transmitted beam (including both co- and cross-polarizations) remains in the same subspace as

$$E_t = b_L|l_1L\rangle + b_R|l_2R\rangle, \tag{5.9}$$

while the local response (equation (5.3)) of the metasurface preserves its form:

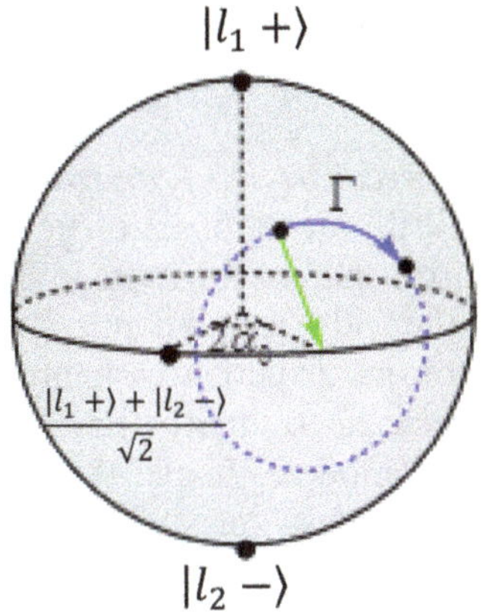

Figure 5.5. Poincaré sphere with north/south poles being $|l_1 + \rangle/|l_2 - \rangle$ for the subspace of the incident beam. The incident beam interacts with a metasurface of topological charge $q = (l_2 - l_1)/2$. '+' ('−') stands for $L(R) - CP$ polarization. The blue/green arrows illustrate the action of a metasurface constructed from identical wave plates with retardation phase Γ/ linear polarizers with principal axis along α_0. These correspond exactly to standard polarization control on the Poincaré sphere, generalized here to higher-order modes.

$$\begin{pmatrix} b_L \\ b_R \end{pmatrix} = \begin{pmatrix} t_{LL}^{(0)} & e^{-i2\alpha_0} t_{LR}^{(0)} \\ e^{i2\alpha_0} t_{RL}^{(0)} & t_{RR}^{(0)} \end{pmatrix} \begin{pmatrix} a_L \\ a_R \end{pmatrix}. \tag{5.10}$$

This summarizes the global response without explicit azimuthal dependence. The usual methods of representing polarization control on the Poincaré sphere (see section A.4) can now be applied to this higher-order case. For example, a metasurface constructed by rotating wave plates with retardation Γ and orientation α (equation (5.6)) rotates the state of the incident beam around the axis at angle $2\alpha_0$ on the equator by angle Γ on the Poincaré sphere (blue arrow in figure 5.5). Similarly, a metasurface made from rotated polarizer elements projects the state to the point at angle $2\alpha_0$ on the equator (green arrow). Thus, $2\alpha_0$ can be interpreted as a higher-order geometric phase associated with the metasurface.

5.3 Metasurface holograms

As metasurfaces are much thinner than a wavelength, diffraction between neighboring meta-atoms is weak. This means that the phase profile imposed across the surface can be controlled with very fine resolution, limited mainly by the size of each individual meta-atom. Such precise control over phase makes metasurfaces particularly powerful for constructing high-resolution holograms.

5.3.1 Metasurface holograms with propagation phase

So far, we have seen that metasurfaces can steer beams or convert spin into orbital angular momentum by controlling the geometric phase. Here, we begin by showing how to design a metasurface hologram that displays two different images depending on the input polarization, either polarized in horizontal or vertical polarizations.

One advantage of using a metasurface approach is the ease of this functionality multiplexing and upgrade.

Step 1—unit cell and resonance-controlled propagation phase
The unit cell, shown schematically in the inset of figure 5.6, is a titanium dioxide (TiO_2) nanobrick placed on a transparent substrate in this example. Its thickness and lattice period are fixed, while the width w and depth d are tunable parameters. Light transmitted through the nanobrick experiences a propagation phase because the nanobrick acts as a short dielectric resonator. By tuning w and d, we shift the resonance conditions of guided modes, change the effective refractive index, and therefore adjust the accumulated phase delay.

For horizontally polarized (H) incident light, both w and d influence the resonance and thus the transmission phase and amplitude. This dependence is shown in figures 5.7(a) and (b). For vertically polarized (V) incident light, the response is obtained by interchanging w and d because of the structural symmetry.

Step 2—building the phase–amplitude library
By sweeping (w, d) in simulations we generate a library of achievable phases and transmission amplitudes. Each entry in the library corresponds to a pair (ϕ_H, ϕ_V). To maintain high efficiency, unit cells with transmission below 90% are excluded. In this way, the device operates close to the phase-only regime.

Step 3—computing the target phase maps with the Gerchberg–Saxton algorithm
To determine the required phase distribution across the metasurface, we use the Gerchberg–Saxton (GS) algorithm. This is an iterative Fourier transform method.

We begin with the desired holographic images in the far field: 'CityU' for H polarization and 'UST' for V polarization. The far field is related to the metasurface plane by a Fourier transform. The algorithm alternates between the two planes:

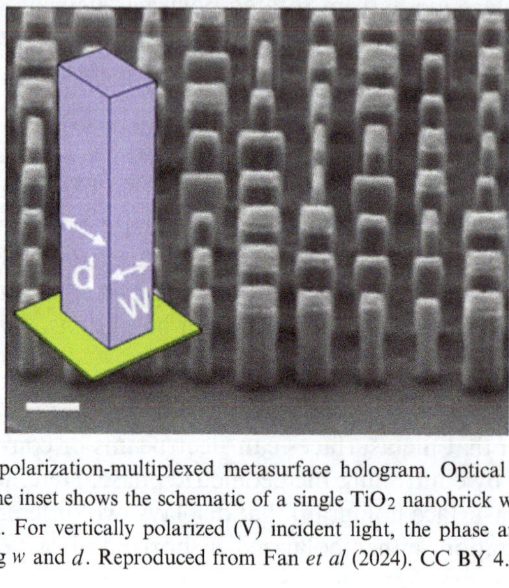

Figure 5.6. Design of a polarization-multiplexed metasurface hologram. Optical microscope image of the fabricated metasurface; the inset shows the schematic of a single TiO_2 nanobrick with width w, depth d, and fixed thickness $h = 1\ \mu m$. For vertically polarized (V) incident light, the phase and transmission maps are obtained by interchanging w and d. Reproduced from Fan *et al* (2024).

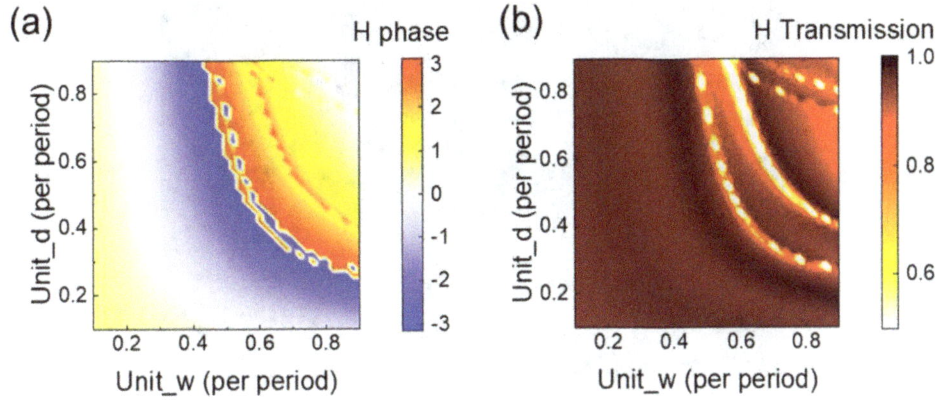

Figure 5.7. Design of a polarization-multiplexed metasurface hologram from Fan *et al* (2024). (a) Simulated transmission phase of the nanobrick unit cell for horizontally polarized (H) incident light as a function of normalized width w/l and depth d/l (l is the length of the nanobrick). (b) Corresponding transmission amplitude map for H polarization.

- In the image plane, we replace the amplitude with $\sqrt{I_{\text{target}}(k_x, k_y)}$ while keeping the phase from the previous step. The phase here is unconstrained, since only intensity is known.
- After inverse Fourier transforming to the metasurface plane, we enforce the amplitude to be unity (phase-only assumption) while keeping the phase profile.

After repeated iterations, the algorithm converges to the required phase maps $\phi_H^0(x, y)$ and $\phi_V^0(x, y)$.

A compact update formula is

$$E^{(n+1)}(x, y) = \exp(i \arg(\mathcal{F}^{-1}[\sqrt{I_{\text{target}}(k_x, k_y)}\ e^{\,i \arg\{\mathcal{F}[E^{(n)}(x, y)]\}}])),$$

where $\mathcal{F}$ denotes the Fourier transform.

Step 4—amplitude constraint at the metasurface plane

In the GS algorithm we usually set the metasurface-plane amplitude to 1, which corresponds to an ideal phase-only hologram. This is justified because metasurfaces are designed to have nearly uniform, high transmission. In practice, the actual transmission amplitude varies with (w, d). These variations are accounted for later during mapping: cells with poor transmission are excluded.

Step 5—dual-channel mapping from target phases to nanobricks

At each pixel we require $\phi_\text{H}(x, y) \approx \phi_\text{H}^0(x, y)$ and $\phi_\text{V}(x, y) \approx \phi_\text{V}^0(x, y)$. We search the phase library to find the (w, d) combination that minimizes the total phase error for both polarizations, for example,

$$\varepsilon(x, y) = \left| e^{i\phi_\text{H}(w,\, d)} - e^{i\phi_\text{H}^0(x,\, y)} \right| + \left| e^{i\phi_\text{V}(w,\, d)} - e^{i\phi_\text{V}^0(x,\, y)} \right|.$$

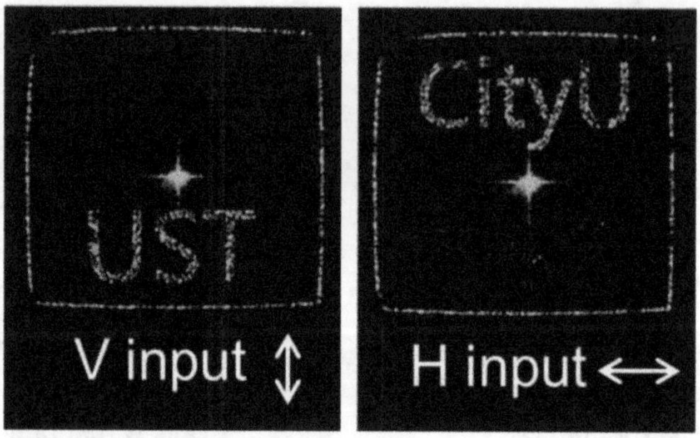

Figure 5.8. Classical holographic reconstruction from the fabricated metasurface. (a) Reconstructed hologram of the letters 'UST' under vertical (V) input polarization. (b) Reconstructed hologram of the letters 'CityU' under horizontal (H) input polarization. Reproduced from Fan *et al* (2024). CC BY 4.0.

Step 6—fabrication and classical verification

After assigning nanobricks across the array, the metasurface is fabricated. Under classical illumination the two holograms are reconstructed: with V (H) polarization, the 'UST' ('CityU') hologram is observed in the left (right) panel of (figure 5.8).

Resonances in the nanobricks determine the propagation phases. Both w and d affect the H-polarization maps in figures 5.7(a,b), while the V channel is obtained by swapping axes. The GS algorithm, based on alternating Fourier and inverse Fourier transforms, computes the target phase maps. The library connects these phase targets to real geometries, with efficiency maintained by excluding low-transmission cells. This workflow serves as a starting example of how to design a polarization-multiplexed metasurface hologram. Very often, the GS algorithm is modified to include additional constraints as functionalities are added.

5.3.2 Jones-matrix holograms

We now take one step further: instead of generating only one function, we would like the same metasurface to generate *different holograms* depending on the different polarization of the incoming and outgoing light, preferably including co-polarization and cross-polarization channels. This is possible because a metasurface can be described by its full Jones matrix. When both the propagation phases (φ, φ') and the geometric phase ($\pm 2\alpha$) are used, we can construct what is called a *Jones-matrix hologram*. From one perspective, we can instruct the hologram to have different polarizations at different locations (Rubin *et al* 2021). The key idea here is that each input–output polarization channel can be used as a separate 'hologram channel' on the same device. Figure 5.1(b) shows the unit cell we use: an amorphous Si nanopillar on a SiO_2 substrate. The pillars have a height of 500 nm and are arranged in a square lattice with periodicity 350 nm. By adjusting the pillar width W, length L, and orientation α, we can tune both the propagation phases and the geometric phase. This is the basic 'toolbox' we use to encode holograms.

In the CP basis (equation (5.3)), the transmission elements are

$$t_{LL} = t_{RR} = \frac{1}{2}(t_{xx}^{(0)} + t_{yy}^{(0)}) = |t|e^{i\varphi},$$
$$\sqrt{t_{LR}t_{RL}} = \frac{1}{2}(t_{xx}^{(0)} - t_{yy}^{(0)}) = |t'|e^{i\varphi'}.$$

The propagation phases φ, φ' come from the resonant response of the nanopillars, while the geometric phase $\pm 2\alpha$ comes from their orientation. Numerical simulations show that as we vary W and L, both φ and φ' can cover the full range 0 to 2π (figure 5.9). This means that by controlling α, W, and L, we can realize three independent phase profiles: φ, $\varphi' + 2\alpha$, and $\varphi' - 2\alpha$. Each of these can be used to form a different hologram.

To illustrate, suppose we want the metasurface to produce three simple holograms: an '=' sign, a '+' sign, and a '×' sign. We assign these holograms to different input–output polarization channels as follows:

$$\begin{aligned}\hat{M}|L\rangle &= \beta_1|L, =\rangle + \beta_2(|R, +\rangle + |R, \times\rangle),\\ \hat{M}|R\rangle &= \beta_1|R, =\rangle + \beta_2(|L, +\rangle - |L, \times\rangle).\end{aligned} \tag{5.11}$$

Here, the operator $\hat{M}$ represents the metasurface. The coefficients β_1, β_2 describe the fractions of light in the co-polarized and cross-polarized channels. For example, if

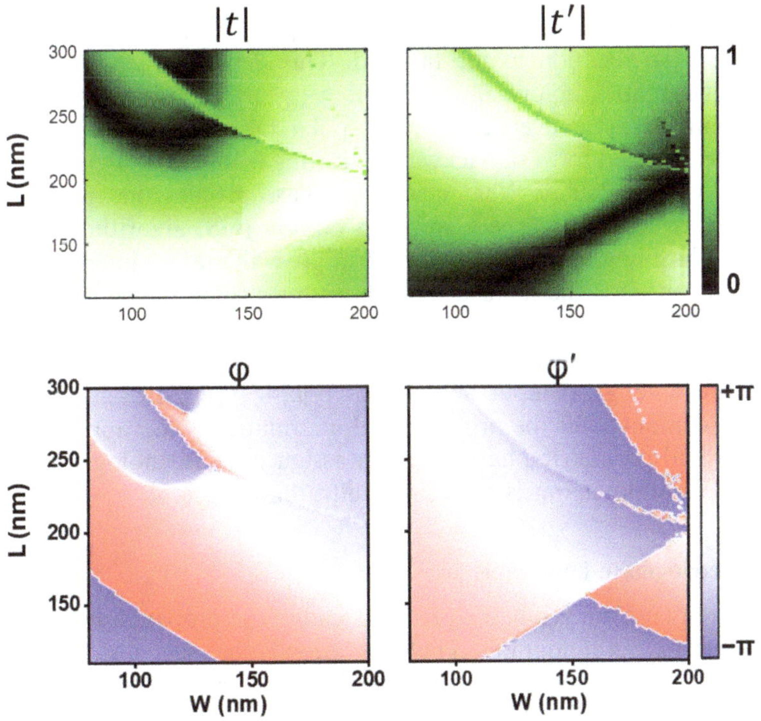

Figure 5.9. Dependence of amplitudes $|t|$, $|t'|$ and phases φ, φ' on nanopillar width W and length L.

an LCP beam is incident, then the co-polarized output (L-to-L) produces the '=' hologram, while the cross-polarized output (L-to-R) produces a combination of the '+' and '×' holograms. Similarly, for RCP incidence, the R-to-R channel again produces the '=' hologram, while the R-to-L channel gives a slightly different combination of '+' and '×'.

The same reasoning can be applied in the linear polarization bases. In the H/V basis we find

$$\begin{aligned}\hat{M}|H\rangle &= \beta_1|H, =\rangle + \beta_2(|H, +\rangle - \mathrm{i}|V, \times\rangle),\\ \hat{M}|V\rangle &= \beta_1|V, =\rangle - \beta_2(|V, +\rangle + \mathrm{i}|H, \times\rangle).\end{aligned}$$

Here, a hologram with a '×' appears when converting between H and V polarizations.

In the D/A basis we obtain

$$\begin{aligned}\hat{M}|D\rangle &= \beta_1|D, =\rangle + \beta_2(|A, +\rangle - \mathrm{i}|D, \times\rangle),\\ \hat{M}|A\rangle &= \beta_1|A, =\rangle + \beta_2(|D, +\rangle + \mathrm{i}|A, \times\rangle).\end{aligned}$$

Thus, holograms with a '+' symbol appear when converting between D and A polarizations.

Altogether, the metasurface generates an '=' hologram for co-polarized channels (L-to-L, R-to-R), a '×' hologram for H-to-V or V-to-H conversion, and a '+' hologram for D-to-A or A-to-D conversion. Other input/output combinations correspond to superpositions of these three holograms, each designed to appear at a different spatial location in the far-field. This demonstrates the multi-channel nature of Jones-matrix holography: a single metasurface can carry many different holographic images, selectively accessed by polarization.

In practice, we need to work backwards: starting from the desired far-field images ('=', '+', '×'), we calculate what near-field phase profiles the metasurface must provide. This is done using the Gerchberg–Saxton (GS) algorithm, which finds a phase-only solution whose Fourier transform gives the target intensity. The results are then mapped onto the metasurface library of structures (figure 5.9), choosing parameters α, W, and L that realize the required phases.

Figure 5.10 shows experimental results: the same metasurface produces different holograms depending on input and output polarizations, in agreement with the design. This demonstrates the principle of Jones-matrix holography: by combining propagation and geometric phases, we can assign different holographic images to different polarization channels on a single ultrathin device.

5.3.3 Geometric-phase holograms

Suppose we only use the geometric phase instead of the propagation phase profile to generate Jones-matrix holograms. The motivation is that such a simplification gives higher robustness against fabrication errors. In this case, we simply use identical metamaterial atoms across the metasurface, with their orientation angle providing the required phase (Huang 2013b; Ni *et al* 2013; Zheng *et al* 2015; Wen 2015). While

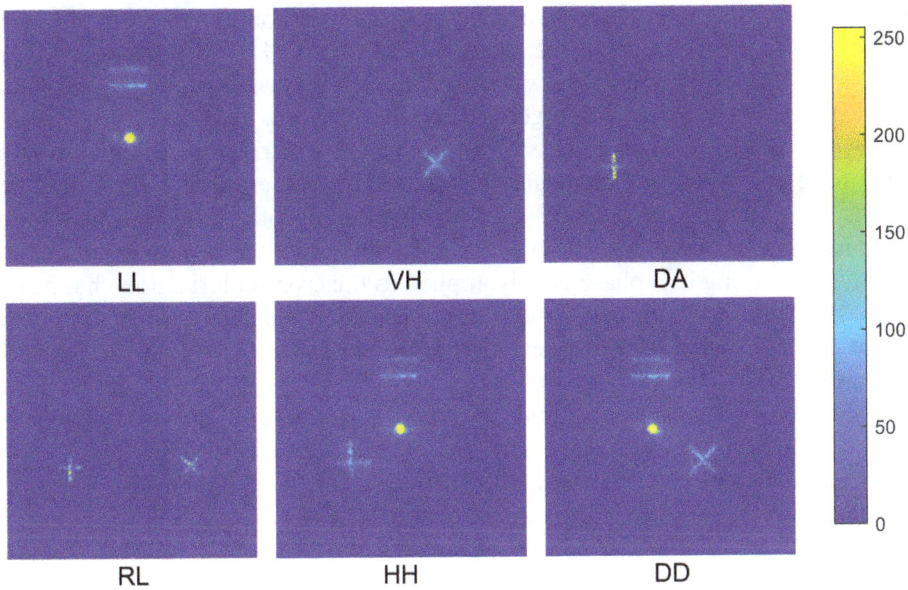

Figure 5.10. Far-field Jones-matrix holograms for different input and output polarizations. The first (second) letter in each label denotes output (input) polarization. Input light has wavelength 810 nm and the far-field images are recorded with a CCD camera.

the geometric phase does not affect the diagonal elements in the Jones matrix J_{CP}, the off-diagonal elements satisfy the important restriction

$$t_{LR} = t_{RL}^*.$$

At first sight, it seems we can only get a single hologram from the Jones matrix, since the L-to-R and R-to-L near-field phase profiles are conjugates of each other. However, instead of forming the holograms in the far-field, we can design the metasurface to form a real image at a chosen distance f from the surface. In this case, it becomes possible to generate two different holograms: hologram ψ_L for L-to-R conversion and hologram ψ_R for R-to-L conversion.

Suppose from the standard GS algorithm we obtain the two phase masks φ_L and φ_R such that

$$FT(\mathrm{e}^{\mathrm{i}\varphi_L}) \cong \psi_L, \qquad FT(\mathrm{e}^{\mathrm{i}\varphi_R}) \cong \psi_R.$$

We now set

$$\arg(t_{RL}) = \arg\left(\mathrm{e}^{\mathrm{i}\varphi_L}\mathrm{e}^{-\mathrm{i}\frac{2\pi}{\lambda}(\sqrt{r^2+f^2}-f)} + \mathrm{e}^{-\mathrm{i}\varphi_R}\mathrm{e}^{\mathrm{i}\frac{2\pi}{\lambda}(\sqrt{r^2+f^2}-f)}\right), \tag{5.12}$$

where λ is the free-space wavelength.

For LCP incidence, the first term carries the converging lens phase factor $\mathrm{e}^{-\mathrm{i}\frac{2\pi}{\lambda}(\sqrt{r^2+f^2}-f)}$, so the hologram ψ_L appears as a real image at the plane $z = f$. The second term, with the diverging lens phase, produces only a virtual image that is not collected at the image plane. For RCP incidence, the condition $t_{LR} = t_{RL}^*$ flips the roles of the two terms, and a real image ψ_R is obtained at distance f. Thus, the metasurface can still generate two different holograms, one for each circular polarization, even when only the geometric phase is used.

This design scheme is summarized in figure 5.11. The rightmost panel shows an SEM image of a fabricated metasurface, with a unit cell period of 0.7 μm. Each cell consists of a geometric-phase grating of width 105 nm, rotated by half the required transmission phase. We note that a residual co-polarization signal is present, but since the converging lens phase is only applied to the cross-polarization channel, the co-polarized part does not focus and is not collected at the image plane. In practice, one can simply place a polarizer before the camera to capture the intended hologram.

To further enrich the variations of holograms one can obtain from such a geometric-phase-only scheme, we can use input polarizations other than circular. Linear and diagonal polarizations, being superpositions of LCP and RCP, produce interference between the two holograms ψ_L and ψ_R. For an input polarization of the form

$$\frac{1}{\sqrt{2}}(|L\rangle - e^{i2\theta}|R\rangle),$$

the generated hologram is proportional to $\psi_L - e^{i2\theta}\psi_R$. Choosing four angles $\theta = 0, \pi/4, \pi/2, 3\pi/4$ corresponds to input polarizations V, A, H, and D, and yields four different holograms proportional to $\psi_L - \psi_R$, $\psi_L - i\psi_R$, $\psi_L + \psi_R$, and $\psi_L + i\psi_R$, respectively.

By further designing the target holograms ψ_L and ψ_R to have equal amplitudes and a phase relation $\arg(\psi_L/\psi_R)$ restricted to one of the four constants $0, \pi/2, \pi, 3\pi/2$ in different regions of the image (figure 5.12), constructive and destructive interference can be controlled locally. In this way, by simply rotating the

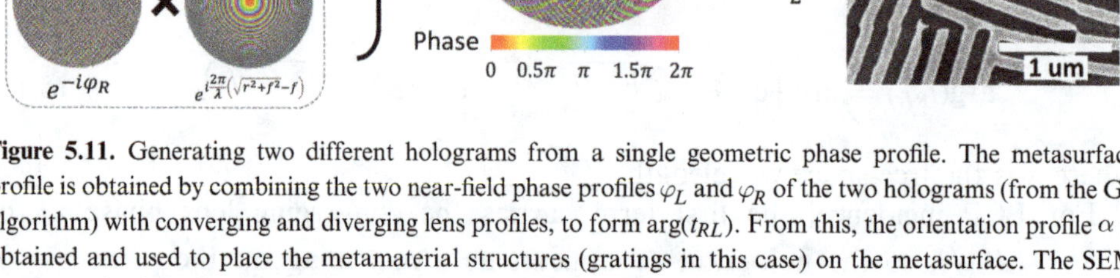

Figure 5.11. Generating two different holograms from a single geometric phase profile. The metasurface profile is obtained by combining the two near-field phase profiles φ_L and φ_R of the two holograms (from the GS algorithm) with converging and diverging lens profiles, to form $\arg(t_{RL})$. From this, the orientation profile α is obtained and used to place the metamaterial structures (gratings in this case) on the metasurface. The SEM image on the right shows a fabricated example. Reproduced from Liang *et al* (2025). CC BY 4.0.

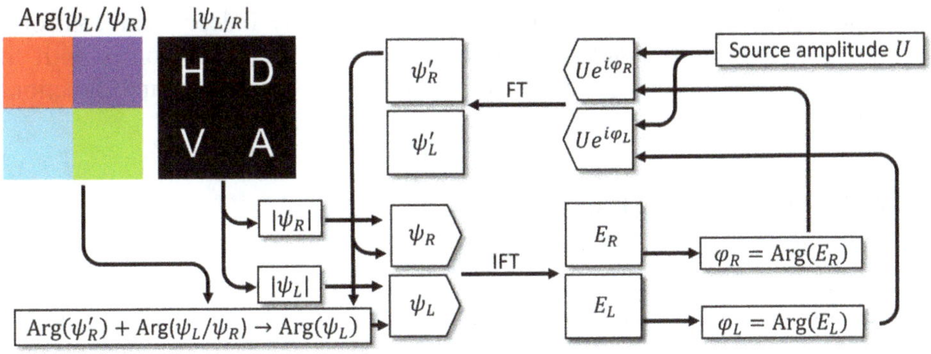

Figure 5.12. The modified GS algorithm to design two phase masks φ_L and φ_R that generate two holograms with a designed phase relationship. The target holograms ψ_L and ψ_R are chosen with equal amplitudes $|\psi_L| = |\psi_R|$ and a controlled phase relation $\arg(\psi_L/\psi_R)$, restricted to four constant values (schematically represented by colors) in different regions of the letters 'H', 'D', 'V', and 'A'. Reproduced from Liang *et al* (2025) CC BY 4.0.

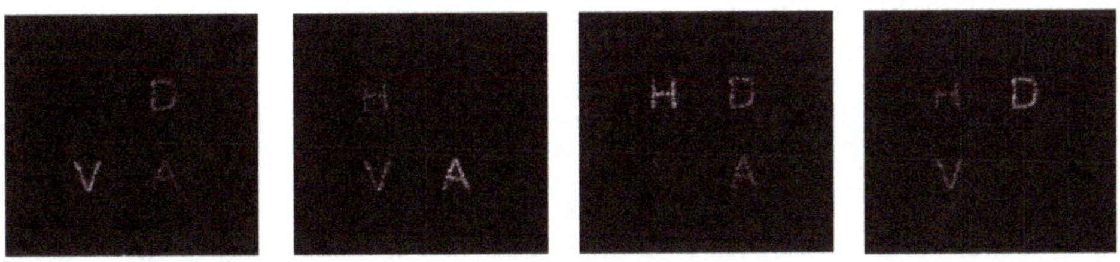

Figure 5.13. Holograms obtained by rotating the input linear polarization in steps of 45 degrees. Reproduced from Liang *et al* (2025) CC BY 4.0.

input linear polarization in steps of 45 degrees, four distinct holograms can be obtained (figure 5.13).

Before closing this section, it is worth mentioning that in addition to combining converging and diverging phase profiles, we may relax the unit-cell design by introducing two types of slots that generate geometric phase with slightly different orientations. As illustrated in figure 5.1(d), the unit cell contains two slot pairs with orientations $\alpha_1 = \alpha_0 + \Delta\alpha/2$ and $\alpha_2 = \alpha_0 - \Delta\alpha/2$. Treating the two contributions with an equal (50:50) weighting, the resulting Jones matrix in the circular-polarization basis is

$$J_{CP} = \begin{pmatrix} t_{LL} & t_{LR} \\ t_{RL} & t_{RR} \end{pmatrix} = \begin{pmatrix} |t|e^{i\varphi} & \cos(\Delta\alpha)\ |t'|e^{i(\varphi\prime-2\alpha_0)} \\ \cos(\Delta\alpha)\ |t'|e^{i(\varphi\prime+2\alpha_0)} & |t|e^{i\varphi} \end{pmatrix}.$$

In this way, we can also control the transmission amplitude which is proportional to $\cos(\Delta\alpha)$, while the geometric phase is still governed by the averaged orientation α_0. Comparing to controlling both propagation and geometric phases from the same structure, it suffers transmission power loss but have a higher robustness in fabrication. This two-angle approach provides genuinely independent control of cross-polarization transmission amplitude and phase, though it comes with the trade-off that the power loss from the cosine factor is not channeled into functional operation.

5.4 Surface wave manipulation

So far, we have seen how metasurfaces can shape light in the far field, producing holograms and other structured beams by carefully tailoring geometric and propagation phases. In many photonic applications, however, we are equally interested in controlling waves bound to a surface, such as surface plasmon polaritons (SPPs) supported at metal–dielectric interfaces. Because SPPs confine light below the diffraction limit and guide it along chip-scale geometries, they provide an important platform for integrated optics.

Metasurfaces offer a natural way to couple light from free space into such surface waves, and to manipulate their phase and direction once launched. In the following sections, we extend the concepts of geometric-phase metasurfaces to the control of SPPs, showing how they can be used for efficient excitation, hologram formation directly on the surface, and unidirectional launching of guided waves.

5.4.1 Geometric-phase matching rule

To have metasurfaces designed to couple free-space light into guided waves such as SPPs, we start with SPP radiation from just one meta-atom. Figure 5.14 illustrates the basic idea, where a nano-slot on a metal surface excites an SPP.

A far-field radiation pattern from an in-plane dipole moment $\mathbf{p}$ can be expressed through its out-of-plane electric field component at a direction $\hat{k}$ and a distance ρ:

$$E_z = C_0 \frac{e^{ik\rho}}{\sqrt{k\rho}} \hat{k} \cdot \mathbf{p} = C_0 \frac{e^{ik\rho}}{\sqrt{k\rho}} (p_x \cos\beta + p_y \sin\beta),$$

where k is the SPP wave number and $\beta = \arg(k_x + ik_y)$ is the azimuthal angle of the radiating direction. C_0 is the coupling coefficient. (The z-dependent scaling factor is omitted here for clarity.)

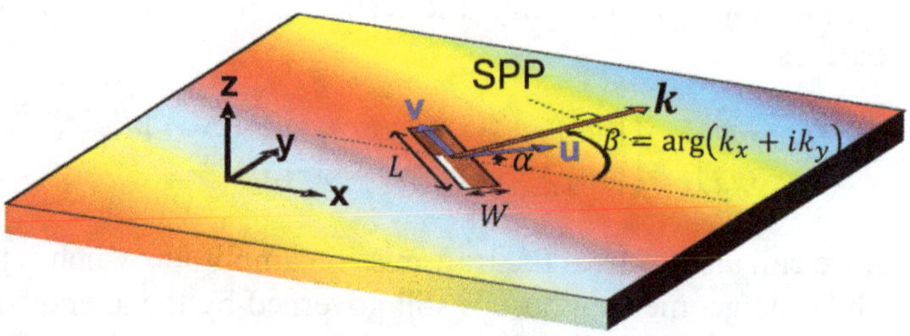

Figure 5.14. Schematic diagram for generating surface plasmon polaritons (SPPs) from nano-slots on a metal surface.

It is convenient to work in the circular polarization (CP) basis for the dipole moment:

$$\mathbf{p} = p_x\hat{x} + p_y\hat{y} = \frac{1}{\sqrt{2}}(p_L(\hat{x} + \mathrm{i}\hat{y}) + p_R(\hat{x} - \mathrm{i}\hat{y})),$$

which gives

$$E_z = C_0 \frac{\mathrm{e}^{\mathrm{i}k\rho}}{\sqrt{2k\rho}} (p_L\mathrm{e}^{\mathrm{i}\beta} + p_R\mathrm{e}^{-\mathrm{i}\beta}). \tag{5.13}$$

This shows that a dipole oriented as $\hat{x} \pm \mathrm{i}\hat{y}$ excites an SPP point source with topological charge ± 1.

Next, we consider how an incident plane wave couples into such a dipole moment. Suppose light propagates along $+z$ with field $\mathbf{E}_0$. It couples to the meta-atom, which has principal axes $\hat{u}$, $\hat{v}$, generating an in-plane dipole response $\mathbf{p}$. We can write this excitation in the familiar 'Jones matrix' form:

$$\begin{pmatrix} p_u \\ p_v \end{pmatrix} = \begin{pmatrix} t_{uu} & t_{uv} \\ t_{vu} & t_{vv} \end{pmatrix}\begin{pmatrix} a_u \\ a_v \end{pmatrix}, \tag{5.14}$$

where the incident field is

$$\mathbf{E}_0 = a_x\hat{x} + a_y\hat{y} = a_u\hat{u} + a_v\hat{v} = \frac{1}{\sqrt{2}}(a_L(\hat{x} + \mathrm{i}\hat{y}) + a_R(\hat{x} - \mathrm{i}\hat{y})).$$

Rotating into the CP basis, with $\hat{x} \pm \mathrm{i}\hat{y} = \mathrm{e}^{\pm\mathrm{i}\alpha}(\hat{u} \pm \mathrm{i}\hat{v})$, gives

$$\begin{pmatrix} p_L \\ p_R \end{pmatrix} = \begin{pmatrix} t_{LL}^{(0)} & \mathrm{e}^{-\mathrm{i}2\alpha}t_{LR}^{(0)} \\ \mathrm{e}^{\mathrm{i}2\alpha}t_{RL}^{(0)} & t_{RR}^{(0)} \end{pmatrix}\begin{pmatrix} a_L \\ a_R \end{pmatrix}, \tag{5.15}$$

where the Jones matrix without rotation ($\alpha = 0$) in CP basis is

$$\begin{pmatrix} t_{LL}^{(0)} & t_{LR}^{(0)} \\ t_{RL}^{(0)} & t_{RR}^{(0)} \end{pmatrix} = \frac{1}{2}\begin{pmatrix} 1 & -\mathrm{i} \\ 1 & \mathrm{i} \end{pmatrix}\begin{pmatrix} t_{uu} & t_{uv} \\ t_{vu} & t_{vv} \end{pmatrix}\begin{pmatrix} 1 & 1 \\ \mathrm{i} & -\mathrm{i} \end{pmatrix}.$$

For a structure with mirror symmetry (about $\hat{u}$ or $\hat{v}$), we have $t_{uv} = t_{vu} = 0$, so $t_{LL}^{(0)} = t_{RR}^{(0)} = (t_{uu} + t_{vv})/2$ and $t_{LR}^{(0)} = t_{RL}^{(0)} = (t_{uu} - t_{vv})/2$. Equation (5.15) then parallels the standard Jones matrix description of far-field optics, with the rotation angle α introducing a geometric phase of $\pm 2\alpha$.

By combining equations (5.13) and (5.15), we obtain

$$E_z \propto \frac{\mathrm{e}^{\mathrm{i}k\rho}}{\sqrt{2k\rho}}\begin{pmatrix} \mathrm{e}^{\mathrm{i}\beta} & \mathrm{e}^{-\mathrm{i}\beta} \end{pmatrix}\begin{pmatrix} t_{LL}^{(0)} & \mathrm{e}^{-2\mathrm{i}\alpha}t_{LR}^{(0)} \\ \mathrm{e}^{2\mathrm{i}\alpha}t_{RL}^{(0)} & t_{RR}^{(0)} \end{pmatrix}\begin{pmatrix} a_L \\ a_R \end{pmatrix}. \tag{5.16}$$

Now suppose we place many such atoms on a surface to generate a desired target SPP profile $E_z^{(\mathrm{tar})}$. Constructive interference among them requires a phase-matching rule. For LCP incidence,

$$E_z \propto \frac{e^{ik\rho}}{\sqrt{2k\rho}}\left(e^{i\beta}t_{LL}^{(0)} + e^{i(2\alpha-\beta)}t_{RL}^{(0)}\right).$$

The first term corresponds to conversion into a topological charge +1 SPP without geometric phase. The second term corresponds to conversion into a topological charge −1 SPP, and it carries the geometric phase 2α. To make use of this tunable phase, we set

$$2\alpha - \beta = \arg E_z^{(\mathrm{tar})} + \text{constant},$$

at each location. Here β specifies the radiation direction. Unlike in free-space optics, where we usually work near the paraxial direction, the direction β varies strongly along an SPP. As an approximation, we take β to follow the local power flow of the target SPP:

$$\beta = \arg(k_x + ik_y) = \arg\left((\partial_x + i\partial_y)E_z^{(\mathrm{tar})}\right) - \arg(E_z^{(\mathrm{tar})}) - \pi/2.$$

This leads to the matching rule for LCP incidence:

$$\alpha = \frac{1}{2}\arg\left((\partial_x + i\partial_y)E_z^{(\mathrm{tar})}\right) + \text{constant}. \tag{5.17}$$

For RCP incidence, the radiation from a single atom at the origin is

$$E_z \propto \frac{e^{ik\rho}}{\sqrt{2k\rho}}\left(e^{-i(2\alpha-\beta)}t_{LR}^{(0)} + e^{-i\beta}t_{RR}^{(0)}\right),$$

which gives the orientation rule

$$\alpha = -\frac{1}{2}\arg\left((\partial_x - i\partial_y)E_z^{(\mathrm{tar})}\right) + \text{constant}. \tag{5.18}$$

In summary, the matching rules for designing the orientation profile of meta-atoms to couple into a target SPP field are

$$\alpha_\pm = \pm\frac{1}{2}\arg\left((\partial_x \pm i\partial_y)E_z^{(\mathrm{tar})}\right) + \text{constant}, \tag{5.19}$$

where the + sign is used for LCP incidence and the − sign for RCP incidence. These rules ensure that the phases of the local scatterers add constructively to form the desired surface wave.

5.4.2 Surface plasmon polaritons holograms

The opposite geometric phases for LCP and RCP incidence often lead to opposite spin-dependent shifts in optical phenomena. Examples include the transverse displacement of a beam center in the optical spin Hall effect, or the spin-dependent changes in linear or orbital angular momentum introduced by a metasurface. The same principle also applies when exciting SPPs. If equation (5.17) is satisfied for LCP incidence with a chosen orientation profile α and a target $E_z^{(\mathrm{tar})}$, then by substituting $E_z^{(\mathrm{tar})*}$ into equation (5.18), one recovers the same orientation profile α (up to a global additive constant) for RCP incidence.

Figure 5.15 shows a practical platform: nano-slots patterned in a ring-shaped region around a central slit-free area. When either LCP or RCP light is incident from the top, the nano-slots reradiate SPPs toward the center, where they interfere to form a hologram with a specified intensity pattern. Importantly, the opposite geometric phases for the two spins do not prevent us from specifying two different holograms in the same central region.

As an example, suppose we want the hologram for LCP incidence to be a single point source, described by a standing-wave profile $J_0(k\rho)$, where k is the SPP wave number and ρ the radial distance. Instead of directly using this standing wave as $E_z^{(\mathrm{tar})}$ in equation (5.17), we decompose it into Hankel functions:

$$J_0(k\rho) = \frac{1}{2}\left(H_0^{(1)}(k\rho) + H_0^{(2)}(k\rho)\right).$$

The inward-radiating part, $H_0^{(2)}(k\rho)$, is taken as the effective target $E_z^{(\mathrm{tar})}$, since only the inward component contributes to the central region. The outward radiation is unimportant as we are not interested in the external field.

This idea lets us generalize the matching rule of equation (5.19). By noting that complex conjugation swaps inward and outward components, we define

$$\alpha = \frac{1}{2}\arg\left((\partial_x + \mathrm{i}\partial_y)(E_z^{(\mathrm{tar}+)} + E_z^{(\mathrm{tar}-)*})\right) + \text{constant}, \tag{5.20}$$

where $E_z^{(\mathrm{tar}+)}$ and $E_z^{(\mathrm{tar}-)}$ denote the inward-radiating parts of the two target holograms for LCP and RCP incidence.

For LCP incidence, only the inward-radiating component of $E_z^{(\mathrm{tar}+)} + E_z^{(\mathrm{tar}-)*}$ contributes, producing the hologram corresponding to $E_z^{(\mathrm{tar}+)}$. For RCP incidence,

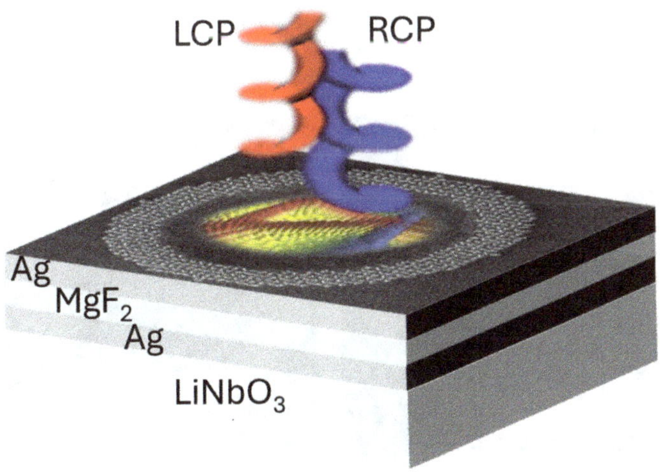

Figure 5.15. Identical nano-slots on a metal surface with orientation profile $\alpha(x, y)$. Light is normally incident from the top, and the orientation profile is designed to generate target SPP holograms in the central region.

the conjugated field $(E_z^{(\mathrm{tar}+)} + E_z^{(\mathrm{tar}-)*})^*$ is relevant, producing the hologram corresponding to $E_z^{(\mathrm{tar}-)}$. The constant in equation (5.20) can be taken as zero for simplicity. Thus, with a single set of nano-slots having orientation profile $\alpha(x, y)$, two different SPP holograms can be generated in the central region, depending on the input polarization.

Figure 5.16 illustrates this independent hologram capability. In the central region, a triangle appears for LCP incidence, while a cross appears for RCP incidence, even though the same metasurface is used. The bright ring surrounding the holograms comes from specular reflection at the nano-slots. This demonstrates how geometric-phase control allows the two spins to be harnessed coherently for flexible and independent generation of SPP holograms (Xiao *et al* 2015).

5.4.3 Dipolar and monopolar SPP excitation

So far, we have mainly considered the simplest case of SPP excitation by a nano-slot or nano-bar. Such structures behave like in-plane dipoles placed close to the surface, which reradiate into SPPs. While this mechanism works, we may ask: can we design the scatterer shape to couple more efficiently into the cross-polarized SPP channel?

To see how this works, recall from equation (5.14) that the cross-polarized SPP is strongest when the co-polarized Jones elements are suppressed, i.e., when

$$|t_{uu} + t_{vv}| \ll |t_{uu} - t_{vv}|.$$

One structure with this property is shown in figures 5.17(a) and (b): a gold bar placed above a gold surface with a thin SiO_2 spacer.

To measure how efficiently such a scatterer excites a particular SPP mode, we look at the field along a circle of radius $R = 800\,\mathrm{nm}$ and define

$$A^{(\upsilon)} = \int_0^{2\pi} E_z(R, \phi)\mathrm{e}^{-\mathrm{i}\sigma\phi}\,\mathrm{d}\phi, \tag{5.21}$$

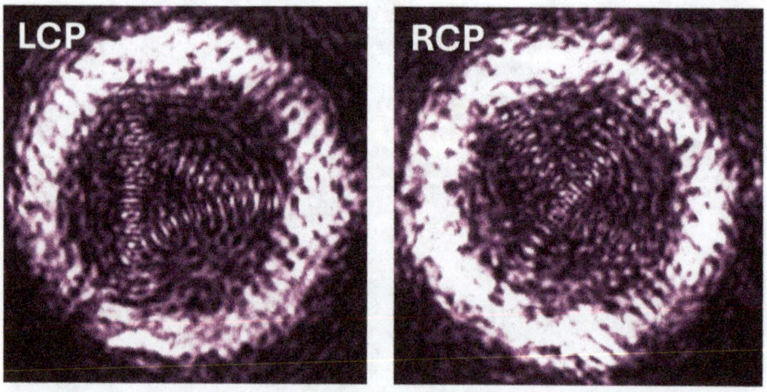

Figure 5.16. Experimental demonstration of independent SPP holograms. For LCP incidence, the hologram forms a triangle; for RCP incidence, the hologram forms a cross. The bright ring around the holograms is the direct specular reflection from the nano-slots. Reproduced from Xiao *et al* (2015). CC BY 4.0.

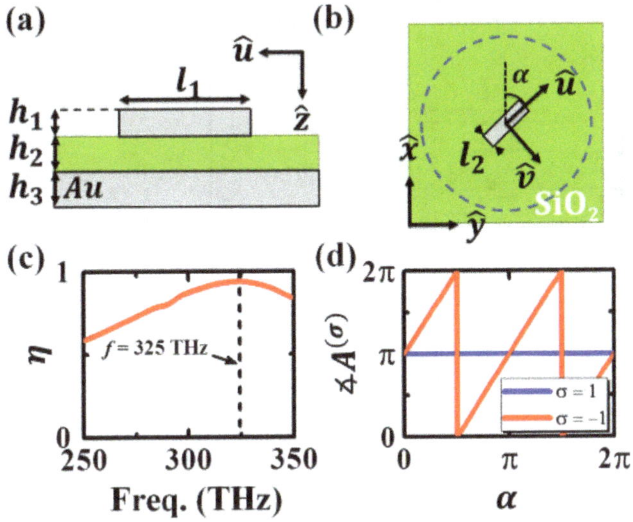

Figure 5.17. (a) and (b) Side and top views of a plasmonic particle on a gold film with SiO_2 spacer. Structural parameters: $h_1 = h_2 = 50\,\text{nm}$, $h_3 = 100\,\text{nm}$, $l_1 = 270\,\text{nm}$, $l_2 = 110\,\text{nm}$. (c) Cross-polarization conversion efficiency versus frequency. (d) Geometric phase $\arg(A^{(\sigma)})$ versus orientation angle α under LCP incidence. Unpublished results from Xiao and Li (2025).

where σ labels the topological charge of the SPP radiation. The efficiency is then given by

$$\eta_\sigma = \frac{|A^{(\sigma)}|^2}{\sum_{\sigma'=-1}^{1} |A^{(\sigma')}|^2}. \tag{5.22}$$

Numerical simulations (COMSOL) show that this design can reach a cross-polarization efficiency as high as $\eta = 0.93$ at 325 THz (figure 5.17(c)). Moreover, the phase of the scattered SPP follows the expected geometric law: $\arg(A^{(-1)}) \approx 2\alpha$, $\arg(A^{(1)}) \approx 0$ (figure 5.17(d)).

So far, we have dealt only with dipolar excitation, coming from in-plane dipoles. But a vertical dipole can also radiate into SPPs. Such radiation is monopolar in character, and together with the dipolar channel it forms a complete basis for building more complex SPP patterns.

To realize monopolar excitation, the dipole structure can be modified by connecting the horizontal bar to the surface with a vertical bar, as shown in figure 5.18(a). The offset l_s breaks mirror symmetry: when $l_s = 0$ the structure is symmetric and no monopolar component is generated, but for non-zero l_s the horizontal dipole partly couples to a vertical dipole. This effect is similar to bianisotropy in split-ring resonators. By tuning l_s, monopolar SPP generation can be maximized. Figure 5.18(b) shows that, after optimization, nearly perfect monopolar efficiency ($\eta_{\sigma=0} \approx 1$) can be achieved around 350 THz. In this case, the geometric phase varies linearly with the orientation angle:

$$\frac{\mathrm{d}}{\mathrm{d}\alpha} \arg(A^{(0)}) = 1.$$

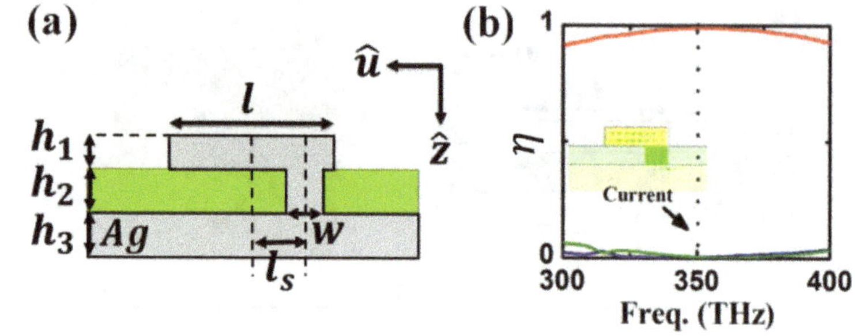

Figure 5.18. (a) An 'L'-shaped meta-atom, consisting of a horizontal metal bar connected to the surface by a vertical bar through a SiO_2 spacer. The offset is chosen as $l_s = (l - w)/2$. (b) Normalized $|A^{(\sigma)}|^2$ for $\sigma = 0, \pm 1$, and absolute values versus frequency for LCP incidence. The inset shows the current distribution. Unpublished results from Xiao and Li (2025).

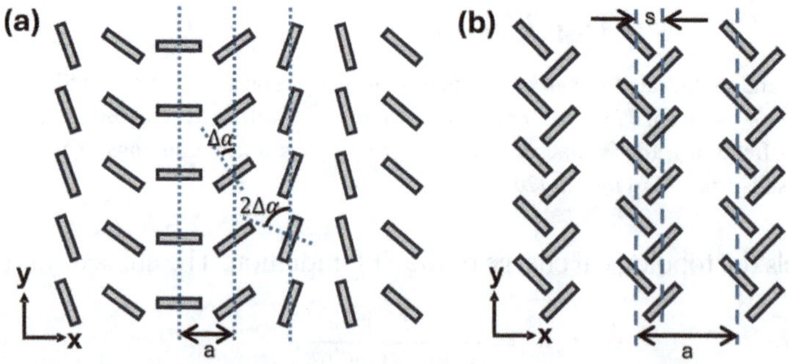

Figure 5.19. Unidirectional SPP propagation to positive x-direction by (a) nanoslots with additional rotation of $\Delta\alpha$ per lattice constant a or (b) periodic arrangement of two columns of nanoslots with $+45°$ and $-45°$ rotations with separation s and an overall lattice constant a. Light is normally incident along the positive z-direction onto the metal surface with the nanoslots. The excited SPP propagates to the right for LCP incidence and to the left for RCP incidence.

In summary, dipolar excitation (from in-plane dipoles) follows the law $\arg(A^{(-1)}) \approx 2\alpha$. Monopolar excitation (from vertical dipoles) follows $\arg(A^{(0)}) \propto \alpha$. By combining both channels, we can in principle design metasurfaces with higher efficiency and richer control of SPP wavefronts.

5.4.4 Unidirectional SPP excitation

To achieve unidirectional propagation, let's say only along the positive x-direction for LCP incidence, of SPP, we can utilize the geometric phase of the dipolar SPP radiation. Figure 5.19(a) shows a typical structure with a periodic arrangement of nano-slots, each with an orientation angle α and an additional rotation of $\Delta\alpha$ per lattice constant a (Huang *et al* 2013a). The geometric-phase matching condition to excite such SPP from the nano-slots can be described by the following equation:

$$2\alpha - \beta = \arg E_z^{(tar)} + \text{constant},$$

where β is the propagation angle of the SPP radiation direction, 0 in this case for a target SPP propagating along the positive x-direction. The target SPP field $E_z^{(tar)}$ is defined as the SPP field profile: $\exp(ikx)$ where k is the wavenumber of the SPP. Therefore, equivalently by directly invoking equation (5.19), we obtain that the orientation profile for the nano-slots should satisfy

$$2\Delta\alpha = ka \ (\text{mod } 2\pi), \tag{5.23}$$

for a linear orientation profile with an additional rotation angle of $\Delta\alpha$ at each distance a. For the radiation towards the negative x-direction, the orientation profile of the nanoslots is not designed to match $\exp(-ikx)$. There will be no radiation to the left and the SPP radiation is unidirectional in nature. The same argument can be applied to RCP incidence. In this case, the geometric phase has an additional minus sign, i.e., the orientation profile is designed to match $\exp(-ikx)$ for the radiation towards the negative x-direction. The SPP radiation is unidirectional in nature as well.

From a bottom-up approach, we can also sum the radiation from each nanoslot to obtain the total SPP radiation. For the ease of discussion, we take the nanoslot as a polarizer with $t_{uu} = 1$ and $t_{uv} = t_{vv} = t_{vu} = 0$ or $t_{LL} = t_{RR} = t_{LR} = t_{RL} = 1/2$. According to equation (5.16), it can be proved that each nanoslot radiates as $e^{i\sigma\alpha}\cos(\alpha - \beta)$, which can be decomposed as $(e^{i\sigma\beta} + e^{i\sigma(2\alpha-\beta)})/2$, where the first term is the 'spin-conserved' component with phase $\sigma\beta$ and the second term is the 'spin-flipped' component with phase $\sigma(2\alpha - \beta)$. A series sum over a number of nanoslots with a periodicity of a in the x-direction should be taken over the two terms separately so that the amplitude of each term does not depend on orientation α. Then, the first term will give a series sum with denominator $(1 - \exp(ika))$, which becomes zero when $ka = 2\pi m$ for integer m. This corresponds to the conventional grating method to excite SPP. For the second term, it can be shown that the series sum gives a denominator $(\exp(2i\sigma\Delta\alpha) - \exp(ika))$. When $\Delta\alpha$ is chosen to satisfy equation (5.23) (e.g., for LCP incidence $\sigma = 1$), the denominator becomes zero. This corresponds to the geometric-phase scheme to excite SPP.

In the above, the geometric-phase term is used to achieve constructive interference in the SPP radiation in one direction while in the opposite direction the geometric-phase matching condition is not satisfied so that SPP radiation is not purposely excited and the radiation dims when we use a larger number of nanoslots in the propagation direction to increase the contrast. On the other hand, it is actually possible to have constructive interference for SPP in one propagating direction and destructive interference in the opposite direction to be satisfied exactly. This can be achieved by using two columns of nanoslots with opposite orientation angles in each unit cell of lattice constant a, as shown in figure 5.19(b) (Lin *et al* 2013).

The total radiation $e^{i\sigma\alpha}\cos(\alpha - \beta)$ from each nanoslot includes both the 'spin-conserved' and 'spin-flipped' components. $\beta = 0, \pi$ for radiation to the right and left, respectively. For each unit cell, the left column has an orientation of $\alpha_1 = \pi/4$ and the right column has an orientation of $\alpha_2 = -\pi/4$. The separation distance between the two columns is denoted as s. The opposite orientation angles of the two columns give the same amplitude for the radiation coming from the two columns, creating the chance for completely constructive and destructive interference. Radiating to the right (positive x-direction), the radiation from the left column has a phase of $\sigma\alpha_1 + ks$, catching up to a phase of $\sigma\alpha_2$ from the right column. Radiating to the left (negative x-direction), the radiation from the right column has a phase of $\sigma\alpha_2 + ks$, catching up to a phase of $\sigma\alpha_1$ from the left column. Then, for LCP incidence, constructive interference for radiation to the left together with a destructive interference for radiation to the right translates to the following condition:

$$\alpha_1 - \alpha_2 = ks = \pi - ks \ (\mathrm{mod}\ 2\pi),$$

which can be satisfied by choosing $ks = \pi/2$ and $\alpha_1 = \pi/4$, $\alpha_2 = -\pi/4$. The same condition results from requesting constructive interference for radiation to the right together with a destructive interference for radiation to the left for RCP incidence. Finally, the two columns of nanoslots are periodically placed with a lattice constant of SPP wavelength λ_{SPP} so that the unidirectional SPP radiation can be enhanced from different unit cells.

5.5 Artificial electromagnetic boundary

A metasurface is a thin layer of metamaterial atoms, it can only be an approximation if we use the effective medium theory developed for bulk metamaterials to describe a metasurface using effective medium parameters (such as permittivity and permeability) with a very thin thickness. Actually, metasurfaces can be classified as penetrable and impenetrable depending on whether transmitted waves are allowed to pass through the metasurface or not. Here, for an impenetrable metasurface, we can describe it using an artificial electromagnetic boundary condition, which can be characterized by a generally anisotropic impedance tensor relating surface electric and magnetic fields.

To describe a reflective metasurface in a simple but powerful way, we can treat it as an *artificial electromagnetic boundary* (AEB). Suppose the metasurface lies in the x–y plane with surface normal $\hat{z}$. Then the tangential electric field is related to the tangential magnetic field by a 2×2 surface impedance matrix Z:

$$\begin{pmatrix} E_x \\ E_y \end{pmatrix} = \begin{pmatrix} Z_{11} & Z_{12} \\ Z_{21} & Z_{22} \end{pmatrix} \begin{pmatrix} -H_y \\ H_x \end{pmatrix}. \tag{5.24}$$

This description is valid as long as the metasurface is made of subwavelength elements, so that the response is local. Suppose the metasurface has principal axes along x- and y-directions and we can focus on their reflection coefficients. At normal incidence, the reflection simply introduces a phase shift on the electric field:

$$r_x = \mathrm{e}^{\mathrm{i}\phi_x}, \qquad r_y = \mathrm{e}^{\mathrm{i}\phi_y}, \tag{5.25}$$

where ϕ_x and ϕ_y are the reflection phases for TM (E-field along x) and TE (E-field along y) waves. By deriving reflection coefficients using equation (5.24) at normal incidence and comparing to equation (5.25), we obtain the impedance matrix

$$Z_t = -\begin{pmatrix} \mathrm{i}\cot\dfrac{\phi_x}{2} & 0 \\ 0 & \mathrm{i}\cot\dfrac{\phi_y}{2} \end{pmatrix}, \tag{5.26}$$

for incidence along the $+z$ direction on metasurface with surface normal $-\hat{z}$, and

$$Z_b = \begin{pmatrix} \mathrm{i}\cot\dfrac{\phi_x}{2} & 0 \\ 0 & \mathrm{i}\cot\dfrac{\phi_y}{2} \end{pmatrix}, \tag{5.27}$$

for incidence along the $-z$ direction on metasurface with surface normal $+\hat{z}$. The Z matrices only differ by a sign as it is the same metasurface just with opposite surface normals. The key point is that the *normal reflection phases act like constitutive parameters* for the metasurface. Once extracted, they can be used to predict reflection not only at normal incidence but also at oblique angles. In other words, the metasurface can be modeled as an effective boundary whose behavior is fully specified by these phases.

Figure 5.20 illustrates a typical reflective metasurface at microwave frequencies, consisting of metallic cross structures on a dielectric spacer backed by a conducting plate. Its simulated response can be compared with the effective boundary condition. The metasurface unit cell is designed to give specific reflection phases at normal incidence. Using these phases, we extract the impedance matrix of the corresponding artificial electromagnetic boundary. Figures 5.20(a) and (b) show the reflection

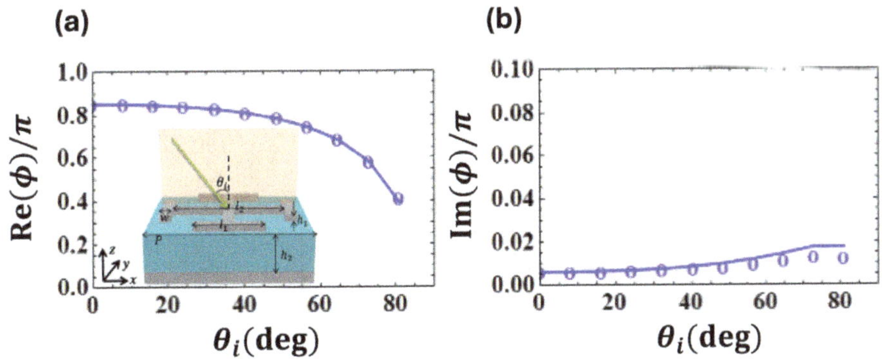

Figure 5.20. (a) and (b) Reflection phase as a function of incident angle at 10 GHz, comparing full-wave simulations (symbols) with the effective boundary condition (lines). Inset: unit cell of a reflective metasurface, consisting of metallic cross structures on a dielectric spacer backed by a conducting plate.

phase as a function of incidence angle at 10 GHz. The results from the metasurface (symbols) and from the boundary condition (solid lines) agree well even at large angles. This confirms that the effective boundary description captures the metasurface response accurately, also indicating the metasurface has a local response.

For a perfectly reflecting boundary (no transmission or absorption), the condition

$$\mathrm{Re}(\hat{z} \cdot \mathbf{E}^* \times \mathbf{H}) = 0$$

requires

$$Z_{11}^* = -Z_{11}, \qquad Z_{22}^* = -Z_{22}, \qquad Z_{21}^* = -Z_{12}. \tag{5.28}$$

5.5.1 Helicity-preserving mirror

5.5.1.1 Cross-circularly polarized mirror

A metasurface can be engineered to realize very different reflection properties by tailoring its effective impedance matrix. Suppose we want a mirror that always converts an incident circular polarization into its opposite handedness, i.e., a *cross-CP mirror*. Imposing this condition together with equation (5.28), we obtain

$$Z_{\text{cross-CP}} = \begin{pmatrix} 0 & -m \\ m & 0 \end{pmatrix}, \tag{5.29}$$

with real parameter m. This impedance is isotropic since $R(-\alpha)Z_{\text{cross-CP}}\,R(\alpha) = Z_{\text{cross-CP}}$ for any rotation $R(\alpha)$. Special cases include the perfect electric conductor (PEC, $m = 0$) and perfect magnetic conductor (PMC, $m \to \infty$), while other m values correspond to the more general perfect electromagnetic conductor (PEMC) class. The reflection coefficients in E-field at normal incidence follow as

$$r_{LR} = -\frac{\mathrm{i} - m}{\mathrm{i} + m}, \qquad r_{RL} = -\frac{\mathrm{i} + m}{\mathrm{i} - m}, \qquad r_{LL} = r_{RR} = 0, \tag{5.30}$$

where the first index labels the output polarization and the second the input.

5.5.1.2 Co-circularly polarized mirror

In contrast, a *co-CP mirror* reflects each circular polarization back to itself. In this case, it can be proved that co-circular total reflection (with surface normal $+\hat{z}$) requires

$$Z_{\text{co-CP}} = R(-\alpha)\begin{pmatrix} \mathrm{i}\cot\dfrac{\phi_u}{2} & 0 \\ 0 & \mathrm{i}\cot\dfrac{\phi_v}{2} \end{pmatrix}R(\alpha), \tag{5.31}$$

with ϕ_u, ϕ_v, α being real and $\phi_u - \phi_v = \pi \pmod{2\pi}$. The principal axes of the surface lie along $\hat{u} = \hat{x}\cos\alpha + \hat{y}\sin\alpha$ and $\hat{v} = -\hat{x}\sin\alpha + \hat{y}\cos\alpha$. The reflection coefficients become

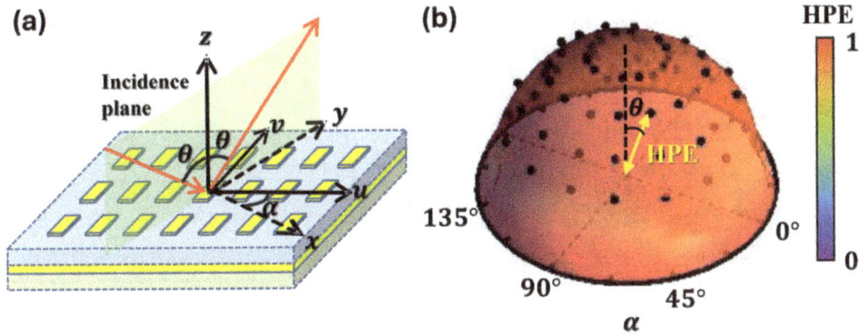

Figure 5.21. (a) Example metasurface implementation: array of gold nanorods on a dielectric spacer backed by gold film. (b) Measured (dots) and simulated (surface) helicity preserving efficiency versus incidence angle and azimuthal angle, showing high co-CP reflection. Xiao *et al* (2016) John Wiley & Sons. © 2016 WILEY-VCH Verlag GmbH & Co. KGaA, Weinheim.

$$
\begin{aligned}
r_{LL} &= \exp\left[2i\tan^{-1}\left(\frac{\tan(\frac{\phi_u}{2} - \alpha)}{\cos\theta}\right)\right], \\
r_{RR} &= \exp\left[2i\tan^{-1}\left(\frac{\tan(\frac{\phi_u}{2} + \alpha)}{\cos\theta}\right)\right], \\
r_{RL} &- r_{LR} = 0,
\end{aligned}
\tag{5.32}
$$

for incidence angle θ in the y–z plane. This condition is reciprocal ($Z_{12} = Z_{21}$) and only requires a π phase difference in reflection between the two orthogonal linear polarizations. At normal incidence, r_{LL} and r_{RR} acquire phases -2α and $+2\alpha$, consistent with the geometric phase obtained by rotating the metasurface elements. One way to realize a co-CP mirror is to place an array of gold nanorods above a gold film with a dielectric spacer, designed to enforce the π phase difference at a target wavelength. Figure 5.21(a) shows such a design at infrared wavelengths. The experimentally measured helicity preserving efficiency, defined as the co-CP reflected power over total reflected power, remains above 91% up to 60° incidence across different azimuthal angles (figure 5.21(b)), in good agreement with simulations. Such a co-CP mirror with wide-angle performance, also called a helicity-preserving mirror (Xiao *et al* (2016); Plum and Zheludev (2015)), can find applications in spin-controlled photonics and imaging systems.

5.5.2 Waveguide with metasurface boundaries

Metasurfaces can also be used to modify the boundary conditions of a parallel-plate waveguide. By describing each plate with an artificial electromagnetic boundary

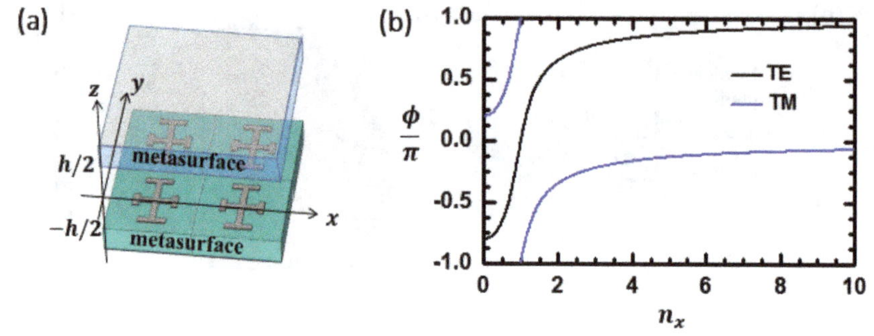

Figure 5.22. (a) A parallel-plate waveguide with metasurfaces as top and bottom boundaries. Their reflection phases ϕ_x and ϕ_y control the boundary conditions. (b) Relationship between reflection phase and mode index n_x for TE (black) and TM (blue) modes, solved for $h = 0.4\lambda$. Reproduced from Wei *et al* (2016).

(AEB), we can directly link the metasurface reflection phase to the dispersion of guided modes. This provides a powerful way to design waveguides that exhibit unusual propagation features, such as asymmetric transmission.

Consider a waveguide of height h, with top and bottom plates replaced by identical metasurfaces, as illustrated in figure 5.22(a). The boundaries at $z = \pm h/2$ can be described by the impedance matrices

$$Z_b = \begin{pmatrix} \mathrm{i}\cot\dfrac{\phi_x}{2} & 0 \\ 0 & \mathrm{i}\cot\dfrac{\phi_y}{2} \end{pmatrix}, \qquad Z_t = -\begin{pmatrix} \mathrm{i}\cot\dfrac{\phi_x}{2} & 0 \\ 0 & \mathrm{i}\cot\dfrac{\phi_y}{2} \end{pmatrix},$$

where ϕ_x and ϕ_y are the normal reflection phases for x-polarized and y-polarized waves, respectively. Because the metasurface thickness is deeply subwavelength, the AEB description remains valid for all incidence angles.

Applying Maxwell's equations with these boundary conditions leads to the following dispersion relations for the fundamental TE (E_y, H_x, H_z) and TM (H_y, E_x, E_z) modes propagating along x:

$$\begin{aligned} \text{TE:} \quad & \cot\frac{\phi_y}{2} = -\frac{k_0}{k_z}\cot\frac{k_z h}{2}, \\ \text{TM:} \quad & \cot\frac{\phi_x}{2} = \frac{k_z}{k_0}\tan\frac{k_z h}{2}, \end{aligned} \tag{5.33}$$

with $k_z = \sqrt{k_0^2 - n_x^2 k_0^2}$, $k_0 = 2\pi/\lambda$ the free-space wavenumber, and n_x the effective modal index. Figure 5.22(b) shows the relation between reflection phase and mode index for a height of waveguide fixed at $h = 0.4\lambda$. In practice, the metasurface geometry can be tuned to sweep the reflection phase ϕ: ϕ_x or ϕ_y across the full $[-\pi, \pi]$ range, allowing almost arbitrary control of the guided mode index.

5.5.3 Asymmetric wave propagation

Beyond uniform boundaries, we can imprint a spatially varying reflection phase along the x direction. Suppose

$$\phi(x) = \phi_0 + \xi(x - x_0),$$

where ξ is the phase gradient. Each reflection then provides an additional in-plane momentum ξ, modifying the reflection law to

$$k_{r,x} - k_{i,x} = \xi,$$

which is the reflective analog of the generalized Snell's law for transmissive metasurfaces. A positive gradient ($\xi > 0$) means that, after each bounce between the plates, the parallel wavevector increases. Eventually $k_{r,x} > k_0$, and the propagating bulk mode is converted into a surface-guided mode localized near the metasurface (figure 5.23). The capability of gradient metasurfaces to convert propagating waves into surface waves has been demonstrated firstly in microwave regime (Sun *et al* 2012).

To see the effect of a phase gradient more clearly, consider the ray picture in figure 5.23. For a forward-going wave (along $+x$), each reflection from the metasurface adds an in-plane momentum ξ. The reflected angle therefore increases step by step, until eventually $k_{r,x}$ exceeds k_0, and the bulk mode is converted into a surface-guided wave localized near the boundary. This behavior is illustrated by the forward ray paths (black lines) in figure 5.23, and can be seen directly in the simulated field distribution of figures 5.24(a) and (c): the wave launched from the left port gradually couples into a surface mode that hugs the metasurface.

The backward-going case (along $-x$) is quite different. Now each reflection subtracts momentum ($k_{r,x} = k_{i,x} - \xi$), so the propagation angle decreases. After multiple reflections, the effective $k_{r,x}$ may even change sign, meaning the wave is forced back toward the $+x$ direction. In other words, the backward input is blocked. This is shown by the green rays in figure 5.23, and confirmed in the simulations of figures 5.24(b) and (d): waves launched from the right cannot pass through, and are reflected back after a short distance.

Together, these results demonstrate that a simple linear phase gradient on the metasurface boundaries is enough to break the symmetry of a waveguide, enabling forward waves to evolve into surface-guided modes while suppressing backward transmission. In the following sections, we will see how the same AEB framework can be extended to design other types of guided-wave devices.

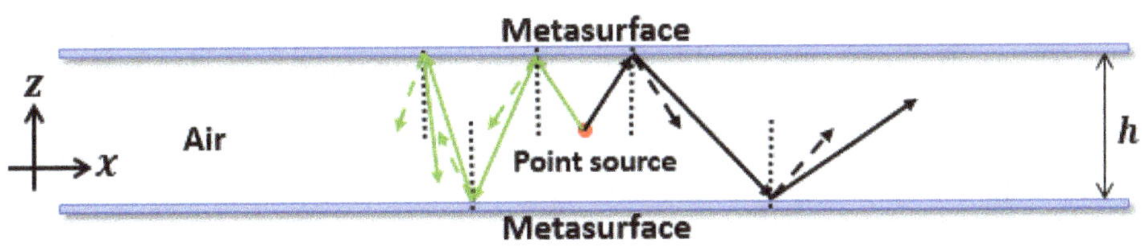

Figure 5.23. Concept of a gradient metasurface waveguide. A phase gradient ξ along x modifies the reflection law and gradually converts a propagating mode into a surface wave.

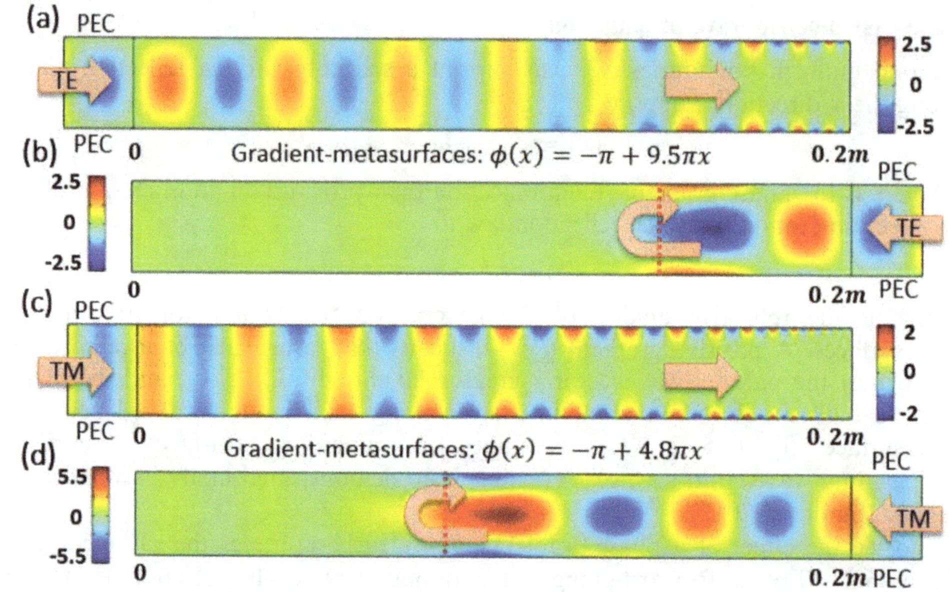

Figure 5.24. Simulated field patterns in a gradient metasurface waveguide at 10.5 GHz. (a,b) TE polarization (E_y). (c,d) TM polarization (H_y). Forward incidence couples into surface waves, while backward incidence is reflected. Results adapted from Xu *et al* (2013).

5.5.4 Zero-index waveguides

The metasurface waveguide approach also allows us to create effective indices that extend below unity, even approaching zero. This is quite different from surface-wave based metasurfaces, where the mode index is always larger than one due to leakage. By placing metasurfaces as both the top and bottom boundaries, the guided mode is fully confined, and its effective index can be tuned continuously from nearly zero to very large values by adjusting the reflection phases. This flexibility is useful for transformation-optics applications, where indices smaller than one can be employed to compensate phase accumulation along elongated paths.

In practice, the required reflection phase for a mode index n_y (for propagation along the y-direction) can be read directly from the same reflection phase–mode index curve used earlier for n_x (figure 5.22). By choosing appropriate normal reflection phases ϕ_x and ϕ_y, we can therefore realize arbitrary anisotropic index pairs (n_x, n_y). For simplicity, we neglect the small imaginary part of the reflection phase in the microwave regime, since it has little influence on the device performance. In experimental design, this imaginary component can be included by extracting it from the reflection of specific metasurface unit cells.

As a demonstration, we design a two-dimensional wave collimator based on a zero-index region. We first compare with the conventional effective-medium picture. Figure 5.25(a) shows a 2D simulation of Maxwell's equations for H_z polarization at free-space wavelength $\lambda = 0.04$ m. A zero-index material fills the space between a

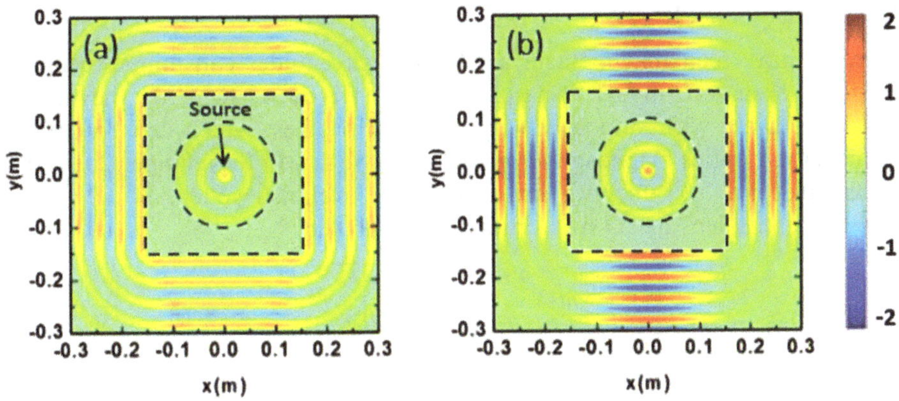

Figure 5.25. Wave collimator based on a zero-index region. (a) Conventional design using a bulk zero-index medium, simulated with 2D Maxwell's equations for H_z polarization at $\lambda = 0.04$ m. (b) Realization with a metasurface-loaded parallel-plate waveguide, simulated in 3D with AEB boundaries. The H_z field pattern at $z = 0$ shows the same collimation effect: four beams with flat wavefronts emerge from the square boundary, in excellent agreement with the 2D model. Reproduced from Wei *et al* (2016).

circular inner boundary and a square outer boundary (dashed lines). A point source placed at the center radiates waves that, due to the vanishing phase delay in the zero-index region, emerge as collimated beams with flat wavefronts aligned to the square boundary.

The same effect can be achieved using metasurfaces. In this realization, the 'free-space' region is implemented with metasurfaces designed for $n_x = n_y = 1$ (corresponding to reflection phase $\phi_x = \phi_y = 0$, i.e., a PMC-like boundary). The zero-index region is implemented with metasurfaces tuned to $n_x = n_y = 0$, corresponding to isotropic reflection phase $\phi_x = \phi_y \approx -0.8\pi$ (read from the TE curve of the reflection phase–mode index relation). A full 3D finite-element simulation (COMSOL Multiphysics) of this metasurface-loaded parallel-plate waveguide, described by its AEB boundaries, is shown in figure 5.25(b). The magnetic field distribution H_z is plotted at $z = 0$. The results confirm the collimation effect: radiation from the point source inside the circular domain is transformed into four beams with flat wavefronts, matching the square output interface. The agreement between the 3D metasurface simulation and the 2D effective-medium design demonstrates the consistency of the AEB approach.

In a conventional parallel-plate waveguide, the modal index is mainly determined by the refractive index of the bulk material filling the core, which places limits on how small or large the index can be. In contrast, a metasurface-loaded waveguide allows the modal index to be tuned directly through the reflection phase at the boundaries. This offers a much wider design space: the effective index can be varied continuously from nearly zero, up to very large values without relying on extreme bulk materials.

5.5.5 Gradient-index lenses

So far, we have seen that metasurface-loaded waveguides can mimic media with uniform or even zero index. A natural next step is to see whether they can also reproduce devices with smoothly varying index profiles, which can be useful for transformation optics (Mencagli *et al* 2014; Quarfoth and Sievenpiper 2013). Here, two examples are the Luneburg lens and the Maxwell fisheye lens.

The Luneburg lens in 2D has a radial index profile

$$n(r) = n_0\sqrt{2 - \left(\frac{r}{R}\right)^2},$$

which means that a plane wave arriving from any direction will be focused to a point on the rim of the lens. The Maxwell fisheye lens has a different profile,

$$n(r) = \frac{2n_0}{1 + \dfrac{r^2}{R^2}},$$

which has the remarkable property that a source placed on the rim will be focused to the opposite side. In both cases, n_0 is the surrounding background index. Here we take $n_0 = 1$ (air), a radius $R = 1$ m, and a working wavelength of $\lambda = 0.4$ m.

Figures 5.27(a) and (c) show what happens if we simulate these two lenses directly as bulk media for E_z polarization. The Luneburg lens indeed focuses a plane wave onto the rim, while the Maxwell fisheye lens transfers a point source on one side to the opposite side. These bulk simulations are useful references.

To realize the same effect with metasurface-loaded waveguides, we make use of the AEB framework. From the TM dispersion curve in figure 5.22(b), we can work out what reflection phase $\phi(r)$ corresponds to the required index $n(r)$. The resulting phase profiles for the Luneburg and fisheye lenses are plotted in figure 5.26.

When we impose these reflection phases on the metasurface boundaries, the 3D waveguide simulations in figures 5.27(b) and (d) show almost identical behavior to the bulk models. The Luneburg lens focuses the incoming plane wave onto the rim,

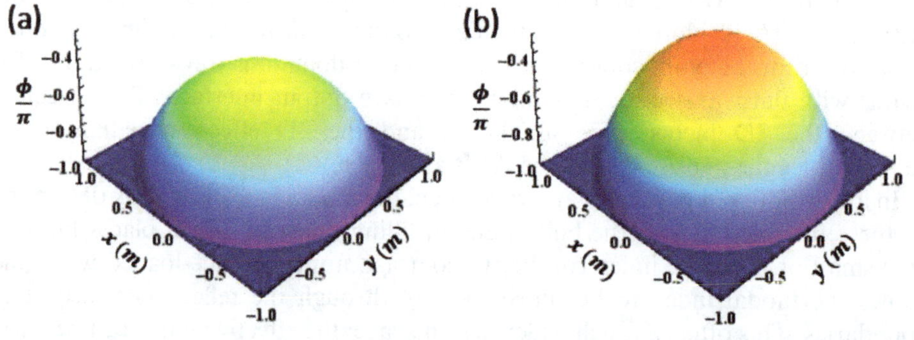

Figure 5.26. Required reflection phase $\phi(r)$ for (a) the Luneburg lens and (b) the Maxwell fisheye lens, obtained from the TM mode dispersion.

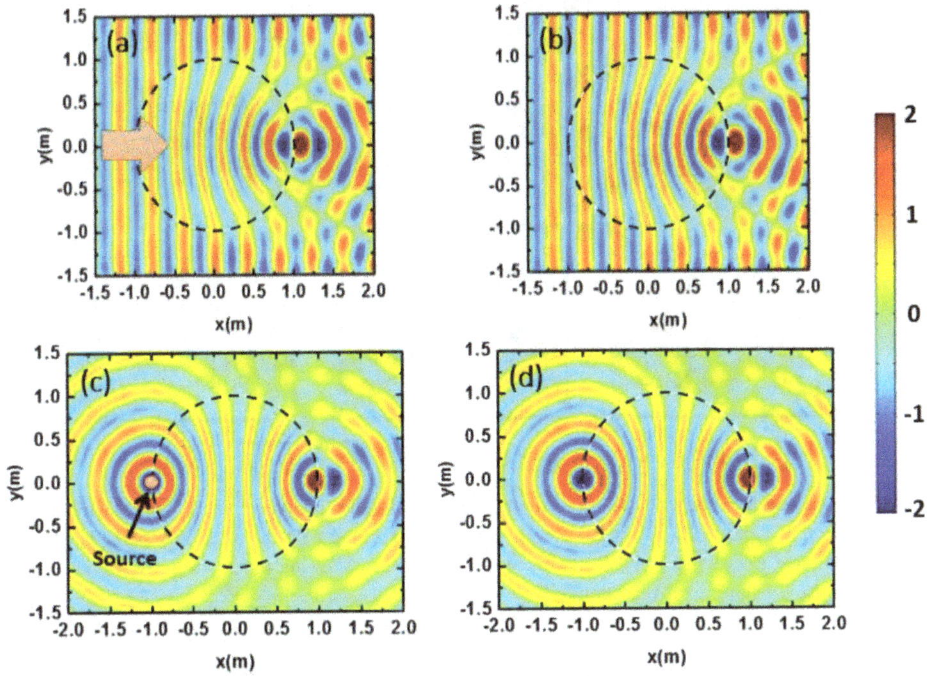

Figure 5.27. Field patterns (E_z) of gradient-index lenses. Top row: Luneburg lens illuminated by a plane wave from the left. Bottom row: Maxwell fisheye lens with a source on the left rim. (a,c) 2D bulk-medium simulations. (b,d) 3D metasurface-loaded waveguide simulations using AEB boundaries. The agreement is excellent. Reproduced from Wei *et al* (2016).

and the fisheye lens carries the source across to the opposite side. This close agreement tells us that gradient-index lenses, normally thought of as bulk optical devices, can also be built from flat metasurface-loaded waveguides. It is interesting to note that when a lens, such as the Eaton lens, has a singularity in its index profile, it is possible to apply a radial transformation to remove the singularity and still yield the same lens function (Tyc and Leonhardt 2008; Ma *et al* 2009).

5.5.6 Cloaking and illusion with a metasurface

So far we have used AEBs to design waveguides and lenses. A final step is to ask: can we design the reflection from a metasurface so that a curved boundary looks flat, or even so that one object appears as another? This is the principle behind metasurface cloaks and metasurface illusions.

Metasurface carpet cloak. Let us start with the simple case of a carpet cloak, where a bump on a reflective surface distorts the wavefront. Our aim is to design a metasurface that reshapes the reflection so it looks as if the bump were not there at all.

Figure 5.28(a) shows the problem: a curved bump on a conducting plane produces scattering different from a flat mirror. To correct this, we ask what

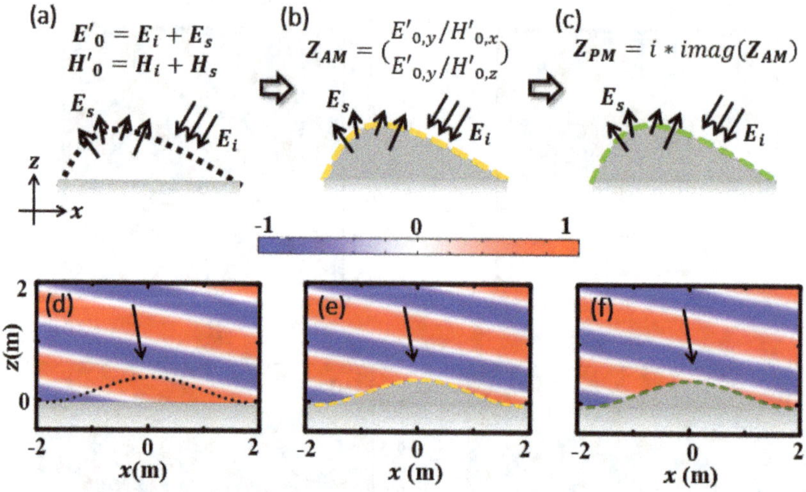

Figure 5.28. Metasurface carpet cloak. (a) Original bump on a reflective plane. (b) Active design with ideal complex surface impedance Z_{AM}. (c) Passive approximation keeping only the imaginary part, Z_{PM}. (d)–(f) Simulated scattered electric field (E_y) for an incident beam at 10°, showing that both the active and passive cloaks reproduce the flat-mirror reflection.

boundary condition would reproduce the fields of the *virtual* flat case, but applied along the actual curved surface. For TE polarization ($E \| y$), this required surface impedance can be written as

$$Z_{\mathrm{AM}} = \begin{pmatrix} E_y'/H_x' \\ E_y'/E_z' \end{pmatrix},$$

where E_y' and $H_{x,\,z}'$ are the electric and magnetic fields of the flat mirror, evaluated along the bump. This gives us the *active* design, since Z_{AM} is in general complex (figure 5.28(b)).

In practice, implementing the real part of this impedance would require gain or loss. A simple but effective approximation is to keep only the imaginary part,

$$Z_{\mathrm{PM}} = \mathrm{i}\ \mathrm{Im}(Z_{\mathrm{AM}}),$$

which corresponds to a purely reactive, hence passive, metasurface (figure 5.28(c)). Such a metasurface carpet cloak was demonstrated in the visible regime (Ni *et al* 2015).

Panels (d)–(f) of figure 5.28 show the simulated E_y field patterns. Both the active and passive cloaks restore the flat-mirror reflection, confirming that the passive approximation works very well in practice. This also highlights an important compromise: by compressing a bulky transformation-optics cloak into a thin sheet, the device only works perfectly at its design angle, with a typical tolerance of about $\pm 10°$.

From cloaks to illusions. Once the principle of 'reshaping reflection' is established, we can push it further for both cloaking and illusion effects (Liu *et al* 2018). Instead of

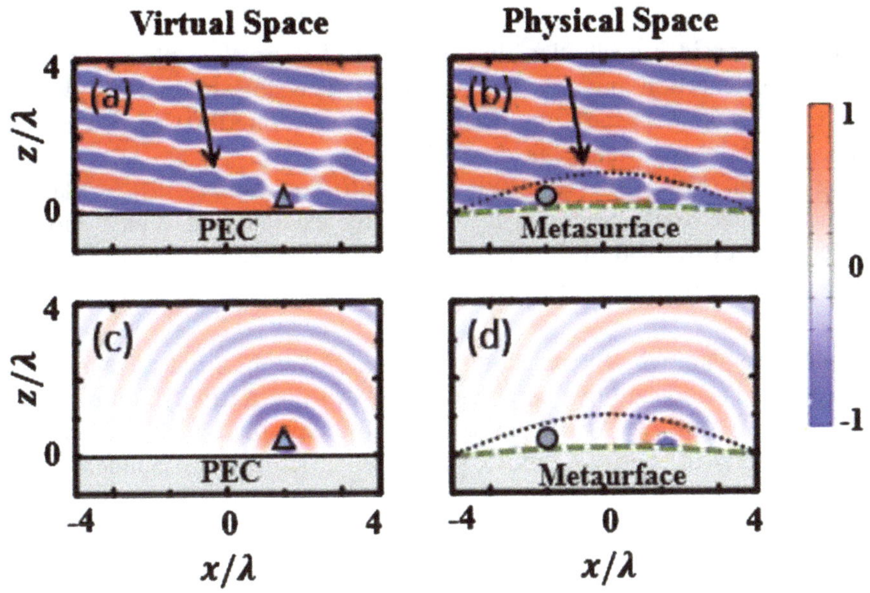

Figure 5.29. Metasurface illusion optics. Numerical demonstration where the scattering from a metallic circle (real object of permittivity −1) is transformed by the metasurface into that of a dielectric triangle (virtual object of permittivity 2.25) at an incident angle of 10°. Results from unpublished work Liu and Li *et al* (2025).

hiding a bump, we can make the scattering of a real object look like that of a different virtual object. Figure 5.29 illustrates such an illusion: the scattering from a metallic circle, when backed by a passive metasurface, is engineered to be indistinguishable from that of a dielectric triangle located elsewhere. To an external observer, the circle has effectively been replaced by the triangle.

Metasurface cloaks and illusions show how far the AEB approach can be extended. By carefully tailoring the reflection phase, a single patterned surface can hide a bump or even make one object appear as another. The cost of this ultra-thin implementation is limited angular bandwidth, but the advantage is striking simplicity.

5.6 Problems

Problem 5.1 For a 1D array of geometric-phase elements, the transmitted cross-polarized phase is $\Phi_{\rm PB}(x) = 2\sigma\,\alpha(x)$, with $\sigma = +1$ for LCP and $\sigma = -1$ for RCP incidence. Show that to realize a focusing lens of focal length f, the required orientation profile is

$$\alpha(x) = \frac{1}{2\sigma}\frac{2\pi}{\lambda}(f - \sqrt{f^2 + x^2}).$$

Using the fixed orientation profile above, determine the cross-polarized phase profile and its corresponding focal length if the handedness of the incident polarization is reversed.

Problem 5.2 Consider a circular ring of identical nano-slots placed on a metal surface with orientation angles to be determined. The slots are oriented with angle $\alpha(\phi)$ that may vary with the azimuthal coordinate ϕ along the ring. We illuminate the metasurface with normally incident LCP light.

Using the matching rules for SPP excitation (equation (5.19)), explain what orientation profile $\alpha(\phi)$ is required if we want the excited SPP to have monopolar character (with cylindrical SPP wavefront) radiating outward from the center. You can also use the formula in Bliokh *et al* (2008).

Problem 5.3 Explain why the reflection phases ϕ_x and ϕ_y of a reflective metasurface can be regarded as constitutive parameters, similar to permittivity and permeability in bulk media. Using the AEB formulation, show how these phases determine the impedance matrix for both top and bottom boundaries of a waveguide.

Problem 5.4 For a reflective metasurface with surface normal $+\hat{z}$ and impedance matrix

$$Z = \begin{pmatrix} \mathrm{i}\cot\dfrac{\Phi_x}{2} & 0 \\ 0 & \mathrm{i}\cot\dfrac{\Phi_x+\pi}{2} \end{pmatrix},$$

prove that the boundary condition can be written in terms of circular polarizations:

$$(\mathbf{E} \pm \mathrm{i}\mathbf{H}) \cdot \left(\hat{x}\sin\frac{\Phi_x}{2} \pm \hat{y}\cos\frac{\Phi_x}{2}\right) = 0.$$

This shows that left- and right-circularly polarized waves reflect back without changing their handedness, unlike reflections from a PEC or PMC surface.

Problem 5.5 Starting from the artificial electromagnetic boundary condition in equation (5.24), derive the TE dispersion relation in equation (5.33) for a parallel-plate waveguide with metasurfaces at $z = \pm h/2$.

Hints:

1. Write down the TE field distribution in the waveguide as a superposition of forward and backward waves in z.

2. Apply the boundary conditions using the impedance matrices Z_t and Z_b for the top and bottom plates.

3. Eliminate the field amplitudes to obtain the condition relating ϕ_y, k_z, and h.

References

Allen L, Beijersbergen M, Spreeuw R and Woerdman J 1992 Orbital angular momentum of light and the transformation of laguerre-gaussian laser modes *Phys. Rev.* **45** 8185–9

Allen L, Lembessis V and Babiker M 1996 Spin-orbit coupling in free-space laguerre-gaussian light beams *Phys. Rev.* **53** R2937–9

Berry M V 1987 The adiabatic phase and pancharatnamas phase for polarized light *J. Mod. Opt.* **34** 1401–7

Bliokh K and Bliokh Y 2004 Modified geometrical optics of a smoothly inhomogeneous isotropic medium: the anisotropy, berry phase, and the optical magnus effect *Phys. Rev.* **70** 026605

Bliokh K, Gorodetski Y, Kleiner V and Hasman E 2008 Coriolis effect in optics: unified geometric phase and spin-hall effect *Phys. Rev. Lett.* **101** 030404

Bomzon Z, Biener G, Kleiner V and Hasman E 2002 Space-variant pancharatnam-berry phase optical elements with computer-generated subwavelength gratings *Opt. Lett.* **27** 1141–3

Fan Y, Liang H, Wang Y, Chen S, Lai F, Chen M K, Xiao S, Li J and Tsai D P 2024 Dual-channel quantum meta-hologram for display *Adv. Photonics Nexus* **3** 016011

Gorodetski Y, Niv A, Kleiner V and Hasman E 2008 Observation of the spin-based plasmonic effect in nanoscale structures *Phys. Rev. Lett.* **101** 043903

Gorodetski Y, Shitrit N, Bretner I, Kleiner V and Hasman E 2009 Observation of optical spin symmetry breaking in nanoapertures *Nano Lett.* **9** 3016–9

Hosten O and Kwiat P 2008 Observation of the spin hall effect of light via weak measurements *Science* **319** 787–90

Huang L, Chen X, Bai B, Tan Q, Jin G, Zentgraf T and Zhang S 2013a Helicity dependent directional surface plasmon polariton excitation using a metasurface with interfacial phase discontinuity *Light: Sci. Appl.* **2** e70

Huang L, Chen X, Mühlenbernd H, Li G, Bai B, Tan Q, Jin G, Zentgraf T and Zhang S 2012 Dispersionless phase discontinuities for controlling light propagation *Nano Lett.* **12** 5750–5

Huang L *et al* 2013b Three-dimensional optical holography using a plasmonic metasurface *Nat. Commun.* **4** 2808

Kang M, Feng T, Wang H and Li J 2012 Wave front engineering from an array of thin aperture antennas *Opt. Express* **20** 15882–90

Karimi E, Schulz S, Leon I D, Qassim H, Upham J and Boyd R 2014 Generating optical orbital angular momentum at visible wavelengths using a plasmonic metasurface *Light: Sci. Appl.* **3** e167

Li G, Kang M, Chen S, Zhang S, Pun E, Cheah K and Li J 2013 Spin-enabled plasmonic metasurfaces for manipulating orbital angular momentum of light *Nano Lett.* **13** 4148–51

Li Y, Liu Y, Ling X, Yi X, Zhou X, Ke Y, Luo H, Wen S and Fan D 2015 Observation of photonic spin hall effect with phase singularity at dielectric metasurfaces *Opt. Express* **23** 1767–74

Liang H, Wong W C, An T and Li J 2025 Metasurface-enabled quantum holograms with hybrid entanglement *Adv. Photon.* **7** 026006

Liberman V and Zel'dovich B 1992 Spin-orbit interaction of a photon in an inhomogeneous medium *Phys. Rev.* **46** 5199–207

Lin J, Mueller J P B, Wang Q, Yuan G, Antoniou N, Yuan X and Capasso F 2013 Polarization-controlled tunable directional coupling of surface plasmon polaritons *Science* **340** 331–4

Liu Y and Li J 2025 Metasurface approach to implementing complementary media for optical illusion (unpublished)

Liu Y, Xu J, Xiao S, Chen X and Li J 2018 Metasurface approach to external cloak and designer cavities *ACS Photon* **5** 1749

Luo W, Xiao S, He Q, Sun S and Zhou L 2015 Photonic spin hall effect with nearly 100% efficiency *Adv. Opt. Mater.* **3** 1102

Ma Y, Ong C K, Tyc T and Leonhardt U 2009 An omnidirectional retroreflector based on the transmutation of dielectric singularities *Nat. Mater.* **8** 639–42

Marrucci L, Manzo C and Paparo D 2006 Optical spin-to-orbital angular momentum conversion in inhomogeneous anisotropic media *Phys. Rev. Lett.* **96** 163905

Mencagli M, Martini E, González-Ovejero D and Maci S 2014 Metasurface transformation optics *J. Opt.* **16** 125106

Ni X, Emani N, Kildishev A, Boltasseva A and Shalaev V 2012 Broadband light bending with plasmonic nanoantennas *Science* **335** 427

Ni X, Kildishev A and Shalaev V 2013 Metasurface holograms for visible light *Nat. Commun.* **4** 2807

Ni X, Wong Z, Mrejen M, Wang Y and Zhang X 2015 An ultrathin invisibility skin cloak for visible light *Science* **349** 1310–4

O'Neil A T, Macvicar I, Allen L and Padgett M 2002 Intrinsic and extrinsic nature of the orbital angular momentum of a light beam *Phys. Rev. Lett.* **88** 053601

Onoda M, Murakami S and Nagaosa N 2004 Hall effect of light *Phys. Rev. Lett.* **93** 083901

Plum E and Zheludev N 2015 Chiral mirrors *Appl. Phys. Lett.* **106** 221901

Quarfoth R and Sievenpiper D 2013 Artificial tensor impedance surface waveguides *IEEE Trans. Antennas Propag.* **61** 3597–606

Rubin N A, Zaidi A, Dorrah A H, Shi Z and Capasso F 2021 Jones matrix holography with metasurfaces *Sci. Adv.* **7** eabg7488

Shitrit N, Bretner I, Gorodetski Y, Kleiner V and Hasman E 2011 Optical spin hall effects in plasmonic chains *Nano Lett.* **11** 2038–42

Sun S, He Q, Xiao S, Xu Q, Li X and Zhou L 2012 Gradient-index meta-surfaces as a bridge linking propagating waves and surface waves *Nat. Mater.* **11** 426–31

Tomita A and Chiao R 1986 Observation of berry's topological phase by use of an optical fiber *Phys. Rev. Lett.* **57** 937–40

Tyc T and Leonhardt U 2008 Transmutation of singularities in optical instruments *New J. Phys.* **10** 115038

Wei P, Xiao S, Xu Y, Chen H, Chu S and Li J 2016 Metasurface-loaded waveguide for transformation optics applications *J. Opt.* **18** 044015

Wen D *et al* 2015 Helicity multiplexed broadband metasurface holograms *Nat. Commun.* **6** 8241

Xiao S, Mühlenbernd H, Li G, Kenney M, Liu F, Zentgraf T, Zhang S and Li J 2016 Helicity-preserving omnidirectional plasmonic mirror *Adv. Opt. Mater.* **4** 654–8

Xiao S and Li J 2025 Dipolar and monopolar SPP excitation from freespace with high efficiency (unpublished)

Xiao S, Zhong F, Liu H, Zhu S and Li J 2015 Flexible coherent control of plasmonic spin-hall effect *Nat. Commun.* **6** 8360

Xu Y, Gu C, Hou B, Lai Y, Li J and Chen H 2013 Broadband asymmetric waveguiding of light without polarization limitations *Nat. Commun.* **4** 2561

Yin X, Ye Z, Rho J, Wang Y and Zhang X 2013 Photonic spin hall effect at metasurfaces *Science* **339** 1405–7

Yu N, Genevet P, Kats M, Aieta F, Tetienne J, Capasso F and Gaburro Z 2011 Light propagation with phase discontinuities: Generalized laws of reflection and refraction *Science* **334** 333–7

Zheng G, Mühlenbernd H, Kenney M, Li G, Zentgraf T and Zhang S 2015 Metasurface holograms reaching 80% efficiency *Nat. Nanotechnol.* **10** 308–12

IOP Publishing

Essential Concepts in Metamaterials

Jensen Li

Chapter 6

Acoustic metamaterials

In this chapter, we will explore the extension of the concepts of metamaterials from electromagnetic waves to acoustic waves. In the macroscopic level, the governing equations of these waves share some features similar to those of electromagnetic waves, but they involve different physical quantities such as pressure and velocity. For example, we will discuss how the principles of field-averaging and transformation optics can be applied to these types of waves. On the other hand, we will need very radically different design strategies to achieve the desired metamaterial properties, as the material parameters of acoustic and metamaterials are not as straightforward as those of electromagnetic metamaterials.

6.1 Acoustic field-averaging method

We now adapt the field-averaging method to acoustic waves. The microscopic fields inside a metamaterial unit cell consist of the pressure $p(\mathbf{r})$ and particle velocity $\mathbf{v}(\mathbf{r})$. In analogy with polarization and magnetization in electromagnetism, the unit cell may contain secondary sources: a scalar monopolarization M (representing local volume change) and a vector dipolar polarization $\mathbf{D}$ (representing dipolar flow). It is also useful to define two auxiliary quantities: the momentum density

$$\boldsymbol{\mu} = \rho_0(\mathbf{v} + \mathbf{D}),$$

and the volumetric strain

$$-\epsilon = \beta_0(p + M),$$

where ρ_0 and β_0 are the density and compressibility of the background medium.

The microscopic wave equations then read

$$\begin{aligned} \nabla p &= \mathrm{i}\omega\boldsymbol{\mu} = \mathrm{i}\omega\rho_0(\mathbf{v} + \mathbf{D}), \\ \nabla \cdot \mathbf{v} &= -\,\mathrm{i}\omega\epsilon = \mathrm{i}\omega\beta_0(p + M). \end{aligned} \tag{6.1}$$

doi:10.1088/978-0-7503-1131-1ch6

Within the unit cell, the fields vary rapidly due to resonances. To homogenize the description, we replace them by effective fields that vary smoothly. By averaging equation (6.1) over a cell volume V and using the divergence theorem, we define the effective strain and momentum density as

$$\begin{aligned} \epsilon_{\text{eff}} &= \frac{1}{V}\int_V \epsilon \, \mathrm{d}V = -\frac{1}{\mathrm{i}\omega V}\oint_{\partial V} \mathbf{v} \cdot d\mathbf{S}, \\ \mu_{\text{eff}} &= \frac{1}{V}\int_V \mu \, \mathrm{d}V = \frac{1}{\mathrm{i}\omega V}\oint_{\partial V} p \, \mathrm{d}\mathbf{S}. \end{aligned} \tag{6.2}$$

These definitions rely only on surface fields, treating the unit cell as a 'black box.'

Inner-product identity
To connect microscopic sources to effective fields, we compare the true solution $(p, \mathbf{v})$ in the inhomogeneous cell with a reference 'local' solution $(p_{\text{loc}}, \mathbf{v}_{\text{loc}})$ in the homogeneous background medium:

$$\nabla p_{\text{loc}} = \mathrm{i}\omega\rho_0 \mathbf{v}_{\text{loc}}, \qquad \nabla \cdot \mathbf{v}_{\text{loc}} = \mathrm{i}\omega\beta_0 p_{\text{loc}}.$$

A straightforward calculation gives

$$\nabla \cdot (p_{\text{loc}}^* \mathbf{v} + p\, \mathbf{v}_{\text{loc}}^*) = i\omega\Big(\beta_0 p_{\text{loc}}^* M + \rho_0 \mathbf{v}_{\text{loc}}^* \cdot \mathbf{D}\Big).$$

Integrating over the unit cell volume and applying the divergence theorem leads to the acoustic analog of the EM inner-product identity:

$$\frac{1}{\mathrm{i}\omega}\oint_{\partial V} (p_{\text{loc}}^* \mathbf{v} + p\, \mathbf{v}_{\text{loc}}^*) \cdot d\mathbf{S} = \int_V \Big(\beta_0 p_{\text{loc}}^* M + \rho_0 \mathbf{v}_{\text{loc}}^* \cdot \mathbf{D}\Big) \, \mathrm{d}V. \tag{6.3}$$

Effective fields in 3D
We first define the background wavenumber and acoustic impedance as

$$k = \frac{\omega}{c} = \omega\sqrt{\rho_0\beta_0}, \qquad \eta_0 = \sqrt{\rho_0/\beta_0}.$$

The local field arriving at the center of the unit cell, up to dipolar expansion, can be expressed using spherical Bessel functions as

$$\begin{aligned} p_{\text{loc}}(\mathbf{r}) &= p_{\text{loc}}(0)\, j_0(kr) + 3\mathrm{i}\,\eta_0\, j_1(kr)\, \mathbf{v}_{\text{loc}}(0) \cdot \hat{r}, \\ \mathbf{v}_{\text{loc}}(\mathbf{r}) &= 3\mathbf{v}_{\text{loc}}(0) \cdot \left(\frac{j_1(kr)}{kr}\overline{\overline{I}} + r\frac{\mathrm{d}}{\mathrm{d}r}\left(\frac{j_1(kr)}{kr}\right)\hat{r}\hat{r}\right) - \mathrm{i}\,\frac{p_{\text{loc}}(0)}{\eta_0}\, j_0'(kr)\, \hat{r}, \end{aligned} \tag{6.4}$$

where j_l is the spherical Bessel function of order l; $\overline{\overline{I}}$ is the identity tensor, and $\hat{r} = \mathbf{r}/r$. Now, we expand the local solution to first spatial order about the cell center:

$$\begin{aligned} p_{\text{loc}}(\mathbf{r}) &\approx p_{\text{loc}}(0) + \mathrm{i}\omega\rho_0 \mathbf{v}_{\text{loc}}(0) \cdot \mathbf{r}, \\ \mathbf{v}_{\text{loc}}(\mathbf{r}) &\approx \mathbf{v}_{\text{loc}}(0) + \frac{1}{3}\mathrm{i}\omega\beta_0 p_{\text{loc}}(0)\mathbf{r}. \end{aligned} \tag{6.5}$$

Substituting into equation (6.3) and collecting coefficients of $p_{\text{loc}}(0)^*$ and $\mathbf{v}_{\text{loc}}(0)^*$ motivates the definitions

$$\begin{aligned}
M_{\text{eff}} &= \frac{1}{V}\int_V \left(M - \frac{1}{3}\mathrm{i}\omega\rho_0\mathbf{r}\cdot\mathbf{D}\right)\mathrm{d}V,\\
\mathbf{D}_{\text{eff}} &= \frac{1}{V}\int_V (\mathbf{D} - \mathrm{i}\omega\beta_0\mathbf{r}\,M)\,\mathrm{d}V,\\
p_{\text{eff}} &= \frac{1}{3V}\oint_{\partial V} p\,\mathbf{r}\cdot\mathrm{d}\mathbf{S},\\
\mathbf{v}_{\text{eff}} &= \frac{1}{V}\oint_{\partial V} \mathbf{r}\,\mathbf{v}\cdot\mathrm{d}\mathbf{S},
\end{aligned} \tag{6.6}$$

so that the inner-product identity becomes the effective constitutive relations which take the form

$$-\epsilon_{\text{eff}} = \beta_0(p_{\text{eff}} + M_{\text{eff}}), \qquad \boldsymbol{\mu}_{\text{eff}} = \rho_0(\mathbf{v}_{\text{eff}} + \mathbf{D}_{\text{eff}}). \tag{6.7}$$

The consideration of field-averaging can highlight the analogy between the electromagnetic and acoustic waves, table 6.1 summarizes the correspondence between electromagnetic and acoustic quantities.

Dimensional reduction to 2D and 1D

Because acoustics is governed by a scalar pressure field, the above derivation simplifies when the problem is reduced from three to lower spatial dimensions. The replacements are systematic:

- The factor $1/3$ in 3D expansions in both local and effective field formulas becomes $1/2$ in 2D and 1 in 1D.
- The position vector $\mathbf{r} = x\hat{x} + y\hat{y} + z\hat{z}$ is replaced by $\mathbf{r}_T = x\hat{x} + y\hat{y}$ (the in-plane vector (x, y)) in 2D, and by $z\hat{z}$ in 1D.

With these replacements, the formulas in equation (6.6) immediately yield the correct definitions of M_{eff}, $\mathbf{D}_{\text{eff}}$, p_{eff}, and $\mathbf{v}_{\text{eff}}$ in reduced dimensions.

Table 6.1. Correspondence between electromagnetic and acoustic quantities.

Concept	Electromagnetism	Acoustics
Primary fields	$\mathbf{E}$, $\mathbf{H}$	p, $\mathbf{v}$
Auxiliary fields	$\mathbf{D} = \epsilon_0\mathbf{E} + \mathbf{P}$, $\mathbf{B} = \mu_0\mathbf{H} + \mathbf{M}$	$-\epsilon = \beta_0(p + M)$, $\boldsymbol{\mu} = \rho_0(\mathbf{v} + \mathbf{D})$
Secondary sources	Polarization $\mathbf{P}$, Magnetization $\mathbf{M}$	Monopolarization M, Dipolar polarization $\mathbf{D}$
Constitutive parameters	ϵ_{eff}, μ_{eff}	β_{eff}, ρ_{eff}
Effective field extraction	Surface integrals of $\mathbf{E}$, $\mathbf{H}$	Surface integrals of p, $\mathbf{v}$
Local-field approximation	Linear polynomial in $\mathbf{r}$	Linear polynomial in $\mathbf{r}$
Effective-local-fields relationship	same for $\mathbf{E}$ and $\mathbf{H}$	different for p and $\mathbf{v}$

For the case of two dimensions with fields independent of z and velocities and dipolar polarizations confined to x–y plane:

$$\begin{aligned} M_{\text{eff}} &= \frac{1}{V}\int_V \left(M - \frac{1}{2}\mathrm{i}\omega\rho_0\mathbf{r}_T\cdot\mathbf{D}\right)\mathrm{d}V, \\ \mathbf{D}_{\text{eff}} &= \frac{1}{V}\int_V (\mathbf{D} - \mathrm{i}\omega\beta_0\mathbf{r}_T\, M)\,\mathrm{d}V, \\ p_{\text{eff}} &= \frac{1}{2V}\oint_{\partial V} p\,\mathbf{r}_T\cdot\mathrm{d}\mathbf{S}, \\ \mathbf{v}_{\text{eff}} &= \frac{1}{V}\oint_{\partial V} \mathbf{r}_T\,\mathbf{v}\cdot\mathrm{d}\mathbf{S}, \end{aligned} \tag{6.8}$$

where the volume V has a cylindrical shape with an arbitrary height as the field and material being invariant along z. The surface integrals are performed on the cylindrical surface and the top and bottom surfaces do not contribute.

For the case of 1D with fields invariant in x, y, velocities and dipolar polarizations parallel to $\hat{z}$:

$$\begin{aligned} M_{\text{eff}} &= \frac{1}{V}\int_V (M - \mathrm{i}\omega\rho_0 z\hat{z}\cdot\mathbf{D})\,\mathrm{d}V, \\ \mathbf{D}_{\text{eff}} &= \frac{1}{V}\int_V (\mathbf{D} - \mathrm{i}\omega\beta_0 z\hat{z}\, M)\,\mathrm{d}V, \\ p_{\text{eff}} &= \frac{1}{V}\oint_{\partial V} p\, z\hat{z}\cdot\mathrm{d}\mathbf{S}, \\ \mathbf{v}_{\text{eff}} &= \frac{1}{V}\oint_{\partial V} z\hat{z}\,\mathbf{v}\cdot\mathrm{d}\mathbf{S}, \end{aligned} \tag{6.9}$$

where the volume V is a slab of thickness a in z. The surface integrals are performed on the two end faces at $z = -a/2$ and $z = a/2$. The side surfaces do not contribute. In the application of extracting effective fields using the inner-product identity, a is chosen as the lattice constant along z.

6.2 Understanding acoustic metamaterials with dipolar model

6.2.1 Acoustic Clausius–Mossotti equation

We now develop an acoustic analog of the Clausius–Mossotti (CM) equation using the field-averaging framework described above. Imagine a periodic array of identical, subwavelength particles embedded in a background medium with density ρ_0 and compressibility β_0. Each particle is illuminated by a local field (p_{loc}, $\mathbf{v}_{\text{loc}}$) and, in response, generates secondary sources: a monopole m and a dipole $\mathbf{d}$.

Total fields around one inclusion. In the long-wavelength multipole expansion, and keeping monopole and dipole terms, the total fields around a particle can be expressed as

$$p(\mathbf{r}) = p_{\text{loc}}(\mathbf{r}) + \frac{ik^3 m}{4\pi} h_0(kr) - \eta_0 \frac{k^3}{4\pi} h_1(kr)\, \mathbf{d} \cdot \hat{r},$$
$$\mathbf{v}(\mathbf{r}) = \mathbf{v}_{\text{loc}}(\mathbf{r}) + \frac{ik^3\mathbf{d}}{4\pi} \cdot \left(\frac{h_1(kr)}{kr} \overline{\overline{I}} + r\frac{\mathrm{d}}{\mathrm{d}r}\left(\frac{h_1(kr)}{kr}\right)\hat{r}\hat{r} \right) + \frac{k^3 m}{4\pi\eta_0} h_0'(kr)\, \hat{r}, \tag{6.10}$$

where h_ℓ are spherical Hankel functions of the first kind of order l, $\overline{\overline{I}}$ is the identity tensor, and $\hat{r} = \mathbf{r}/r$.

When these expressions are substituted into the surface-integral formulas for effective fields (equation (6.6)) and expanded in powers of k, keeping the lowest non-vanishing real and imaginary terms, we obtain

$$p_{\text{eff}} = p_{\text{loc}}(0) + \frac{\mathrm{i}\, k^3}{4\pi} m,$$
$$\mathbf{v}_{\text{eff}} = \mathbf{v}_{\text{loc}}(0) - \frac{2}{3V}\mathbf{d} + \frac{\mathrm{i}\, k^3}{12\pi}\mathbf{d}, \tag{6.11}$$

where V is the unit-cell volume. The additional terms proportional to $\mathrm{i}k^3$ represent radiative corrections in 3D.

From polarizabilities to effective parameters. We now assume the particle response is isotropic and describe it using monopolar and dipolar polarizabilities:

$$m = \alpha_0\, p_{\text{loc}}(0), \qquad \mathbf{d} = \alpha_1\, \mathbf{v}_{\text{loc}}(0). \tag{6.12}$$

Using the effective constitutive relations (equation (6.7)) with $M_{\text{eff}} = m/V$ and $\mathbf{D}_{\text{eff}} = \mathbf{d}/V$, we write

$$(\beta_{\text{eff}} - \beta_0)\, p_{\text{eff}} = \beta_0 \frac{m}{V}, \qquad (\rho_{\text{eff}} - \rho_0)\, \mathbf{v}_{\text{eff}} = \rho_0 \frac{\mathbf{d}}{V}. \tag{6.13}$$

By eliminating the local and effective fields between equations (6.11)–(6.13), we arrive at the acoustic CM-type relations:

$$\boxed{\frac{\beta_0}{\beta_{\text{eff}} - \beta_0} = \frac{V}{\alpha_0} + \frac{\mathrm{i}\, k^3 V}{4\pi}}, \qquad \boxed{\frac{1}{3}\,\frac{2\rho_{\text{eff}} + \rho_0}{\rho_{\text{eff}} - \rho_0} = \frac{V}{\alpha_1} + \frac{\mathrm{i}\, k^3 V}{12\pi}}. \tag{6.14}$$

The first relation governs the effective compressibility, which originates from monopolar response. The second governs the effective density, which originates from dipolar response. As in the electromagnetic case, the radiative corrections guarantee that the polarizabilities become complex at finite frequency, even if the inclusions themselves are lossless.

Homogeneous spherical inclusions

To make the CM equation concrete, let us consider the case where each particle is a homogeneous sphere of radius a_s and volume $V_s = 4\pi a_s^3/3$, with compressibility β_s and density ρ_s. Inside the sphere, the wavenumber is $k_s = \omega/c_s = \omega\sqrt{\rho_s\beta_s}$. Enforcing continuity of pressure and normal velocity at the boundary $r = a_s$ leads to the Mie scattering coefficient in scattering $j_l(kr)$ into $h_l(kr)$:

$$s_l = -\frac{\rho_s k j_l'(ka_s) j_l(k_s a_s) - \rho_0 k_s j_l(ka_s) j_l'(k_s a_s)}{\rho_s k h_l'(ka_s) j_l(k_s a_s) - \rho_0 k_s h_l(ka_s) j_l'(k_s a_s)}. \tag{6.15}$$

With the normalization factors obtained from equations (6.4) and (6.10), the monopolar and dipolar polarizabilities are

$$\alpha_0 = \frac{4\pi}{\mathrm{i}k^3} s_0, \qquad \alpha_1 = \frac{12\pi}{\mathrm{i}k^3} s_1. \tag{6.16}$$

These expressions are exact and include possible resonances of the sphere. They can be directly substituted into the acoustic CM relations (equation (6.14)) to compute $\beta_{\text{eff}}(\omega)$ and $\rho_{\text{eff}}(\omega)$.

In the simpler limit $ka_s \ll 1$, where no resonance is excited inside the sphere, we can expand $1/s_l$ in powers of k to obtain the quasistatic polarizabilities:

$$\frac{1}{\alpha_0} \approx \frac{\beta_0}{V_s(\beta_s - \beta_0)} - \frac{\mathrm{i}\,k^3}{4\pi}, \qquad \frac{1}{\alpha_1} \approx \frac{1}{3V_s}\frac{2\rho_s + \rho_0}{\rho_s - \rho_0} - \frac{\mathrm{i}\,k^3}{12\pi}. \tag{6.17}$$

These expressions resemble the CM equations, and this is no accident. They can also be derived by applying the field-averaging procedure directly to a single sphere in free space, where the enclosing surface is taken just outside the sphere. In that case, the effective parameters must equal those of the sphere itself, so equation (6.14) reduces to equation (6.17) once we replace $(\beta_{\text{eff}}, \rho_{\text{eff}}, V)$ with (β_s, ρ_s, V_s).

Acoustic Maxwell–Garnett equation

Finally, in the case of non-resonant spheres at long wavelengths, we can eliminate α_0 and α_1 between equations (6.14) and (6.17). This gives the Maxwell–Garnett type mixing formulas for a suspension of spheres with filling fraction $f = V_s/V$:

$$\boxed{\beta_{\text{eff}} = f\,\beta_s + (1-f)\,\beta_0}, \qquad \boxed{\frac{\rho_{\text{eff}} - \rho_0}{2\rho_{\text{eff}} + \rho_0} = f\,\frac{\rho_s - \rho_0}{2\rho_s + \rho_0}}. \tag{6.18}$$

The compressibility channel is always given by the volume average, which reflects the scalar nature of the pressure field. This remains true in 2D and 1D and does not depend on the particle shape. The density channel, in contrast, depends on dipolar motion and can therefore inherit anisotropy from particle shape or lattice symmetry. In such cases, both α_1 and ρ_{eff} must be treated as tensors, and excitations in different directions need to be considered separately.

This difference highlights a key point: in acoustics, the CM and Maxwell–Garnett equations treat p and $\mathbf{v}$ differently, whereas in electromagnetism electric and magnetic fields appear symmetrically due to the duality of Maxwell's equations. Another important observation is that the effective density in fluids does not reduce to a simple volume average, but instead follows the more complicated CM-type mixing formula (Mei *et al* 2007). The reason is that the fluid constituents are free to flow around each other without friction as the idealized situation. By contrast, in solids—for example, when solid inclusions are embedded in a solid matrix under elastic wave propagation—the constituents are locked together by the shear modulus. In this case, the effective density is simply the volume average, independent of the inclusion shape or the system's dimensionality.

6.2.2 Negative index from soft spheres in water

The acoustic CM equation (equation (6.14)) can be used to design metamaterials with unusual effective parameters. By substituting equation (6.16) into equation (6.14), we obtain a compact form of the CM equation written directly in terms of the monopolar and dipolar scattering coefficients:

$$\frac{\beta_0}{\beta_{\text{eff}} - \beta_0} = \frac{\mathrm{i}k^3V}{4\pi}\left(\frac{1}{s_0} + 1\right), \qquad \frac{2\rho_{\text{eff}} + \rho_0}{\rho_{\text{eff}} - \rho_0} = \frac{\mathrm{i}k^3V}{4\pi}\left(\frac{1}{s_1} + 1\right), \tag{6.19}$$

where s_0 and s_1 are the monopolar and dipolar Mie scattering coefficients of the spheres, given in equation (6.15).

As an example, let us consider soft spheres (such as rubber) suspended in water (Li and Chan 2004). These spheres have a much lower sound speed than water due to a larger compressibility, and therefore support strong monopolar and dipolar Mie resonances. With suitable design of the sphere size and filling fraction, the resonances can occur in the subwavelength regime and can simultaneously drive the effective compressibility and effective density negative in a narrow frequency range.

Figure 6.1 shows the effective parameters calculated using equation (6.19) for porous rubber spheres in water. The sound speed inside the rubber is taken to be as low as 158 m s^{-1}, roughly ten times slower than in water. The chosen radius is $a_s = 1$ cm, and the filling fraction is $f = 0.3$. The results show that the real parts of both β_{eff} and ρ_{eff} become negative near 6.5 kHz, giving rise to a negative effective refractive index. Such a double-negative acoustic metamaterial can support backward-wave propagation and negative refraction, in close analogy with electromagnetic metamaterials.

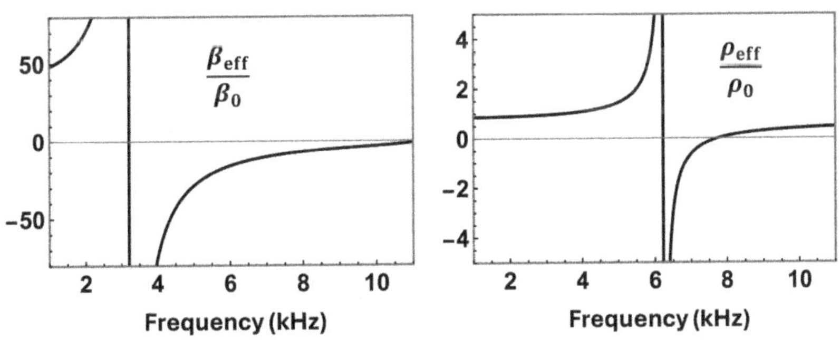

Figure 6.1. Effective medium parameters for rubber spheres in water, with $\rho_s = 600$ kg m^{-3}, $c_s = 158$ m s^{-1}, radius $a_s = 1$ cm, and filling fraction $f = 0.3$. Water has $\rho_0 = 1000$ kg m^{-3} and $c_0 = 1500$ m s^{-1}. Plotted are the real parts of the effective compressibility β_{eff} and density ρ_{eff}, each normalized to the background values.

It should be emphasized, however, that the filling fraction $f = 0.3$ is relatively large and close to the limit of validity of the CM approximation. In this regime, lattice effects (from lattice sums) must be considered, which can shift the resonance frequencies. Moreover, in real porous rubbers with low sound speeds, as reported for example in Brunet *et al* (2015) in achieving an acoustic negative index, material losses are significant. These losses tend to smear out the resonance features, particularly the dipolar resonance, and must be accounted for practical designs.

6.2.3 Sound absorption at low frequencies

In a bulk 3D medium, both monopolar and dipolar resonances can be supported by scatterers whose internal components can deform to generate oscillations. The rubber spheres in water, discussed in the last section, provide a clear example: the large compressibility contrast between rubber and water allows both types of resonances to occur. By deliberately overlapping the monopolar and dipolar resonances, a negative refractive index can be achieved. Physically, the monopolar resonance corresponds to a breathing mode in which the sphere volume oscillates, while the dipolar resonance corresponds to a back-and-forth motion of the spheres in water, much like pushing a spring–mass system.

In lower-dimensional systems, however, there is additional freedom. In 2D or 1D wave propagation with the metamaterials atoms placed in a 2D or 1D array fashion, scatterers are not confined by a fixed 3D volume. Extra volume can be hidden in the remaining dimension, enabling resonances—particularly monopolar ones—to be excited at very low frequencies, while still keeping the unit-cell size subwavelength (in the the wave propagation directions) and maintaining an effective-medium description.

To disentangle the monopolar and dipolar responses and highlight the advantages of reduced dimensions, in this section we study them in isolation within a 1D waveguide. Helmholtz resonators (HRs) act as nearly pure monopolar elements that support breathing-type oscillations, while clamped membranes act as dipolar elements that support vibrating modes. This 1D setting has often been employed as a practical route to demonstrate double negativity: by combining HRs and membranes in the same waveguide, one obtains negative compressibility and negative density simultaneously, and hence a negative index for guided acoustic modes. Beyond this traditional use, however, the same resonances can also be harnessed for other functionalities, such as enhancing absorption at low frequencies, using resonators with a very small size comparing to the free-space wavelength.

One way to illustrate the control of sound absorption and the distinct roles of monopolar and dipolar resonances—especially at low frequencies—is through the phenomenon of coherent perfect absorption (CPA). In CPA, two coherent input waves are completely absorbed when the scattering matrix acquires a zero eigenvalue. This mechanism provides a powerful means to achieve and tune perfect absorption at very low frequencies.

Principles of CPA. Let (a_1, b_1) and (a_2, b_2) denote the pair of complex amplitudes of forward and backward propagating waves on the left (port 1) and right (port 2) sides

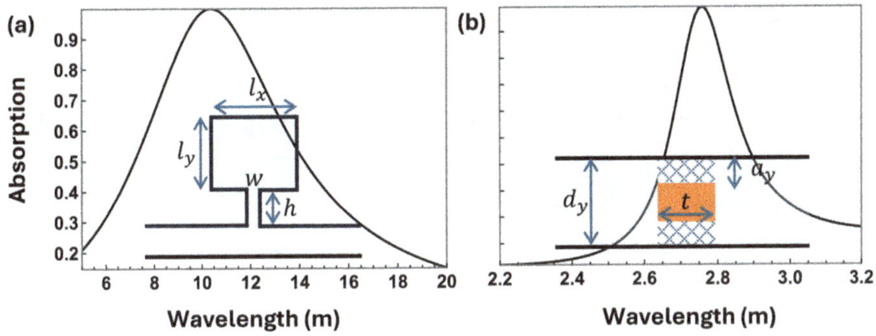

Figure 6.2. Coherent perfect absorption in a 1D waveguide. (a) Helmholtz resonator: schematic with cavity dimensions (l_x, l_y), neck width w, height h, and absorption spectrum under symmetrical excitation (monopolar CPA). (b) Membrane resonator: schematic with baffle opening d_y, membrane thickness t, diameter, and absorption spectrum under anti-symmetrical excitation (dipolar CPA).

of a scatterer within a 1D waveguide, respectively. Figure 6.2 shows the schematic for the case of (a) a HR attached to a waveguide filled with air or (b) a membrane attached to a waveguide filled with water. The scattering matrix S (see appendix A.1 for the details in defining these complex amplitudes) relates these as

$$\begin{pmatrix} a_2 \\ b_1 \end{pmatrix} = S\begin{pmatrix} a_1 \\ b_2 \end{pmatrix} = \begin{pmatrix} t & r \\ r & t \end{pmatrix}\begin{pmatrix} a_1 \\ b_2 \end{pmatrix},$$

where t and r are the transmission and reflection coefficients for a sample with mirror-symmetry along the propagation direction. Any input can be decomposed into *symmetrical* ($a_1 = b_2$) and *anti-symmetrical* ($a_1 = -b_2$) channels, which are eigenvectors of S with eigenvalues

$$s_{\text{sym}} = t + r, \qquad s_{\text{asym}} = t - r.$$

Perfect absorption occurs when one eigenvalue vanishes (Wei *et al* 2014), giving the CPA conditions

$$t + r = 0 \quad \text{(symmetrical CPA)}, \qquad t - r = 0 \quad \text{(anti-symmetrical CPA)}.$$

Physically, symmetrical excitation drives a *monopolar* response (uniform compression/expansion), whereas anti-symmetrical excitation drives a *dipolar* response (opposite particle velocities on the two sides), to the extent that there is no outgoing waves from the scatterer, i.e. $b_1 = a_2 = 0$, meaning a perfect absorption.

Case I: Helmholtz resonator (monopolar CPA). An HR consists of a rigid cavity (volume V_c) connected to the outside (the waveguide) by a narrow neck (cross-section A_n, effective length h_{eff}). The neck behaves as an oscillating mass, and the cavity as a spring, giving the resonance at a radial frequency

$$\omega_0 \approx c\sqrt{\frac{A_n}{h_{\text{eff}} V_c}},$$

where c is the sound speed in the background fluid (chosen as air). To realize CPA, damping must be tuned in addition to ω_0. The dominant loss arises from viscous/thermal boundary layers in the neck; a useful attenuation model in simulation is to tune the imaginary part of the constitutive parameters in the neck to yield an attenuation coefficient α qualitatively satisfying

$$\alpha = 2\mathrm{Im}(k) \propto \sqrt{\eta}/A_n,$$

with η the dynamic viscosity of air. Here, in figure 6.2(a), we are performing simulations for the HR in air. For simplicity, it is a 2D simulation with field invariance in the out-of-plane direction. The neck is a slit of width w and height h, and the cavity is a rectangle of dimensions (l_x, l_y). The parameters controlling the resonance frequency and damping are:

- cavity dimensions $(l_x, l_y) \Rightarrow$ set V_c and tune ω_0;
- neck height $h \Rightarrow$ sets h_{eff} and shifts ω_0;
- neck width $w \Rightarrow$ controls viscous damping to reach critical coupling.

As an example, with $l_x = 40$ cm, $l_y = 30$ cm, $h = 2$ cm, $w = 1$ mm, the monopolar resonance occurs near a wavelength of 10.4 m, and the damping is sufficient to satisfy $t + r = 0$. The corresponding absorption spectrum is shown in figure 6.2(a).

Case II: Membrane resonator (dipolar CPA). A thin elastic membrane clamped in a rigid baffle produces opposite pressure on either side when the membrane vibrates and therefore exhibits a dipolar response, naturally coupling to anti-symmetrical inputs. The resonance frequency is governed by the membrane width and thickness. Dissipation originates from viscoelastic losses and can be modeled by a small loss tangent in the elastic modulus. By tuning these parameters, one can satisfy $t - r = 0$ at resonance. A representative design (baffle opening of width 50 cm with membrane) in water achieves nearly perfect absorption near a wavelength of 2.76 m; see figure 6.2(b).

6.3 Space-coiling acoustic metamaterials

In the last section, we developed the dipolar model for acoustic metamaterials and showed how resonances of soft spheres can generate negative effective parameters, and also Helmholtz resonators and soft membranes can generate resonating monopolar and dipolar responses for perfect absorption. However, achieving a resonant effective density requires exciting a higher-order Mie resonance than that for compressibility. As a result, the density resonance is typically narrowband and highly susceptible to material losses. This motivates the search for alternative design strategies. Importantly, because acoustics is governed by a scalar pressure field, acoustic metamaterials are not limited to direct analogs of electromagnetic designs. Distinct approaches can be employed to achieve broader bandwidths with reduced loss. For airborne acoustics in particular, the strong contrast in density (or acoustic impedance) between air and solids offers a powerful handle to trap scalar sound waves within the air region to have comparatively lower loss.

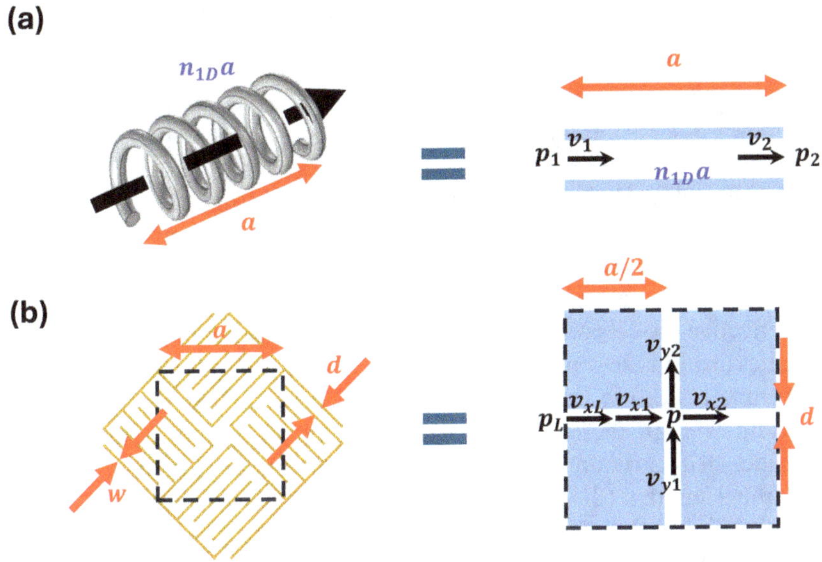

Figure 6.3. (a) Coiling up space so that sound propagates through a curled channel bounded by rigid walls. The elongated path produces a high effective refractive index n_{1D}, equivalent to a straight channel of the same length. (b) Arrangement of curled channels into a 2D network, forming a unit cell of an acoustic metamaterial.

Inspired by how a dielectric slows down light through multiple scatterings and phase delays, we can engineer a high effective refractive index for sound by forcing it to propagate along slits or curled channels (Liang and Li 2012). Consider air as the background medium, with sound speed $c = \frac{1}{\sqrt{\rho_0}\sqrt{\beta_0}}$, and impedance $\eta_0 = \frac{\sqrt{\rho_0}}{\sqrt{\beta_0}}$, where ρ_0 and β_0 are the density and compressibility of air. Figure 6.3(a) illustrates the principle: a curled channel of total length L_{eff} embedded in a rigid solid guides sound across a straight-line distance a, but with an elongated path length. The elongation factor defines an effective refractive index $n_{1\text{D}} = \frac{L_{\text{eff}}}{a}$, which can greatly exceed unity. For such an effective channel (right panel of figure 6.3(a)), we label the acoustic fields at the left and right ports with subscripts 1 and 2, respectively. Using forward (p_{1+}) and backward (p_{1-}) propagating pressure components, the pressure and velocity at each port are written as

$$
\begin{aligned}
p_1 &= p_{1+} + p_{1-}, \quad v_1 = \frac{p_{1+} - p_{1-}}{\eta_0}, \\
p_2 &= p_{1+}e^{i\phi_0} + p_{1-}e^{-i\phi_0}, \quad v_2 = \frac{p_{1+}e^{i\phi_0} - p_{1-}e^{-i\phi_0}}{\eta_0}.
\end{aligned}
\tag{6.20}
$$

We define the phase delay as $\phi_0 = k_0 n_{1\text{D}} a = \frac{2\pi n_{1\text{D}} a}{\lambda}$, where λ is the wavelength in air and $k_0 = \omega/c$ is the free-space wavenumber. Thus, ϕ_0 represents $n_{1\text{D}}$ times the free-space phase accumulation across a distance a. The pressure and velocity at the two ports can then be expressed through the two-port impedance matrix Z_1 of the effective 1D channel:

$$\begin{pmatrix} p_1 \\ p_2 \end{pmatrix} = Z_1 \begin{pmatrix} v_1 \\ v_2 \end{pmatrix}, \qquad Z_1 = \mathrm{i}\eta_0 \begin{pmatrix} \cot\phi_0 & -\csc\phi_0 \\ \csc\phi_0 & -\cot\phi_0 \end{pmatrix}. \tag{6.21}$$

By assembling such curled channels into a 2D network (figure 6.3(b)), we establish the basis for constructing acoustic metamaterials of higher dimensions. This network formulation allows the dispersion and effective refractive indices to be tuned flexibly, operating from the long-wavelength regime to the diffraction regime.

For an effective model composed of the 1D channels, we label the pressure and the four velocities around the central junction, as shown in the right panel of figure 6.3(b). The solid (blue region) is treated as a rigid boundary of thickness w, while the air channel width is d. In the equivalent model, the channel width remains d, but the effective index $n_{1\mathrm{D}}$ accounts for the elongated acoustic path compressed into a straight channel length a (the lattice constant). For a unit cell of equivalent path length a, the actual path length is determined by the air-filled area, $(\frac{a}{\sqrt{2}})^2(1-f_w)$, divided by the channel width d, where f_w is the solid-wall area filling fraction (solid area per unit cell). Thus, the elongation factor $n_{1\mathrm{D}}$ can be approximated as

$$n_{1\mathrm{D}} \simeq \frac{\left(\frac{a}{\sqrt{2}}\right)^2 (1-f_w)}{d\ a} = \frac{a}{2d}(1-f_w). \tag{6.22}$$

To analyze the propagation modes, we introduce Bloch phase factors ϕ_x and ϕ_y along the x- and y-directions, respectively. For the channel to the right of the center in the equivalent model (figure 6.3(b)), the pressure and velocity satisfy

$$\begin{pmatrix} p \\ p\mathrm{e}^{\mathrm{i}\phi_x} \end{pmatrix} = Z_1 \begin{pmatrix} v_{x2} \\ v_{x1}\mathrm{e}^{\mathrm{i}\phi_x} \end{pmatrix}. \tag{6.23}$$

For the channel above the center, we similarly have

$$\begin{pmatrix} p \\ p\mathrm{e}^{\mathrm{i}\phi_y} \end{pmatrix} = Z_1 \begin{pmatrix} v_{y2} \\ v_{y1}\mathrm{e}^{\mathrm{i}\phi_y} \end{pmatrix}. \tag{6.24}$$

At the junction, continuity of volume flux requires

$$v_{x1} + v_{y1} = v_{x2} + v_{y2}. \tag{6.25}$$

Combining equations (6.23)–(6.25) yields the secular equation, giving the dispersion relation for non-trivial solutions:

$$2\cos\phi_0 = \cos\phi_x + \cos\phi_y. \tag{6.26}$$

This secular equation yields a Dirac-like conical dispersion at the zone center (Γ), which underpins the near-zero-index behavior at $\omega = \omega_c$ and the sign change of n across the cone, as shown in figure 6.4(a).

In solving the secular equation, in addition to the band structure we obtained, the Bloch mode is described by the non-trivial solution (p, v_{x1}, v_{x2}, v_{y1}, v_{y2}). From this, the pressure and velocity at the interface of the medium can be derived. Consider the unit cell in figure 6.3(b), with its left interface opened to free-space (air). By taking half of a channel length, we can 'translate' the pressure and velocity values to this interface, denoted by p_L and v_{xL}, respectively. The velocity v_{xL} is evaluated just inside the channel. Since pressure is continuous across the interface while volume flux (rather than velocity) must be continuous, we define the effective fields as $p_{\text{eff}} = p_L$, $v_{x,\text{eff}} = \frac{d}{a} v_{xL}$, where d is the channel width and a is the lattice constant. Relating (p, v_{x1}) to (p_L, v_{xL}) and then to (p_{eff}, $v_{x,\text{eff}}$), the effective impedance of the medium (for $\phi_y = 0$) becomes

$$\eta = \frac{p_{\text{eff}}}{v_{x,\text{eff}}} = \frac{a}{d}\,\eta_0\,\frac{\tan(\phi_0/2)}{\tan(\phi_x/2)}. \tag{6.27}$$

In the long-wavelength limit near the zone center, we set

$$\phi_x = k_x a, \qquad \phi_y = k_y a, \qquad \phi_0 = n_{1\text{D}} a k_0.$$

Let ω_c be the cone-center angular frequency; then $k_{0,c} = \omega_c/c$ is the corresponding free-space wavenumber. We define the detuning

$$\Delta k_0 = k_0 - k_{0,c}.$$

On the first (acoustic-like) band near Γ we then have $\phi_0 \approx n_{1\text{D}} a\,\Delta k_0$, while on the cone bands we write $\phi_0 \approx 2\pi + n_{1\text{D}} a\,\Delta k_0$.

The secular equation then reduces to an isotropic dispersion relation:

$$k_x^2 + k_y^2 \simeq 2(n_{1\text{D}})^2(\Delta k_0)^2. \tag{6.28}$$

Defining an effective index by $k_x^2 + k_y^2 = n^2 k_0^2$, we find

$$n \simeq \sqrt{2}\,n_{1\text{D}}\,\frac{\Delta k_0}{k_0} = \frac{a}{\sqrt{2}d}\,(1 - f_w)\,\frac{\Delta k_0}{k_0}. \tag{6.29}$$

The sign of n is chosen according to Δk_0: positive in the first and third bands, and negative in the second band. Explicitly,

$$\text{sgn}(n) = \text{sgn}(\Delta k_0).$$

The impedance becomes

$$\eta \simeq \frac{a}{d}\,\eta_0\,\frac{n_{1\text{D}}}{n}\frac{\Delta k_0}{k_0} = \frac{a}{\sqrt{2}d}\,\eta_0, \tag{6.30}$$

which is valid for all three bands near the zone center. The analytic expressions for compressibility and density are

$$\frac{\beta}{\beta_0} = \frac{n}{\eta/\eta_0} = \frac{2n_{1D}d}{a}\frac{\Delta k_0}{k_0} = (1 - f_w)\frac{\Delta k_0}{k_0},$$
$$\frac{\rho}{\rho_0} = \frac{n\eta}{\eta_0} = \frac{n_{1D}a}{d}\frac{\Delta k_0}{k_0} = \frac{a^2}{2d^2}(1 - f_w)\frac{\Delta k_0}{k_0}. \tag{6.31}$$

In the long-wavelength limit (first band near the zone center), $\Delta k_0 = k_0$, yielding

$$\frac{\beta}{\beta_0} = 1 - f_w, \qquad \frac{\rho}{\rho_0} = \frac{a^2}{2d^2}(1 - f_w). \tag{6.32}$$

At the conical dispersion, we have $\frac{\Delta k_0}{k_0} = 1 - \frac{\omega_c}{\omega}$, so that

$$\frac{\beta}{\beta_0} = (1 - f_w)\left(1 - \frac{\omega_c}{\omega}\right), \qquad \frac{\rho}{\rho_0} = \frac{a^2}{2d^2}(1 - f_w)\left(1 - \frac{\omega_c}{\omega}\right), \tag{6.33}$$

where ω_c is the frequency at the cone center ($\phi_x = \phi_y = 0$, $\phi_0 = 2\pi$). In practice, finite channel width introduces a small bandgap at ω_c, so the frequencies at which $\beta = 0$ and $\rho = 0$ are slightly shifted apart. This causes significant fluctuations in the impedance η near the conical point.

Figure 6.4(b) shows a slab of the space-coiling metamaterial (with lattice constant $a = 2.33$ cm) of subwavelength thickness, operating at signal frequencies near the conical dispersion (Liang *et al* 2013). When the slab is placed inside a 2D waveguide,

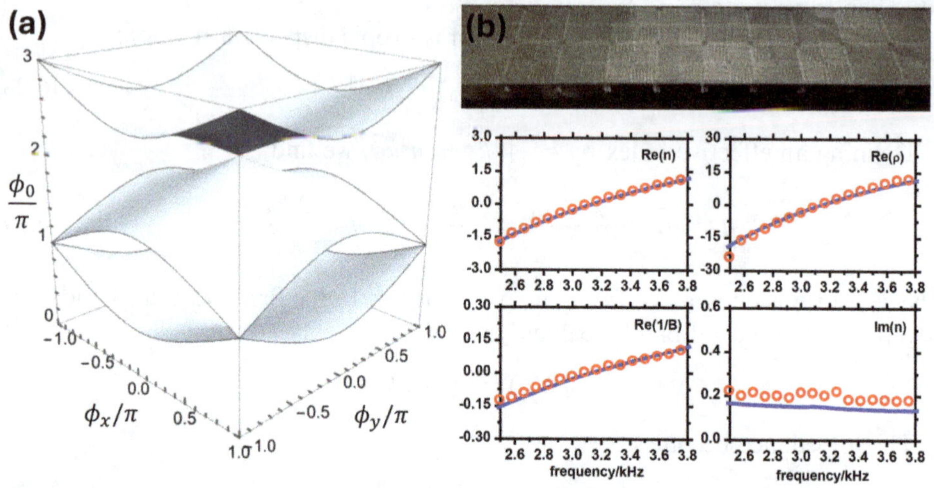

Figure 6.4. (a) Band structure of the space-coiling metamaterial. (b) A sample slab of the metamaterial. The four panels show the extracted effective medium parameters: the real and imaginary parts of the refractive index n, the effective density ρ (normalized to air), and the effective compressibility $\beta = 1/B$, where B is the bulk modulus (also normalized to air). Symbols denote experimental data, while solid lines represent simulation results. Reproduced from Liang *et al* 2013, with permission from Springer Nature.

the complex transmission (t) and reflection (r) coefficients can be measured. From these, the effective medium parameters are extracted using equation (2.15), with the substitutions of permittivity → compressibility and permeability → density (all normalized to their values in free space). The retrieved parameters are also shown in figure 6.4(b). The real parts of the density, compressibility (inverse bulk modulus), and refractive index follow the dependence $\propto (1 - \frac{\omega_c}{\omega})$, where $\omega_c \simeq 2\pi \times 3.12$ kHz is the center frequency of the conical dispersion. At frequencies below the cone, both the effective density and compressibility become negative, yielding a *double-negative* medium with a negative refractive index. Conversely, for applications requiring a near-zero index, the medium can be approximated at the cone center frequency $\omega \approx \omega_c$.

Here, we have formulated how to determine the effective medium of a space-coiling metamaterial, both from analytic models and from simulations. The analytic model was originally derived for a fixed interface or incident angle, but if the effective medium description is valid, it should also apply more generally in 2D. Figure 6.5 illustrates this point using the *coherent potential approximation*, a technique commonly employed to verify the validity of an effective medium (e.g., in disordered systems). At a frequency slightly below the conical dispersion, the

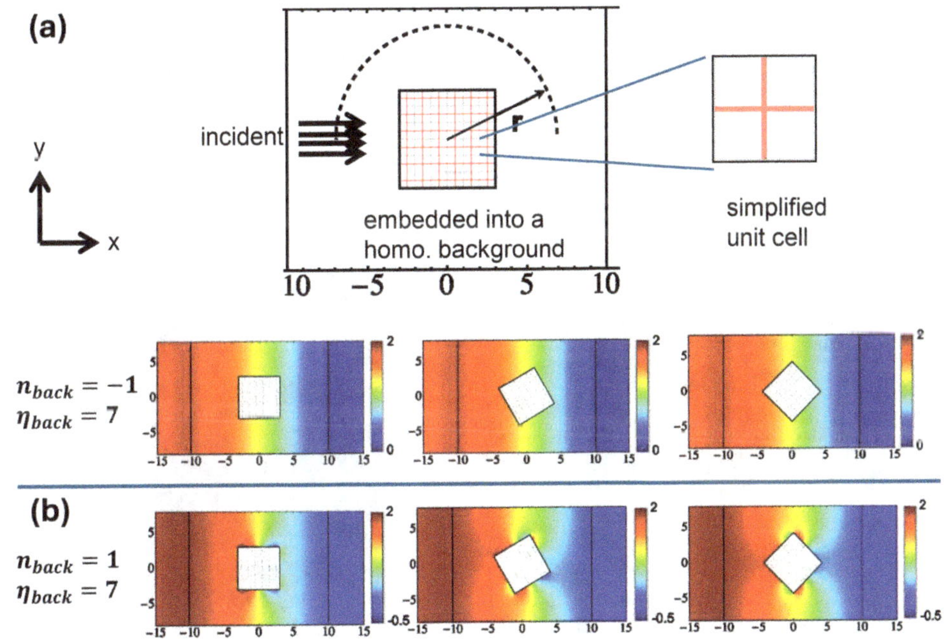

Figure 6.5. (a) Embedding the space-coiling metamaterial into a homogeneous background medium with refractive index $n_{\text{back}} = -1$ and impedance $\eta_{\text{back}} = 7\eta_0$, chosen to match the metamaterial's effective medium at a fixed frequency. Simulations are performed for a square-shaped sample under plane-wave incidence from the left, with various sample orientations. (b) The corresponding simulations when the background refractive index is replaced by $n_{\text{back}} = 1$.

metamaterial exhibits an effective refractive index $n = -1$ and impedance $\eta = 7\eta_0$. For ease of simulation, we replace the detailed microstructure with its equivalent channel model. A square-shaped sample of the metamaterial is then embedded into a homogeneous, isotropic background medium with the same effective parameters ($n = -1, \eta = 7\eta_0$). When illuminated by plane waves from the left at various incident angles, the sample produces almost no scattering, confirming that the effective medium accurately describes the metamaterial beyond normal incidence. In contrast, figure 6.5(b) shows the case when the background index is changed from $n = -1$ to $n = +1$. In this case, strong scattering is observed, highlighting the importance of index matching.

At the first band, the effective index is $n = \sqrt{2}\ n_{1D} = \frac{a}{\sqrt{2}d}\ (1 - f_w)$, which can be significantly larger than unity. This provides a powerful mechanism for designing airborne acoustic metamaterials. Unlike in optics, where refractive indices of solids are typically larger than unity, sound generally propagates faster in solids than in air, giving solids a refractive index usually much below unity. This mismatch leads to large impedance discontinuities when interfacing acoustic devices with air. The space-coiling approach overcomes this limitation by enabling a continuous range of effective indices greater than one. Although some impedance mismatch remains, it is far smaller than that arising from direct air–solid interfaces, making space-coiling an effective strategy for practical acoustic metamaterial design.

Figure 6.6(a) shows an example of negative refraction using a prism made of space-coiling metamaterials. The prism is immersed in air and illuminated by a Gaussian beam at a frequency slightly below the conical dispersion, where the

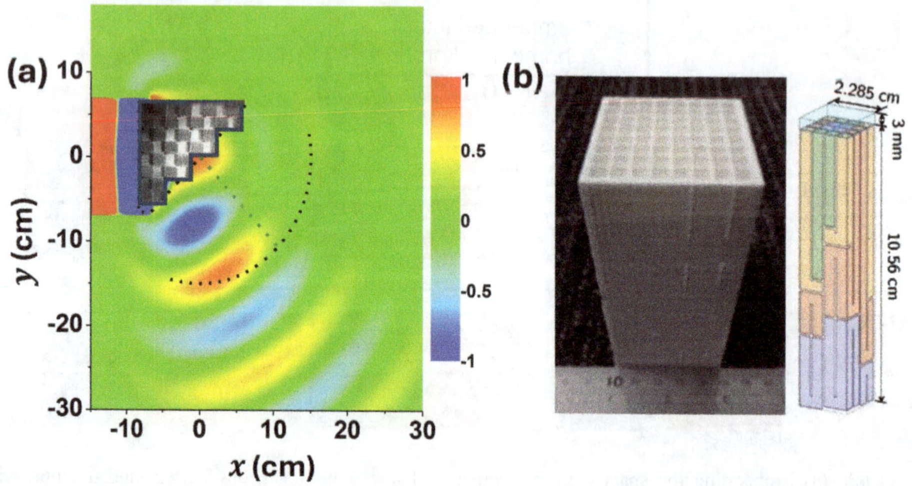

Figure 6.6. (a) Negative refraction demonstrated using a prism made of space-coiling metamaterials, operated at a frequency slightly below the conical center frequency, where the effective index becomes negative. (b) Curled channels with different lengths, designed to tune the effective indices to match successive Fabry–Perot resonances, thereby achieving broadband absorption. Adapted from Yang *et al* (2016) with permission from the Royal Society of Chemistry.

effective index is negative. The simulation shows that the refracted beam bends to the same side of the normal as the incident beam, confirming negative refraction.

Compared with locally resonant metamaterials, space-coiling structures provide a broader operational bandwidth and generally lower losses. The remaining loss arises from a different mechanism, namely viscous and frictional effects along the solid boundaries of the curled channels. For instance, in figure 6.4(b), the extracted refractive index shows an imaginary part of about 0.2. This level of attenuation is sufficiently small that waves can still propagate through a slab without serious decay, and negative refraction remains observable in a prism configuration (figure 6.6(a)). As long as the channel surface area is not excessively large, such boundary-induced losses stay moderate.

At the same time, the same mechanism can be deliberately exploited if absorption is desired. Loss can be enhanced by increasing the surface area of the channels, for example by extending the design into 3D, which increases the area available for viscous interaction. Another strategy is illustrated in figure 6.6(b), where curled channels of different lengths are introduced within a fixed volume (Yang *et al* 2016). By tuning these channels to successive Fabry–Perot resonances, strong broadband absorption across a wide frequency range can be achieved.

6.4 Programmable acoustic response

6.4.1 Acoustic artificial atom with digital feedback

To have a programmable acoustic response, we can adopt feedback circuits to existing structures, such as a Helmholtz resonator, to form a modified metamaterial atom, or directly to an acoustic waveguide, to provide synthetic secondary radiation sources through virtualized artificial atoms. The feedback can be either analog or digital. While analog feedback is more straightforward and consumes less power, digital feedback is more flexible and programmable. As the feedback control allows pumping power into the system to have active response, a digital feedback enjoys larger tunability to achieve stability, which is often crucial in the active regime.

Active metamaterials can be realized by incorporating feedback circuits into artificial atoms, enabling the tuning of their resonant responses (Popa *et al* 2015; He *et al* 2021; Wang and Wang 2025). We begin with the case of using a feedback circuit to modify a Helmholtz resonator, a simple acoustic cavity with a neck that opens to free space. The resonance frequency of the resonator is determined by the geometry of its cavity and neck. This resonance can be tuned by introducing feedback, implemented using a microphone, a speaker, and a microcontroller for digital signal processing, as shown in figure 6.7(a). For the Helmholtz resonator without feedback, the response can be modeled as

$$p = g_0 S_{\text{ext}} = \frac{a_0}{\omega_{res}^{(0)} - \omega} S_{\text{ext}}, \tag{6.34}$$

where the pressure p inside the cavity is measured by microphone D_{ext}. Here, a_0 is the resonance strength, and $\omega_{\text{res}}^{(0)}$ is the complex resonance frequency of the cavity without feedback. Its imaginary part is negative in a passive system. Equation (6.34)

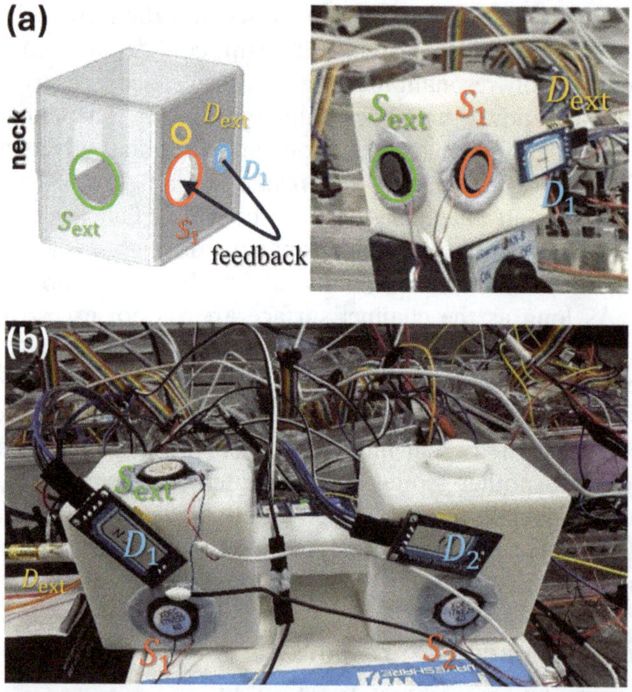

Figure 6.7. (a) A Helmholtz resonator with digital feedback. Microphone D_1 detects the pressure signal inside the cavity, which is processed by a microcontroller (not shown) and fed back to speaker S_1. Together with the Helmholtz resonator, they form an artificial atom with a digital feedback circuit. S_{ext} and D_{ext} are the external speaker and microphone used to excite and measure the acoustic response of the artificial atom. (b) A pair of artificial atoms with digital feedback circuits, connected by a pipe that couples the two cavities.

therefore represents the *passive* response of the resonator. Microphone D_1 detects the same pressure signal p, which is then fed back to speaker S_1 according to

$$S_1 = -\mathrm{i}B_0 \mathrm{e}^{\mathrm{i}\omega\Delta t} p, \tag{6.35}$$

where B_0 is the feedback amplitude and Δt is the delay time. The additional factor $-i$ is a convention chosen so that $\Delta t = 0$ absorbs all other hardware delays, including those from the speaker and microphone responses. The physical cavity now responds to the combined excitation from both the external drive and the feedback:

$$p = gS_{\text{ext}} = g_0(S_{\text{ext}} + S_1). \tag{6.36}$$

Substituting equation (6.35) into equation (6.36) gives

$$g = \frac{a_0}{\omega_{\text{res}} - \omega}, \tag{6.37}$$

which mirrors the passive form in equation (6.34) but with an *effective* resonance frequency ω_{res}. The shift due to feedback is made explicit in

$$\omega_{\text{res}} = \omega_{\text{res}}^{(0)} + \text{i}a_0 B_0 \text{e}^{\text{i}\omega\Delta t}. \tag{6.38}$$

If the resonance is sharp, the relevant frequency range lies close to $\omega_{\text{res}}^{(0)}$. In this case, we can approximate

$$\omega_{\text{res}} \approx \omega_{\text{res}}^{(0)} + \text{i}a_0 B_0 \text{e}^{\text{i}\omega_{\text{res}}^{(0)}\Delta t}.$$

Figure 6.8 shows acoustic response spectra for various feedback delays Δt at a fixed feedback amplitude B_0. The dashed lines are experimental measurements, while the solid lines are model fits based on equation (6.37). The inset plots the extracted complex resonance frequencies $f_{\text{res}} = \omega_{\text{res}}/(2\pi)$ as colored dots, which approximately lie on a circle as predicted by equation (6.38). In other words, tuning the feedback delay rotates the effective resonance frequency along this circle. Finally, figure 6.7(b) illustrates a possible extension to multiple cavities coupled by hollow pipes, where feedback can be individually applied to each cavity. Such coupling enables the study of collective behaviors that emerge when multiple artificial atoms with digital feedback circuits interact.

6.4.2 Virtualized atoms with programmable scattering response

Now, we shift to another approach to apply the digital feedback directly to an acoustic waveguide. Figure 6.9 shows several virtualized atoms coupled to a 1D waveguide along the x-direction (Cho *et al* 2020). Each atom, labeled by index i at position x_i, consists of a microcontroller interconnecting a speaker (S_i) and a

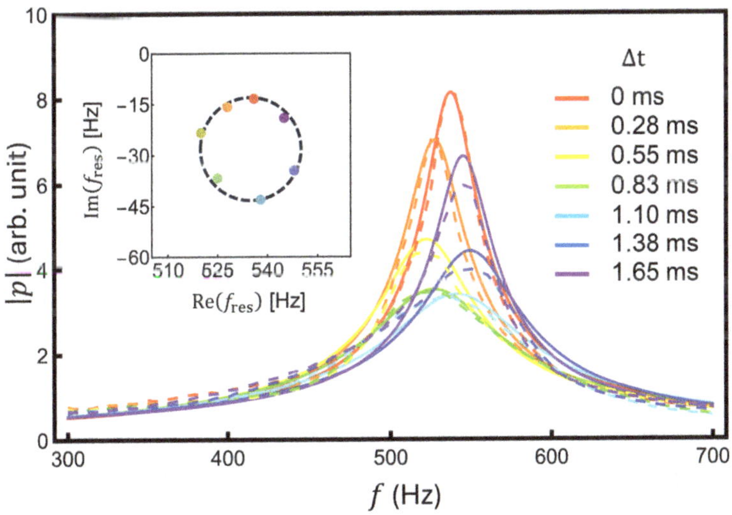

Figure 6.8. Acoustic response spectra for different feedback delays Δt at a fixed feedback amplitude. Dashed lines: experimental results. Solid lines: model fits using equation (6.37). Inset: fitted complex resonance frequencies at different Δt, plotted as colored dots. The points approximately lie on a circle, as expected from equation (6.38). In other words, varying the feedback delay rotates the effective resonance frequency along this circle in the complex plane.

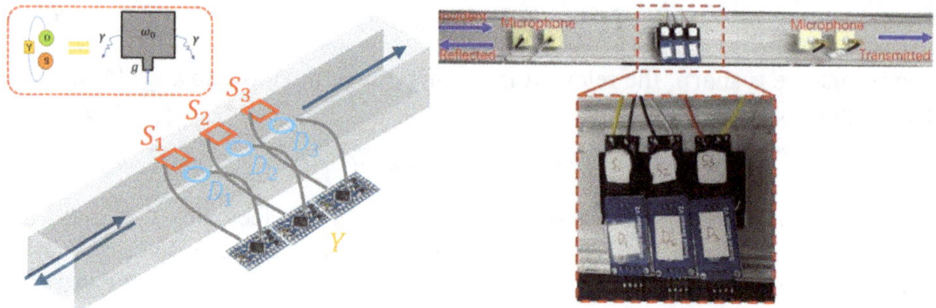

Figure 6.9. A waveguide with three virtualized atoms. Each atom consists of a speaker and a microphone, with the feedback controlled by a microcontroller to generate a monopolar secondary radiation at each location. The waveguide is excited by an external speaker (not shown) at one end.

microphone (D_i) placed at the same location along the propagating x-direction. This arrangement allows the system to sense the local monopolar incoming wave and to generate a monopolar secondary radiation within the waveguide. Unlike the Helmholtz-based artificial atoms described earlier, here the resonant response of each atom does not originate from a physical cavity. Instead, it is *virtualized* by the microcontroller, which reproduces a Lorentzian resonance by digitally integrating the pressure signal in time. The acoustic wave propagation in the waveguide is governed by the equations:

$$\begin{aligned} \partial_x p(x, t) + \rho_0 \partial_t (v(x, t) + D(x, t)) = 0, \\ \partial_x v(x, t) + \beta_0 \partial_t (p(x, t) + M(x, t)) = 0, \end{aligned} \tag{6.39}$$

where ρ_0 and β_0 are the density and compressibility of air, and $p(x, t)$ and $v(x, t)$ are the pressure and velocity fields. D represents the dipolar secondary radiation, which is zero in the present case, and M represents the monopolar secondary radiation (hereafter called monopolar polarization for simplicity), given by

$$M(x, t) = \Sigma_i m_i(t) \delta(x - x_i). \tag{6.40}$$

It is generated by detecting the local pressure at atom i through microphone D_i, and numerically propagating the following differential equation in time t, implemented by the microcontroller:

$$\partial_t^2 m_i + 2\gamma \partial_t m_i + \omega_{\text{res}}^2 m_i = aLp(x_i), \tag{6.41}$$

with implicit t-dependence. Here a, $\omega_{\text{res}}/(2\pi)$, and γ are the resonant strength, resonant frequency, and linewidth associated with atom i, respectively, and L is the distance between neighboring atoms. The m_i obtained from equation (6.41) is fed back to speaker S_i to generate monopolar scattering waves.

This virtualized approach provides a convenient testbed equivalent to a physical resonator but with programmable properties and high flexibility. Figure 6.10 shows the use of these atoms to realize a metamaterial (with subwavelength separation between atoms $kL \ll 1$) exhibiting a Lorentzian-shaped compressibility. In configuration A, the

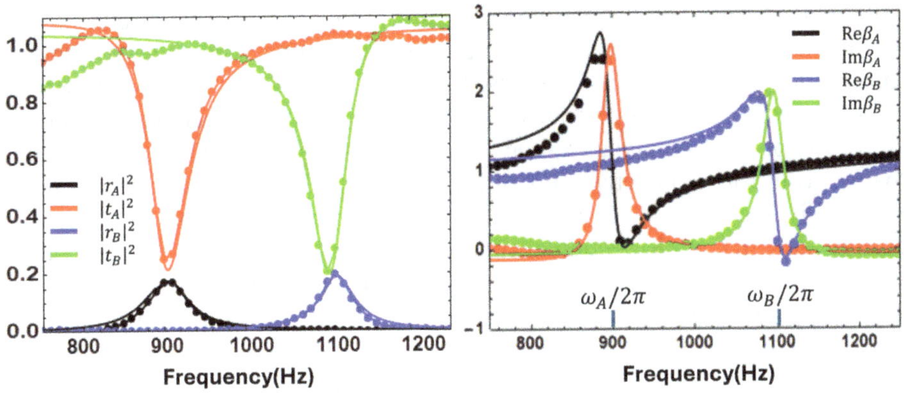

Figure 6.10. Programmable compressibility β (normalized by β_0 in air) implemented by virtualized atoms, with resonant frequency $\omega_A/2\pi = 0.9$ kHz (configuration A) and $\omega_B/2\pi = 1.1$ kHz (configuration B), while the linewidth γ and strength a are kept fixed. The compressibility results (right panel) are extracted from the transmission and reflection coefficients (left panel). Lines are model predictions and dots are experimental results.

resonant frequency is set to $\omega_A/2\pi = 0.9$ kHz. In configuration B, it is changed to $\omega_B/2\pi = 1.1$ kHz by modifying the parameters in the microcontroller program. The compressibility is extracted from the complex transmission and reflection coefficients (t and r) across the three atoms. As shown, the Lorentzian form (both real and imaginary parts) is faithfully reproduced, with good agreement between model and experiment. Equation (6.44) shows explicitly that the virtualized atoms reproduce the Lorentzian frequency dispersion of compressibility, directly analogous to conventional resonators.

To see how the compressibility arises from the virtualized atoms, equations (6.40) and (6.41) can be homogenized from discrete monopoles to a continuous monopolar polarization along x, yielding

$$\partial_t^2 M + 2\gamma\partial_t M + \omega_{res}^2 M = ap, \tag{6.42}$$

where both M and p are now functions of x and t. Combining equation (6.42) with the constitutive relation for monopolar response,

$$\beta p = \beta_0(p + M), \tag{6.43}$$

and applying $\partial_t \to -i\omega$ for steady-state responses, we obtain the Lorentzian form of compressibility:

$$\beta = \beta_0\left(1 + \frac{a}{\omega_{res}^2 - \omega^2 - 2i\gamma\omega}\right). \tag{6.44}$$

Another constitutive relation for the dipolar response is

$$\rho v = \rho_0(v + D), \tag{6.45}$$

which reduces to $\rho = \rho_0$ since $D = 0$ for the present structural arrangement of speakers and microphones.

The structural arrangement of each virtualized atom can be generalized to produce more complex scattering responses, for example including dipolar terms, and the implementation of equation (6.41) can also be extended to multiple resonances, without the need to design physical resonators case by case. On the other hand, electronic virtualization of the Lorentzian response requires careful attention to stability, since feedback may introduce effective gain. Indeed, a closer look at the imaginary part of the compressibility curves in figure 6.10 reveals a small negative contribution. This arises because, in the actual implementation, additional realistic factors must be incorporated, such as time delay in the microcontroller program, the responses of speakers and microphones, and inhomogeneity among atoms. Thus, the active implementation cannot guarantee that the atoms remain strictly passive. At the same time, the same mechanism opens opportunities to realize effective material gain in such systems.

6.5 Time-varying constitutive parameters

The dynamical tunability of the programmable acoustic metamaterials can be immediately used for constructing time-varying acoustic metamaterials as a new dimension of control of wave propagation. An immediate application of such dynamical tunability is to achieve a so-called temporal effective medium if the periodic modulation frequency of material parameters is much faster than the signal frequency.

A static atomic response and its associated compressibility are given by equations (6.42) and (6.44). If we modulate one or more coefficients in the atomic temporal model, equation (6.42), much faster than the signal period (for example, 0.14 ms for modulation period compared with a 1 kHz signal), the modulation produces higher harmonics. These harmonics fall outside the frequency range of interest and, in a damped system, their amplitudes are very small. Therefore, within the frequency band we care about, the medium can still be treated as time-invariant. We call this the *temporal effective medium*.

To model this, we propagate equations (6.39) and (6.42) in time for a continuous infinite medium. In a time-varying but spatially uniform medium, momentum k is conserved (in contrast to a static but spatially varying medium, where frequency is conserved). Replacing $\partial_x \to ik$, the propagation takes the matrix form

$$\mathrm{i}\partial_t \psi = \hat{\omega}\psi, \quad \hat{\omega} = \begin{pmatrix} 0 & \dfrac{k}{\beta_0} & 0 & \mathrm{i} \\ \dfrac{k}{\rho_0} & 0 & 0 & 0 \\ 0 & 0 & 0 & -\mathrm{i} \\ -\mathrm{i}a & 0 & \mathrm{i}\omega_{res}^2 & -i2\gamma \end{pmatrix}, \tag{6.46}$$

where $\psi = (p, v, M, -\partial_t M)^T$.

If the modulation period T_m is very small, such that $\omega_{res} T_m \ll 1$ and $\omega T_m \ll 1$, the total transfer matrix over one modulation cycle is approximately

$$\exp(-\mathrm{i}\hat{\omega}_{\text{eff}} T_m),$$

with $\hat{\omega}_{\text{eff}} \approx \langle\hat{\omega}\rangle$. Here $\langle\cdot\rangle$ means a temporal average. This is the lowest-order approximation for fast modulation.

For a single monopolar resonance, the average $\langle\hat{\omega}\rangle$ leads to a temporal effective medium with

$$\beta_{\text{eff}} = \beta_0\left(1 + \frac{\langle a\rangle}{\langle\omega_{\text{res}}^2\rangle - \omega^2 - 2\mathrm{i}\omega\langle\gamma\rangle}\right), \tag{6.47}$$

and $\rho_{\text{eff}} = \rho_0$.

For comparison, the instantaneous static response (with the medium 'frozen' at a given moment) is

$$\beta = \beta_0\left(1 + \frac{a}{\omega_{\text{res}}^2 - \omega^2 - 2\mathrm{i}\omega\gamma}\right).$$

In some special cases, β_{eff} can be written directly in terms of β. If ω_{res} and γ are constant,

$$\frac{\beta_{\text{eff}}}{\beta_0} = \left\langle 1 + \frac{a}{\omega_{\text{res}}^2 - \omega^2 - 2i\omega\gamma}\right\rangle = \left\langle\frac{\beta}{\beta_0}\right\rangle. \tag{6.48}$$

If instead the resonant strength a is constant,

$$\frac{1}{\beta_{eff}/\beta_0 - 1} = \left\langle\frac{\omega_{\text{res}}^2 - \omega^2 - 2\mathrm{i}\omega\gamma}{a}\right\rangle = \left\langle\frac{1}{\beta/\beta_0 - 1}\right\rangle. \tag{6.49}$$

As an example, consider switching the resonant frequency between $\omega_A/2\pi = 0.9$ kHz and $\omega_B/2\pi = 1.1$ kHz (the corresponding static responses are shown in figure 6.10). The duty cycle is 50% and the modulation frequency is 7.2 kHz. The predictions from equation (6.49) are shown in figure 6.11 as black (real part) and red (imaginary part) lines for the effective compressibility. Experimental data, obtained from transmission and reflection through three atoms, are plotted as symbols. The agreement between model and experiment is good.

6.6 Temporal modulation for frequency conversion

In the last section, the modulation frequency is much faster than the signal frequency, and the modulation is effectively homogenized to give a temporal effective medium while the frequency converted components are ignored. On the other hand, if the modulation frequency is comparable to the signal frequency, the modulation can directly couple different frequency components of the signal to have additional phenomena.

In a static periodic medium (periodic in space), certain frequencies cannot propagate (a frequency band gap). In a medium that is uniform in space but periodically modulated in time, certain wave numbers (momenta) cannot propagate. This is a momentum band gap (or k-gap). Inside a k-gap, one Floquet mode grows exponentially in time and the other decays.

As an example, we discuss a k-gap for flexural waves in a beam, which has been realized experimentally in proving such modulation is feasible (Trainiti *et al* 2019).

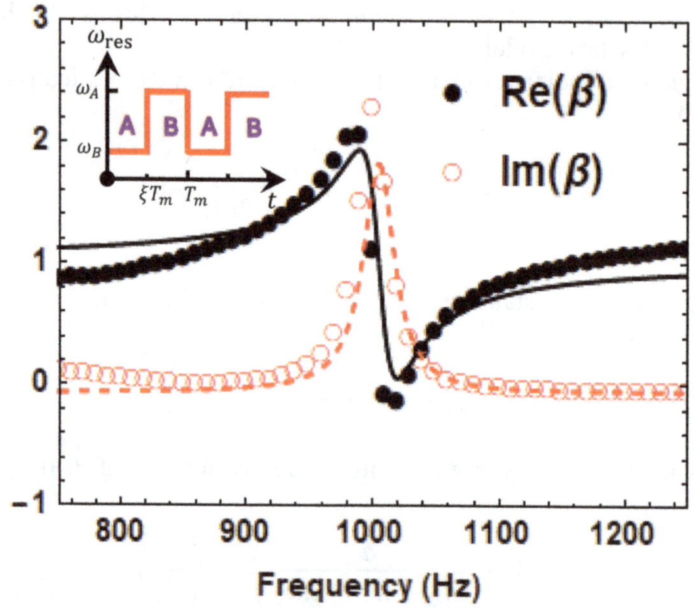

Figure 6.11. Temporal effective medium obtained by switching between two Lorentzian static compressibilities [equation (6.44)], with resonant frequencies $\omega_A/2\pi = 0.9$ kHz and $\omega_B/2\pi = 1.1$ kHz, duty cycle $\xi = 0.5$, and modulation frequency 7.2 kHz. The resulting effective compressibility is plotted in real (black) and imaginary (red) parts (Wen *et al* 2021). Lines show model predictions, and symbols show experimental results extracted from transmission and reflection. Results taken from Zhu *et al* (2023).

We consider transverse waves in a 1D elastic beam whose bending stiffness is time-modulated:

$$D(t)\,\partial_x^4 w(x,\,t) + m\,\partial_t^2 w(x,\,t) = 0, \qquad D(t) = D_0[1 + \alpha\cos(\Omega t)], \tag{6.50}$$

where D_0 is the unmodulated stiffness, m the mass per unit length, α the (small) modulation depth, and Ω the modulation angular frequency. When $\alpha = 0$, the dispersion is $\omega_0(k) = \pm\sqrt{\gamma}\,k^2$ with $\gamma = D_0/m$.

Because the medium is periodic in time, expand the field in temporal Floquet harmonics:

$$w(x,\,t) = \mathrm{e}^{\mathrm{i}(\omega t - kx)}\sum\nolimits_{n\in\mathbb{Z}}\hat{w}_n\,\mathrm{e}^{\mathrm{i}n\Omega t}. \tag{6.51}$$

Near $\omega \approx \Omega/2$, the dominant coupling is between the $n = 0$ and $n = -1$ harmonics (a standard two-mode truncation). Substituting equation (6.51) and using $\cos(\Omega t) = \frac{1}{2}(\mathrm{e}^{\mathrm{i}\Omega t} + \mathrm{e}^{-\mathrm{i}\Omega t})$, the coupled algebraic system for $(\hat{w}_0,\,\hat{w}_{-1})^T$ is

$$\begin{pmatrix} \omega^2 - \gamma k^4 & -\dfrac{\alpha}{2}\gamma k^4 \\ -\dfrac{\alpha}{2}\gamma k^4 & (\omega - \Omega)^2 - \gamma k^4 \end{pmatrix}\begin{pmatrix} \hat{w}_0 \\ \hat{w}_{-1} \end{pmatrix} = \mathbf{0}. \tag{6.52}$$

Non-trivial solutions require the determinant to vanish, giving the *secular equation*

$$(\omega^2 - \gamma k^4)((\omega - \Omega)^2 - \gamma k^4) - \left(\frac{\alpha}{2}\gamma k^4\right)^2 = 0. \tag{6.53}$$

When $\alpha = 0$, the two uncoupled branches cross at

$$\omega^* = \frac{\Omega}{2}, \qquad k^* = \left(\frac{\Omega}{2\sqrt{\gamma}}\right)^{1/2}.$$

For $\alpha \neq 0$, the coupling term opens a finite-k gap: the two branches *merge* over a wave-number interval $k \in [k_-, k_+]$ in which the frequency becomes complex:

$$\omega(k) = \omega_r(k) + \mathrm{i}\,\omega_i(k), \qquad \omega_r(k) \equiv \frac{\Omega}{2}, \quad \omega_i(k) \neq 0. \tag{6.54}$$

The real part is pinned at $\Omega/2$ (a flat band), while the imaginary part gives temporal growth (for $\omega_i > 0$) or decay (for $\omega_i < 0$). Solving (6.53), one obtains the exact roots

$$\omega = \frac{1}{2}\left[\Omega \pm \sqrt{\Omega^2 + 4\gamma k^4 - 2\sqrt{\gamma k^4(4\Omega^2 + \alpha^2\gamma k^4)}}\ \right], \tag{6.55}$$

and the k-gap boundaries (where ω turns complex) are

$$k_\pm = \left[\frac{\Omega^2}{2\gamma\,(2 \mp \alpha)}\right]^{1/4} = k^*\left(1 \pm \frac{\alpha}{4} + O(\alpha^2)\right). \tag{6.56}$$

Thus the gap is centered at k^* and its width scales linearly with the modulation depth α for small α.

Figure 6.12(a) shows the band structure computed from the exact multi-band secular equation for $\alpha = 0.5, \Omega = 1, \gamma = 1$. Several k-gaps appear at $k = k^*$ and at

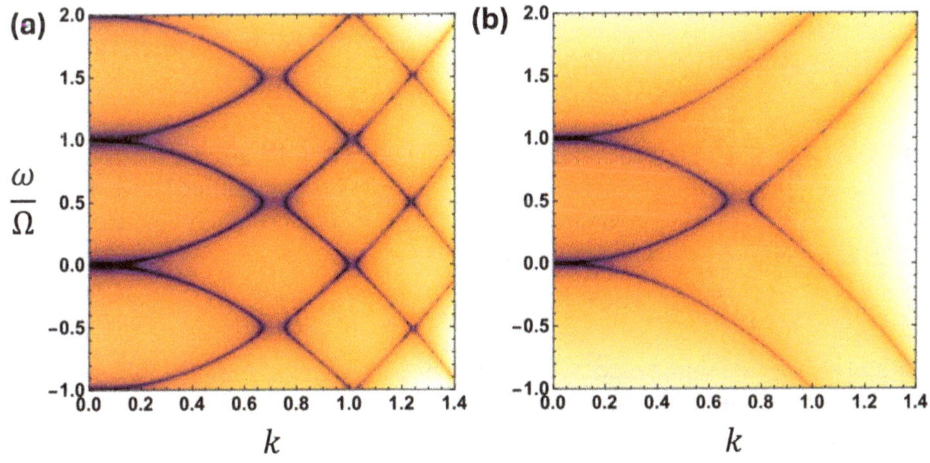

Figure 6.12. (a) Band structure for flexural waves of the time-modulated beam.(b) Band structure when only $n = 0$ and $n = -1$ branches are retained for the two-mode approximation.

higher multiples by adding integral multiples of Ω to $\Omega/2$. Figure 6.12(b) shows the two-mode approximation (6.53), which captures the essential features of the k-gap near $k = k^*$ and $\omega = \Omega/2$.

An elastic-beam realization with time-periodic stiffness modulation uses piezoelectric patches shunted by switchable negative-capacitance circuits to implement $D(t)$. Experimentally, the dispersion branches merge at $\Omega/2$ over a finite k-interval, producing a flat real frequency and a non-zero imaginary part (parametric amplification). At an interface, a broadband input is reflected as a narrowband output centered at $\Omega/2$ (a selective filter set by the modulation frequency) (Trainiti *et al* 2018).

6.7 Transformation approach

Let us develop the theory of the transformation approach for other kinds of classical waves. We take acoustic waves as an example.

The acoustic equations of mass and momentum conservation are

$$\begin{aligned} \nabla \cdot \mathbf{v} &= \mathrm{i}\omega\frac{p}{\kappa}, \\ \nabla p &= \mathrm{i}\omega\overline{\overline{\rho}} \cdot \mathbf{v}, \end{aligned} \tag{6.57}$$

where $\overline{\overline{\rho}}$, κ, $\mathbf{v}$, and p are the mass density tensor, the bulk modulus (or reciprocal of compressibility β), the particle velocity, and the acoustic pressure, respectively. The first equation represents mass conservation: a divergence of velocity corresponds to a change in volumetric strain related to the pressure. The second equation expresses momentum conservation: pressure gradients drive motion through the density tensor.

To simplify the notation and to highlight the analogy with electromagnetism, we define two additional variables. The momentum density is

$$\boldsymbol{\mu} = \overline{\overline{\rho}} \cdot \mathbf{v},$$

and the volumetric strain is

$$\epsilon = -\frac{p}{\kappa}.$$

In this way, $\boldsymbol{\mu}$ and ϵ play roles analogous to $\mathbf{D}$ and $\mathbf{B}$ in electromagnetism.

By integrating the first equation of equation (6.57) over a volume V, we obtain the integral form of mass conservation. Using the divergence theorem, the velocity divergence becomes a flux through the surface:

$$\iiint_V \nabla \cdot \mathbf{v}\, \mathrm{d}V = \iint_S \mathbf{v} \cdot d\mathbf{S} = -i\omega \iiint_V \epsilon\, \mathrm{d}V.$$

This shows that the total velocity flux through the surface equals the time-harmonic change of the strain inside the element.

For the momentum conservation law, consider the gradient of pressure along a chosen axis $i = x, y, z$. Integrating across the element in that direction gives

$$\Delta p = \mathrm{i}\omega \int \mu_i dl_i$$

where μ_i is the ith component of momentum density. This is the discrete analog of Newton's law: a pressure difference drives momentum accumulation along that line element.

Now suppose we have two configurations where the vertices of the finite elements are at different coordinates. One is spanned by $\mathbf{a}$, $\mathbf{b}$, $\mathbf{c}$, and the other by $\mathbf{a}'$, $\mathbf{b}'$, $\mathbf{c}'$. These two configurations are said to be computationally equivalent if the matrix representations of $\overline{\overline{\rho}}$ and κ in the two cases are completely the same. If this holds for all elements, then only one mesh needs to be simulated, because the result for the other follows from simple field transformations.

To preserve equivalence, certain quantities must be conserved. Pressure itself is invariant. Strain must scale with the cell volume. Momentum density must project consistently along line elements, and velocity must be conserved through face fluxes. In equations:

$$p = p', \quad (\mathbf{a} \cdot \mathbf{b} \times \mathbf{c})\, \epsilon = (\mathbf{a}' \cdot \mathbf{b}' \times \mathbf{c}')\, \epsilon', \quad \mathbf{a} \cdot \boldsymbol{\mu} = \mathbf{a}' \cdot \boldsymbol{\mu}', \quad \mathbf{b} \times \mathbf{c} \cdot \mathbf{v} = \mathbf{b}' \times \mathbf{c}' \cdot \mathbf{v}'.$$

Scanning through all three line elements, we can define the transformation matrix that relates the momentum densities in the two configurations as

$$[\boldsymbol{\mu}' \to \boldsymbol{\mu}] = \begin{pmatrix} a_x & a_y & a_z \\ b_x & b_y & b_z \\ c_x & c_y & c_z \end{pmatrix}^{-1} \begin{pmatrix} a_x' & a_y' & a_z' \\ b_x' & b_y' & b_z' \\ c_x' & c_y' & c_z' \end{pmatrix}.$$

For the velocity field $\mathbf{v}$, which is conserved at the flux (face) level, the corresponding matrix is

$$\begin{aligned} [\mathbf{v} \to \mathbf{v}'] &= \begin{pmatrix} (\mathbf{b}' \times \mathbf{c}')_x & (\mathbf{b}' \times \mathbf{c}')_y & (\mathbf{b}' \times \mathbf{c}')_z \\ (\mathbf{c}' \times \mathbf{a}')_x & (\mathbf{c}' \times \mathbf{a}')_y & (\mathbf{c}' \times \mathbf{a}')_z \\ (\mathbf{a}' \times \mathbf{b}')_x & (\mathbf{a}' \times \mathbf{b}')_y & (\mathbf{a}' \times \mathbf{b}')_z \end{pmatrix}^{-1} \begin{pmatrix} (\mathbf{b} \times \mathbf{c})_x & (\mathbf{b} \times \mathbf{c})_y & (\mathbf{b} \times \mathbf{c})_z \\ (\mathbf{c} \times \mathbf{a})_x & (\mathbf{c} \times \mathbf{a})_y & (\mathbf{c} \times \mathbf{a})_z \\ (\mathbf{a} \times \mathbf{b})_x & (\mathbf{a} \times \mathbf{b})_y & (\mathbf{a} \times \mathbf{b})_z \end{pmatrix} \\ &= \frac{\mathbf{a} \cdot (\mathbf{b} \times \mathbf{c})}{\mathbf{a}' \cdot (\mathbf{b}' \times \mathbf{c}')} [\boldsymbol{\mu}' \to \boldsymbol{\mu}]^T. \end{aligned}$$

This consistency ensures that the constitutive relation between momentum density and velocity transforms properly. In particular, the density tensor must satisfy

$$\left[\frac{1}{\overline{\overline{\rho}}'}\right] = [\mathbf{v} \to \mathbf{v}'] \left[\frac{1}{\overline{\overline{\rho}}}\right] [\boldsymbol{\mu}' \to \boldsymbol{\mu}]. \tag{6.58}$$

The matrix $[\boldsymbol{\mu}' \to \boldsymbol{\mu}]$ can be identified as the transpose of the Jacobian matrix $[\Lambda]$ that relates the two coordinate systems when the transformation is carried out differentially. With this identification, equation (6.58) can be expressed in the compact Jacobian form

$$\left[\frac{1}{\overline{\overline{\rho}}'}\right] = \frac{1}{\det[\Lambda]} [\Lambda] \left[\frac{1}{\overline{\overline{\rho}}}\right] [\Lambda]^T. \tag{6.59}$$

Finally, for the bulk modulus we return to the relation $p = -\kappa\epsilon$. Since pressure is invariant, while strain scales with the volume change between the two cells, we have

$$[p \to p'] = 1, \qquad [\epsilon' \to \epsilon] = \frac{\mathbf{a}' \cdot (\mathbf{b}' \times \mathbf{c}')}{\mathbf{a} \cdot (\mathbf{b} \times \mathbf{c})} = \det[\Lambda].$$

Thus the transformed bulk modulus is

$$\kappa' = \kappa \det[\Lambda]. \tag{6.60}$$

To summarize, pressure is invariant, strain scales with the transformed volume, momentum density is conserved along line elements, and velocity is conserved through face fluxes. From these invariants, the transformation rules for the density tensor and the bulk modulus follow. Again, this conservation of invariants is the key principle underlying transformation media (Cummer *et al* 2008) and it complies with results based on tensor calculus with coordinate transformation (Cummer and Schurig 2007; Chen and Chan 2010). This structure directly parallels the electromagnetic case: in both settings, the computational invariants are line- and face-projected quantities, and the Jacobian matrix Λ governs how the constitutive parameters must transform.

Moreover, the analogy with electromagnetism should be understood at a structural level rather than as a one-to-one identification of fields. In electromagnetism, the natural invariants under coordinate transformations are edge-projected electric and magnetic fields (**E**, **H**) and face-flux fields (**D**, **B**). In acoustics, the corresponding roles are played by area-projected velocity and edge-projected momentum density ($\mathbf{v}$, $\boldsymbol{\mu}$), together with the scalar quantities of pressure and strain (p, ϵ), which control flux and volume balance. The transformation rules are not identical—they follow from different governing equations—but the geometric principle is the same: once the correct invariants are identified, the Jacobian of the coordinate deformation dictates how the constitutive parameters ($\overline{\overline{\epsilon}}$, $\overline{\overline{\mu}}$) in electromagnetism or ($\overline{\overline{\rho}}$, κ) in acoustics must transform.

Let us now consider a simple example to design a beam shifter that has an intuitive implementation while the transformation formulas can be easily checked.

In section 4.4, we have discussed how to design a linear map to achieve a beam shifter that shifts a horizontally propagating beam by a constant amount in the vertical direction. The linear map is given by

$$x = x', \quad y = y' + \frac{h}{w}x', \quad z = z', \tag{6.61}$$

where w is the width of the device and h is the amount of vertical shifting. The transformed medium (applied to free space) has an anisotropic density with an in-plane tensor given by

$$\frac{1}{\overline{\overline{\rho}}} = \begin{pmatrix} 1 & \tan\theta \\ \tan\theta & 1 + (\tan\theta)^2 \end{pmatrix}, \tag{6.62}$$

where the shifting angle θ is defined by $\tan\theta = h/w$. The isotropic bulk modulus remains unchanged because the area under transformation is invariant.

For implementation, we can seek further simplification by replacing free space (the virtual medium before transformation) with another effective medium. Suppose we work in airborne acoustics. Horizontal slits running in the x-direction (periodic in the y-direction) have an anisotropic density tensor described by

$$\frac{1}{\overline{\overline{\rho'}}} = \begin{pmatrix} \rho_l^{-1} & 0 \\ 0 & \rho_n^{-1} \end{pmatrix} = \begin{pmatrix} \dfrac{1-f}{\rho_0} + \dfrac{f}{\rho_s} & 0 \\ 0 & \dfrac{1}{f\rho_s + (1-f)\rho_0} \end{pmatrix}, \tag{6.63}$$

where f is the filling fraction of the solid material. Because the density of the solid, ρ_s, is much larger than the air density, ρ_0, the effective densities reduce to

$$\rho_l \approx \frac{\rho_0}{1-f}, \quad \rho_n \approx f\rho_s.$$

Here we have assumed the solid is sufficiently dense that even after averaging, ρ_n is still much larger than ρ_0. With this approximation, equation (6.62) can be rewritten as

$$\frac{1}{\overline{\overline{\rho}}} = \rho_l^{-1}(\cos\theta)^2 \begin{pmatrix} 1 & \tan\theta \\ \tan\theta & (\tan\theta)^2 \end{pmatrix} \cong R(-\theta)\frac{1}{\overline{\overline{\rho'}}}R(\theta), \tag{6.64}$$

where $R(\theta)$ is the 2×2 rotation matrix. This shows that the transformed density corresponds approximately to the reciprocal density tensor of the slit medium, physically rotated by angle θ. Figure 6.13 illustrates both simulation and experimental results for such a beam shifter. The device consists of PMMA

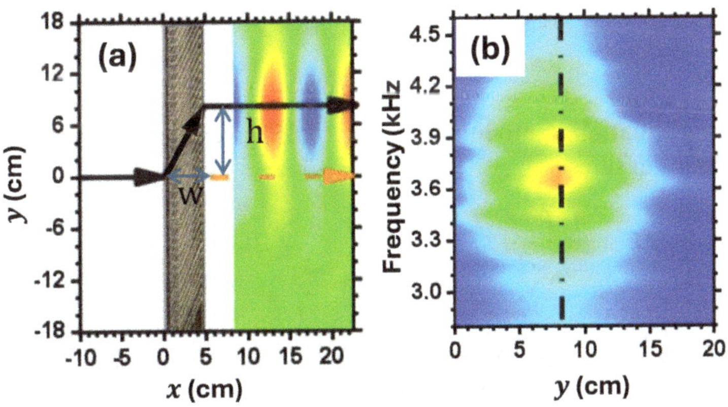

Figure 6.13. (a) Beam-shifter sample and simulated pressure fields showing a horizontal beam entering at $y = 0$ and exiting at $y = h$. (b) Experimental exit beam profiles at different frequencies. Reproduced from Wei *et al* (2015).

plates forming rotated slits. A beam propagating horizontally from left to right is shifted vertically by a constant amount h. The beam profiles at the exit are plotted at different frequencies, demonstrating a vertical displacement of about 8 cm. The shift remains nearly constant over a broad frequency range, while the transmission efficiency peaks around 3.7 kHz due to a Fabry–Pérot resonance along the slits.

6.8 Field transformation for non-reciprocal responses

The usual transformation approach is based on coordinate transformations, which lead to reciprocal constitutive relations. By contrast, applying a transformation directly to the fields—pressure and velocity in the case of acoustics—results in non-reciprocal constitutive relations. This approach can be used to manipulate wave propagation in a non-reciprocal manner or to enhance sensing applications by scaling up the fields arriving at a sensor.

The acoustic equations, including monopole and dipole source terms generated by a metamaterial, are

$$\begin{aligned} \frac{\partial p}{\partial x} &= \mathrm{i}\omega(\rho_0 v_x + D_x), \\ \frac{\partial p}{\partial y} &= \mathrm{i}\omega(\rho_0 v_y + D_y), \\ \frac{\partial v_x}{\partial x} + \frac{\partial v_y}{\partial y} &= \mathrm{i}\omega(\beta_0 p + M), \end{aligned} \tag{6.65}$$

where D_x, D_y are the components of the acoustic dipole density, and M is the monopole density. β_0 and ρ_0 are the compressibility and density of air, respectively. Here, we have absorbed the free-space parameters into the definitions of the polarizations, compared with equation (6.1).

On the other hand, suppose we transform the free-space (air) fields in the virtual space (p_0, v_{x0}, v_{y0}) by a prescribed scalar function $f(x, y)$ into the physical space as

$$p = f(x, y)\, p_0, \quad v_x = f(x, y)\, v_{x0}, \quad v_y = f(x, y)\, v_{y0}. \tag{6.66}$$

Substituting equation (6.66) into equation (6.65) gives the required monopole and dipole moment densities

$$M = \frac{1}{\mathrm{i}\omega}\left(\frac{\partial f}{\partial x} v_{x0} + \frac{\partial f}{\partial y} v_{y0}\right), \qquad \mathbf{D} = \frac{\nabla f}{\mathrm{i}\omega} p_0. \tag{6.67}$$

These are the required sources to implement the desired fields. Next, instead of treating them as impressed sources, we aim to realize them as secondary sources using a metamaterial. In this way, the sources no longer need to be implemented case by case for different incident fields, but can instead be realized in a self-adapted manner. To achieve this, we express these sources directly in terms of the actual fields (p, v_x, v_y) by rewriting equation (6.67) with the help of equation (6.66), yielding

$$M = \frac{i}{c}\,\boldsymbol{\tau} \cdot \mathbf{v}, \qquad \mathbf{D} = \frac{i}{c}\,\boldsymbol{\tau}'\, p, \tag{6.68}$$

where $\boldsymbol{\tau}$ and $\boldsymbol{\tau}'$, being called the Willis coupling terms, are equal in this case and satisfy

$$\boldsymbol{\tau} = \boldsymbol{\tau}' = -\frac{\nabla f}{k_0 f}, \tag{6.69}$$

with $k_0 = \omega/c$ being the wavenumber in air. For a conventional acoustic medium, M and $\mathbf{D}$ couple to the pressure and velocity fields, respectively. Here, however, the monopole density couples to the velocity field, while the dipole density couples to the pressure field. This coupling is known as Willis coupling. Moreover, in a reciprocal medium the Willis coupling coefficients must satisfy $\boldsymbol{\tau} = -\boldsymbol{\tau}'$. In contrast, here we have $\boldsymbol{\tau} = \boldsymbol{\tau}'$, indicating that the medium is non-reciprocal. From equation (6.68), $\boldsymbol{\tau}$, equivalently $\boldsymbol{\tau}'$, behaves like a vector. Equations (6.65) and (6.68) can be recast as the macroscopic wave equations:

$$\begin{aligned} \nabla p &= i\omega\boldsymbol{\mu}, \\ \nabla \cdot \mathbf{v} &= -\, i\omega\epsilon, \end{aligned} \tag{6.70}$$

together with the constitutive relation between the volumetric strain $-\epsilon$, the momentum density components (μ_x, μ_y), and the fields (p, v_x, v_y):

$$\begin{pmatrix} -\,\epsilon \\ \mu_x \\ \mu_y \end{pmatrix} = \begin{pmatrix} \beta & i\tau_x/c & i\tau_y/c \\ i\tau'_x/c & \rho & 0 \\ i\tau'_y/c & 0 & \rho \end{pmatrix} \begin{pmatrix} p \\ v_x \\ v_y \end{pmatrix}, \tag{6.71}$$

where the compressibility $\beta = \beta_0$ and density $\rho = \rho_0$ are the same as in free space, while the off-diagonal terms of the constitutive matrix represent the non-reciprocal Willis coupling defined in equation (6.69).

As an illustrative example, figure 6.14 shows a bean-shaped shell. We set $f(x, y) = f_2 = 1$ in the region outside the shell and $f(x, y) = f_1 = 1.5$ in the closed

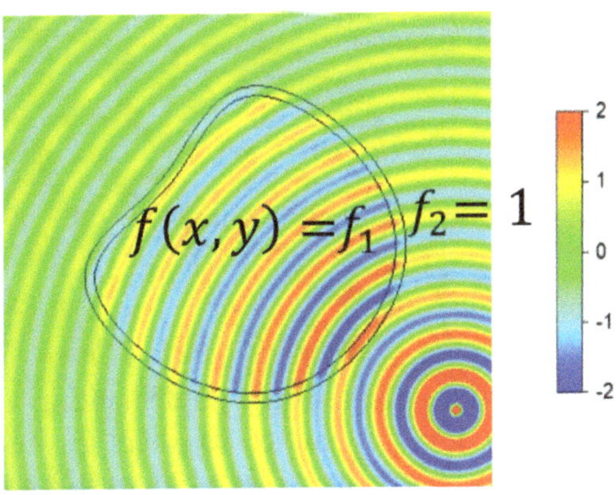

Figure 6.14. Scaling both pressure and velocity fields by a factor $f_2 = 1$ from the outside to $f_1 = 1.5$ inside a bean-shaped shell. Reproduced from Wen *et al* (2024). CC BY 4.0.

region inside the shell. In both regions where f is constant, the required monopole and dipole sources vanish, meaning that the medium is simply air. The shell region between the two boundaries is the non-reciprocal Willis metamaterial. A point source is placed outside the shell to generate incident cylindrical waves. The real part of the pressure field is plotted in the figure, obtained from a full-wave simulation with the constitutive relation in equation (6.71) applied within the shell (Wen *et al* (2024)). The simulation shows that the wavefronts are gradually amplified across the boundaries of the shell, while the fields inside are uniformly scaled up by a factor of 1.5, compared to the fields that would propagate to the same location in free space. Importantly, no scattering is observed from the shell, as indicated by the smooth wavefronts outside, a direct consequence of impedance matching ensured by the transformation approach. Another key feature of the transformation approach is that this field-scaling device is self-adaptive to incident fields. For example, whether a plane wave or a point source at a different location excites the system, the fields inside the shell are amplified in the same way.

In an actual experiment, a circular shell is more convenient. Suppose it has inner and outer radii R_1 and R_2, respectively, and the transformation function $f(r)$ depends only on the radial coordinate r. In this case, the constitutive relations can be simplified in cylindrical coordinates (r, θ) as:

$$\begin{pmatrix} -\epsilon \\ \mu_r \\ \mu_\theta \end{pmatrix} = \begin{pmatrix} \beta & \mathrm{i}\tau/c & 0 \\ \mathrm{i}\tau'/c & \rho & 0 \\ 0 & 0 & \rho \end{pmatrix} \begin{pmatrix} p \\ v_r \\ v_\theta \end{pmatrix}, \tag{6.72}$$

where

$$\beta/\beta_0 = \rho/\rho_0 = 1, \qquad \tau = \tau' = -\frac{f'(r)}{k_0 f(r)}. \tag{6.73}$$

For simplicity, if the transformation function is chosen such that

$$\ln f(r) = \frac{R_2 - r}{R_2 - R_1}\ln f_1 + \frac{r - R_1}{R_2 - R_1}\ln f_2, \tag{6.74}$$

then the Willis coupling is homogeneous:

$$\tau = -\frac{\ln(f_2/f_1)}{k_0(R_2 - R_1)}, \tag{6.75}$$

and the amplification ratio inside the shell becomes

$$\frac{f_1}{f_2} = \mathrm{e}^{\tau k_0(R_2 - R_1)}. \tag{6.76}$$

Figure 6.15 shows the experimental platform implementing a circular shell composed of 24 units, each containing a pair of speakers (S_1 and S_2) and a pair of microphones (D_1 and D_2). Both pairs are aligned along the radial direction, since only the Willis coupling that generates a dipole moment along the radial direction is

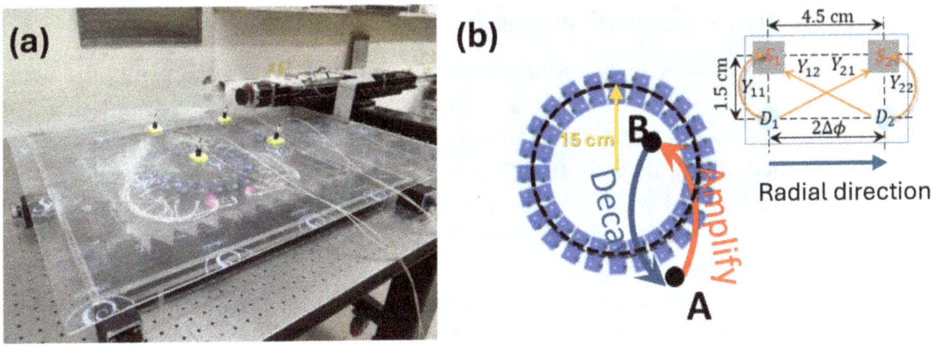

Figure 6.15. (a) Experimental platform of a circular shell of Willis metamaterial of radius 0.15 m. (b) Schematic of the circular shell of 24 unit cells. The inset shows one unit cell containing a pair of speakers and a pair of microphones interconnected through a feedback circuit programmed by a microcontroller. It is aligned in the radial direction to realize Willis coupling in the radial direction. Reproduced from Wen *et al* (2024). CC BY 4.0.

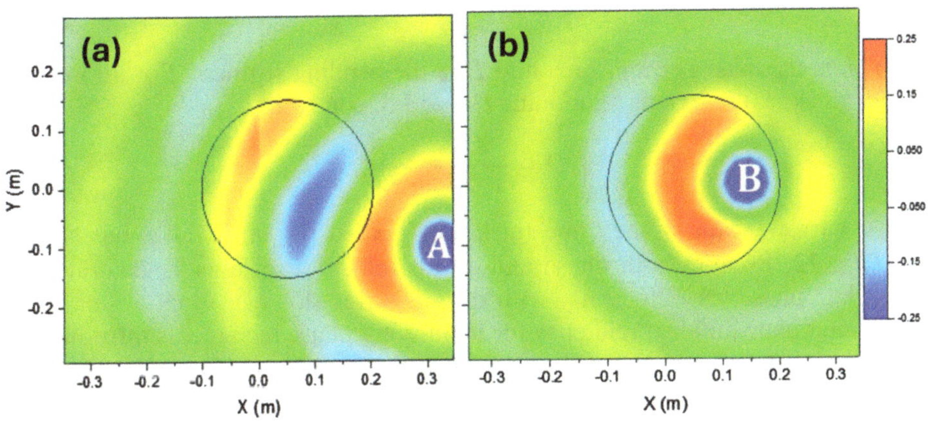

Figure 6.16. Experimental field patterns at a point source situated (a) outside and (b) inside the circular metamaterial shell. Reproduced from Wen *et al* (2024). CC BY 4.0.

required (see equation (6.69)). The microphones and speakers are interconnected through a feedback circuit programmed by a microcontroller to realize the required non-reciprocal Willis coupling.

Figure 6.16 shows the measured acoustic fields at 1.37 kHz when a point source is placed outside (A) and inside (B) the shell. In case (A), the fields inside the shell are amplified by a factor of about 1.5, as expected. In case (B), the field transformation rule should still hold, i.e., the fields outside the shell should be reduced by a factor of 1.5. The experimental results confirm this prediction, demonstrating the one-way field transformation functionality of the non-reciprocal Willis metamaterial shell.

6.9 Extension to flexural waves in elastic plates and beams

While the acoustic waves discussed in the previous sections are scalar pressure waves in air, elastic waves in solids possess a vectorial displacement field and therefore support a variety of polarizations, including longitudinal (P), in-plane shear (SH), out-of-plane shear (SV), as well as different modes of guided waves. In this section we focus on the flexural wave (A_0 Lamb mode), which describes the out-of-plane displacement $w(x, y, t)$ of a plate lying in the x–y plane, or equivalently the bending displacement of a beam along x.

A useful simplification arises from symmetry considerations. In plates with mid-plane symmetry ($z \to -z$) together with reciprocity, the antisymmetric A_0 mode does not couple to the S_0 or the SH_0 modes. Thus, the A_0 mode propagating on the plate can be treated, to an excellent approximation, as a scalar field, allowing many principles developed for airborne acoustics to carry over directly to elastic plates.

At the frequencies considered here, the A_0 mode satisfies a fourth-order plate (or beam) equation with dispersion relation

$$\omega^2 = \frac{D}{\rho h} k^4,$$

where ρh is the mass per unit area and D is the bending stiffness. For an isotropic plate of thickness h, Young's modulus E and Poisson ratio ν,

$$D = \frac{Eh^3}{12(1 - \nu^2)},$$

while a narrow beam of the same thickness (treated per unit width) has stiffness $D = Eh^3/12$. Because both phase and group velocities scale as k, flexural waves accumulate phase very efficiently with distance; this slow-wave nature makes the background plate an ideal reference for identifying phase increments introduced by metamaterial structures.

To adapt the space-coiling concept from airborne acoustics (figure 6.3) to flexural waves, we use the space-coiling unit cell shown in figure 6.17(a). A narrow curved beam of width d is folded into repeated turns of height H_t and equivalent path length l inside a plate of constant thickness. Although compact in footprint, the channel forces the A_0 mode to follow a substantially longer trajectory than in the surrounding plate, leading to a noticeably larger phase advance when projected onto the horizontal x-direction. The resulting behaviour is equivalent to an effective high-index response: the accumulated phase reflects the true propagation length, not the geometric footprint, directly mirroring the space-coiling mechanism used previously for airborne sound.

Figure 6.17(b) illustrates how this curled path enables controllable phase shifts. Increasing the turn height H_t increases the physical path and therefore the transmission phase discontinuity $\Delta\phi$, while full-wave simulations (performed with periodicity $H = 5.9$ mm by repeating the structure in the y-direction) show that the transmission amplitude remains high across a broad band (8–9 kHz). The unexpectedly good transmission arises from the scattering characteristics at the plate–beam interface.

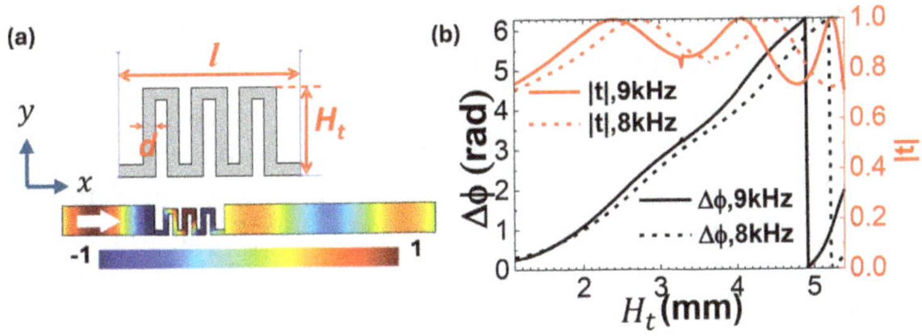

Figure 6.17. (a) Geometry of the space-coiling unit cell for flexural waves, showing the beam width d, turn height H_t, total curved length l, and the plate thickness $h = 3$ mm in the z direction. The example displacement field illustrates how the A_0 mode accumulates a larger phase advance inside the curled channel. (b) Simulated transmission amplitude $|t|$ and phase discontinuity $\Delta\phi = \arg t$ as functions of the turning height H_t at 8 kHz and 9 kHz, demonstrating broadband high transmission and monotonic phase control. Reprinted with permission from Liu *et al* (2017). Copyright (2017) by the American Physical Society.

Because flexural waves obey a fourth-order equation, the field near the plate–beam junction contains two propagating and two evanescent components. Imposing continuity of displacement, slope, bending moment and shear force across the interface between medium 1 (plate) and medium 2 (narrow beam of width d and periodicity H) yields a 4×4 relation among all four mode amplitudes. Eliminating the evanescent components produces an effective 2×2 scattering matrix for the propagating waves. For an A_0 wave incident from the plate, the reflection coefficient can be written as

$$|r| = \left| \frac{Z - 1}{Z + 1} \right|,$$

where the effective flexural impedance is

$$Z = \frac{(1 + F)^2}{4F},$$

and the bending-stiffness ratio

$$F = \frac{D_2 d}{D_1 H} \simeq \frac{d}{H}\left(1 - \nu^2\right),$$

approximates the ratio between the stiffness of the narrow beam and that of the plate. Using $d = 1$ mm, $H = 5.9$ mm and $\nu = 0.38$ gives

$$F \approx 0.15, \qquad Z \approx \frac{(1 + 0.15)^2}{4 \cdot 0.15} \approx 2.3.$$

This value is much closer to unity than one would infer from a purely acoustic analogy, where the impedance mismatch scales with the geometric area ratio $H/d \approx 5.9$. The key difference is that the evanescent flexural components at the interface effectively soften the mismatch and suppress reflection. Consequently, large phase delays can be introduced by the space-coiling geometry without significant amplitude loss. This combination of controllable phase shift and high transmission underpins the use of curled channels for constructing efficient elastic metasurfaces.

By arranging such coiling elements with different geometries along a boundary, we can construct a metasurface—or more precisely, a 'meta-line' in a two-dimensional plate—that shapes flexural wavefronts. To realise a source shifter as an example, we specify a target transmission phase profile $\Delta\phi(\theta)$ on a ring of radius $R \approx 51$ mm, designed such that the outgoing wavefronts match those of a virtual source displaced horizontally by ΔS. For a cylindrical A_0 wave, the original and target phases at the ring are $\phi_{\text{ori}}(R, \theta) = k_0 R$ and $\phi_{\text{tar}}(R, \theta) = k_0\sqrt{R^2 + \Delta S^2 + 2R\Delta S\cos\theta}$, giving the required transmission phase

$$\Delta\phi(\theta) = k_0\left(\sqrt{R^2 + \Delta S^2 + 2R\Delta S\cos\theta} - R\right).$$

For a unit cell located at angle θ on the ring, the required transmission phase is therefore $\Delta\phi(\theta)$, which is implemented by choosing the corresponding turning height $H_t(\theta)$ using the monotonic relation between H_t and arg$t$ in figure 6.17(b).

Figure 6.18 shows the experimentally measured flexural wavefield produced by a ring of space-coiling elements designed using the phase profile above. A physical point source is placed at the centre of the ring, while each unit cell imposes the required local phase shift $\Delta\phi(\theta)$ through its chosen turning height H_t. As a result, the

Figure 6.18. Measured flexural wavefield generated by a ring of space-coiling unit cells designed to shift the apparent position of a point source. The physical source is located at the centre, while the imposed phase profile produces wavefronts that match those of a virtual source displaced horizontally by ΔS (dashed contours).

outgoing A_0 wavefronts no longer appear to radiate from the true source position, but instead from a virtual source displaced horizontally by $\Delta S = 25$ mm. The dashed circular contours superimposed on the measured field indicate the target cylindrical wavefronts corresponding to this virtual source position, demonstrating good agreement with the designed transformation. Inside the ring, the field retains the original circular symmetry, whereas outside it, the wavefronts are shifted as required. This verifies that a ring of space-coiling elastic elements can implement the prescribed phase discontinuity and thereby realise a source shifter using flexural waves in a plate.

6.10 Problems

Problem 6.1 Starting from the secular equation of the space-coiling metamaterial,

$$2\cos\phi_0 = \cos\phi_x + \cos\phi_y,$$

the Bloch mode is described by the non-trivial solution $(p, v_{x1}, v_{x2}, v_{y1}, v_{y2})$.

Assume that the left interface of the unit cell is opened to free space (air). (a) By considering half a channel length, express the pressure p_L and velocity v_{xL} at the left surface in terms of (p, v_{x1}) (see figure 6.3). (b) Define the effective pressure and velocity at the interface as

$$p_{\text{eff}} = p_L, \qquad v_{x,\text{eff}} = \frac{d}{a} v_{xL},$$

where d is the channel width and a is the lattice constant. (c) Show that the effective impedance of the medium,

$$\eta = \frac{p_{\text{eff}}}{v_{x,\text{eff}}},$$

is given by

$$\eta = \frac{a}{d}\eta_0 \frac{\tan(\phi_0/2)}{\tan(\phi_x/2)}.$$

Problem 6.2 Starting from the effective medium formulas, equation (6.33), find the expression for the constant group velocity $v_g = \nabla_{\mathbf{k}}\ \omega$ near the conical dispersion. To describe the effect of finite channel width, assume that the cone center frequency ω_c is slightly shifted to $\omega_c - \delta\omega_g/2$ for the density to become zero, while it is slightly shifted to $\omega_c + \delta\omega_g/2$ for the compressibility to become zero, thereby opening a tiny bandgap of width $\delta\omega_g$. Discuss how this shift affects the group velocity, refractive index, and impedance.

Problem 6.3 Consider a 1D Willis medium along x whose macroscopic fields obey

$$\partial_x\begin{pmatrix} v \\ p \end{pmatrix} = \mathrm{i}\omega\begin{pmatrix} \beta & \mathrm{i}\tau/c \\ \mathrm{i}\tau'/c & \rho \end{pmatrix}\begin{pmatrix} p \\ v \end{pmatrix}.$$

(a) Suppose the medium has a small thickness L in air. We also define the free-space elapsed phase by $\phi_0 = \omega L/c \ll 1$ where c is the speed of sound in air. Derive the scattering matrix of the thin slab (relating forward/backward outgoing waves to incoming waves) and show that, to leading order in ϕ_0,

$$\begin{pmatrix} t_\mathrm{f} & r_b \\ r_\mathrm{f} & t_b \end{pmatrix} \simeq \begin{pmatrix} 1 + \dfrac{\mathrm{i}\phi_0}{2}\left(\beta/\beta_0 + \rho/\rho_0 + 2\mathrm{i}\ \tau_{nr}\right) & \dfrac{\mathrm{i}\phi_0}{2}\left(\beta/\beta_0 - \rho/\rho_0 - 2\mathrm{i}\ \tau_r\right) \\ \dfrac{\mathrm{i}\phi_0}{2}\left(\beta/\beta_0 - \rho/\rho_0 + 2\mathrm{i}\ \tau_r\right) & 1 + \dfrac{\mathrm{i}\phi_0}{2}\left(\beta/\beta_0 + \rho/\rho_0 - 2\mathrm{i}\ \tau_{nr}\right) \end{pmatrix},$$

where the Willis couplings are decomposed into reciprocal and non-reciprocal parts

$$\tau = \tau_r + \tau_{nr}, \qquad \tau' = -\ \tau_r + \tau_{nr}.$$

The transmission coefficients t_f and t_b are for forward and backward incidence, respectively, while r_f and r_b are the corresponding reflection coefficients.

(b) Assume $\beta = \beta_0$ and $\rho = \rho_0$ (same compressibility and density as air) and purely non-reciprocal Willis coupling, i.e., $\tau = \tau' = \tau_{nr}$ (so $\tau_r = 0$). Using your result from (a), prove that the thin slab is reflectionless from both sides to $O(\phi_0)$, i.e. $r_f = r_b = 0$. Argue that cascading N identical, reflectionless thin sections gives

$$t_f\ e^{+\tau_{nr}\phi_0} = t_b\ e^{-\tau_{nr}\phi_0} = e^{i\phi_0}, \qquad r_f = r_b = 0,$$

i.e., exponential amplification in one direction and exponential attenuation in the reverse, with zero reflection (the 'amplifying-diode' behavior).

References

Brunet T, Merlin A, Mascaro B, Zimny K, Leng J, Poncelet O, Aristégui C and Mondain-Monval O 2015 Soft 3d acoustic metamaterial with negative index *Nat. Mater.* **14** 384–8

Chen H and Chan C T 2010 Acoustic cloaking and transformation acoustics *J. Phys. D: Appl. Phys.* **43** 113001

Cho C, Wen X, Park N and Li J 2020 Digitally virtualized atoms for acoustic metamaterials *Nat. Commun.* **11** 251

Cummer S, Rahm M and Schurig D 2008 Material parameters and vector scaling in transformation acoustics *New J. Phys.* **10** 115025

Cummer S and Schurig D 2007 One path to acoustic cloaking *New J. Phys.* **9** 45

He Z, Wang Y-Z, Wang Y-S and Wang Y-S 2021 Active feedback control of effective mass density and sound transmission on elastic wave metamaterials *Int. J. Mech. Sci.* **195** 106221

Li J and Chan C 2004 Double-negative acoustic metamaterial *Phys. Rev.* **70** 055602

Liang Z, Feng T, Lok S, Liu F, Ng K, Chan C, Wang J, Han S, Lee S and Li J 2013 Space-coiling metamaterials with double negativity and conical dispersion *Sci. Rep.* **3** 1614

Liang Z and Li J 2012 Extreme acoustic metamaterial by coiling up space *Phys. Rev. Lett.* **108** 114301

Liu Y, Liang Z, Liu F, Diba O, Lamb A and Li J 2017 Source illusion devices for flexural Lamb waves using elastic metasurfaces *Phys. Rev. Lett.* **119** 034301

Mei J, Liu Z, Wen W and Sheng P 2007 Effective dynamic mass density of composites *Phys. Rev.* B **76** 134205

Popa B, Shinde D, Konneker A and Cummer S 2015 Active acoustic metamaterials reconfigurable in real-time *Phys. Rev.* B **91** 220303

Trainiti G, Xia Y, Marconi J, Cazzulani G, Erturk A and Ruzzene M 2019 Time-periodic stiffness modulation in elastic metamaterials for selective wave filtering: Theory and experiment *Phys. Rev. Lett.* **122** 124301

Wang T and Wang Y-Z 2025 Active feedback control on sound transmission loss of metamaterial shell *AIAA J.* **63** 1541–59

Wei P, Croënne C, Chu S and Li J 2014 Symmetrical and anti-symmetrical coherent perfect absorption for acoustic waves *Appl. Phys. Lett.* **104** 121902

Wei P, Liu F, Liang Z, Xu Y, Chu S T and Li J 2015 An acoustic beam shifter with enhanced transmission using perforated metamaterials *Europhys. Lett.* **109** 14004

Wen X, Cho C, Zhu X, Park N and Li J 2024 Nonreciprocal field transformation with active acoustic metasurfaces *Sci. Adv.* **10** eadm9673

Wen X, Zhu X-F, Wu H-W and Li J 2021 Realizing spatiotemporal effective media for acoustic metamaterials *Phys. Rev.* B **104** L060304

Yang M, Chen S, Fu C and Sheng P 2016 Optimal sound-absorbing structures *Mater. Horiz* **4** 673–80

Zhu X, Wu H-W, Zhuo Y, Liu Z and Li J 2023 Effective medium for time-varying frequency-dispersive acoustic metamaterials *Phys. Rev. B* **108** 104303

IOP Publishing

Essential Concepts in Metamaterials

Jensen Li

Appendix A

A.1 Transfer matrix method

1D wave equation

For 1D systems with stratified layers—such as dielectric layers for electromagnetic wave propagation, transmission lines for electromagnetic power, or sound waves propagating through layers of different rock densities—the transfer matrix method (TMM) provides a convenient framework for both analytical analysis and minimal numerical computation.

A transmission line is a typical example. The wave equation (with harmonic time dependence $e^{-i\omega t}$) for the propagating potential V and current I can be written as

$$\Pi\ \partial_z\begin{pmatrix} V \\ I \end{pmatrix} = i\omega\begin{pmatrix} C & 0 \\ 0 & L \end{pmatrix}\begin{pmatrix} V \\ I \end{pmatrix}, \tag{A.1}$$

where C and L are the capacitance and inductance per unit length of the transmission line. They play roles analogous to permittivity and permeability, respectively, and together form the 2×2 constitutive matrix. The off-diagonal elements are zero for a conventional transmission line. The operator Π is defined as

$$\Pi = \begin{pmatrix} 0 & 1 \\ 1 & 0 \end{pmatrix},$$

which has the operational meaning of extracting the time-averaged power flux:

$$P = \frac{1}{2}\,\mathrm{Re}(V^* I) = \frac{1}{4}\begin{pmatrix} V^* & I^* \end{pmatrix}\Pi\begin{pmatrix} V \\ I \end{pmatrix}. \tag{A.2}$$

Thus Π can be regarded as a 'Poynting operator' because of its relation to power, or more generally as an inner-product matrix defining a conserved quantity along z. Indeed, $\partial_z P = 0$ if the constitutive matrix is Hermitian ($C = C^*$, $L = L^*$).

doi:10.1088/978-0-7503-1131-1ch7

For convenience, we multiply I by the freespace impedance (vacuum in this case) η_0 with the same unit of resistance and normalize the constitutive parameters (in matrix $\mathcal{C}$) to dimensionless numbers:

$$\Pi\ \partial_z\begin{pmatrix} V \\ \eta_0 I \end{pmatrix} = \mathrm{i}k_0\mathcal{C}^{-1}\begin{pmatrix} V \\ \eta_0 I \end{pmatrix} = \mathrm{i}k_0\begin{pmatrix} \epsilon & 0 \\ 0 & \mu \end{pmatrix}\begin{pmatrix} V \\ \eta_0 I \end{pmatrix}, \tag{A.3}$$

where $k_0 = \omega/c$ is the free-space wavenumber. The vacuum speed of light c and impedance η_0 are related to vacuum permittivity ϵ_0 and permeability μ_0 as

$$c = \frac{1}{\sqrt{\epsilon_0}\ \sqrt{\mu_0}}, \qquad \eta_0 = \frac{\sqrt{\mu_0}}{\sqrt{\epsilon_0}}.$$

We have defined the relative permittivity ϵ and relative permeability μ for the transmission-line problem by

$$C = \epsilon\epsilon_0, \qquad L = \mu\mu_0. \tag{A.4}$$

The constitutive matrix $\mathcal{C}$ is Hermitian in the lossless case and antisymmetric if the system is reciprocal.

This form (equation (A.3)) is in fact universal and applies to many other wave systems. For instance, for electromagnetic waves normally incident on stratified media (with ϵ and μ) along the z-axis, the decoupled polarizations satisfy

$$\Pi\ \partial_z\begin{pmatrix} E_x \\ H_y \end{pmatrix} = \mathrm{i}k_0\mathcal{C}^{-1}\begin{pmatrix} E_x \\ H_y \end{pmatrix} = \mathrm{i}k_0\begin{pmatrix} \epsilon & 0 \\ 0 & \mu \end{pmatrix}\begin{pmatrix} E_x \\ H_y \end{pmatrix}, \tag{A.5}$$

and

$$\Pi\ \partial_z\begin{pmatrix} E_y \\ -H_x \end{pmatrix} = \mathrm{i}k_0\mathcal{C}^{-1}\begin{pmatrix} E_y \\ -H_x \end{pmatrix} = \mathrm{i}k_0\begin{pmatrix} \epsilon & 0 \\ 0 & \mu \end{pmatrix}\begin{pmatrix} E_y \\ -H_x \end{pmatrix}. \tag{A.6}$$

For scalar acoustic waves propagating normally to stratified layers (compressibility β and density ρ) along z,

$$\Pi\ \partial_z\begin{pmatrix} p \\ \eta_0 v_z \end{pmatrix} = \mathrm{i}k_0\mathcal{C}^{-1}\begin{pmatrix} p \\ \eta_0 v_z \end{pmatrix} = \mathrm{i}k_0\begin{pmatrix} \beta/\beta_0 & 0 \\ 0 & \rho/\rho_0 \end{pmatrix}\begin{pmatrix} p \\ \eta_0 v_z \end{pmatrix}, \tag{A.7}$$

where β_0, ρ_0, c, and $\eta_0 = \rho_0 c$ are the compressibility, density, sound speed, and intrinsic impedance of the background medium (e.g., air). Here p is the pressure and v_z the velocity along z. Hence, once formulated in this universal form, the transfer matrix method applies equally well to other 1D wave propagation systems.

Normal modes

Consider a homogeneous medium with constant ϵ and μ, which supports waves $\propto e^{\pm ikz}$. This corresponds to solving the eigenvalue problem of the matrix $k_0\Pi^{-1}\mathcal{C}^{-1}$. One eigenmode propagating in the $+z$ direction can be written as

$$V = a\mathrm{e}^{ikz}, \qquad I = \frac{a}{\eta}\mathrm{e}^{ikz}, \tag{A.8}$$

up to an arbitrary complex constant a, where the propagation constant and impedance are

$$k = k_0\sqrt{\epsilon}\sqrt{\mu}, \qquad \eta = \eta_0\frac{\sqrt{\mu}}{\sqrt{\epsilon}}. \tag{A.9}$$

The forward wave power is then

$$P = \frac{|a|^2}{2|\eta|^2}\,\mathrm{Re}(\eta)\,\mathrm{e}^{-2\,\mathrm{Im}(k)\,z}.$$

Similarly, the backward mode is

$$V = b\mathrm{e}^{-\mathrm{i}kz}, \qquad I = -\frac{b}{\eta}\mathrm{e}^{-\mathrm{i}kz}, \tag{A.10}$$

with power

$$P = -\frac{|b|^2}{2|\eta|^2}\,\mathrm{Re}(\eta)\,\mathrm{e}^{2\,\mathrm{Im}(k)\,z}.$$

For a passive system ($\mathrm{Im}\,\epsilon \geqslant 0$, $\mathrm{Im}\,\mu \geqslant 0$), we have $\mathrm{Im}\,k \geqslant 0$ and $\mathrm{Re}\,\eta \geqslant 0$. Thus the forward mode decays as $z \to +\infty$ with power flux along $+z$, while the backward mode decays as $z \to -\infty$ with power flux along $-z$.

Transfer matrix method

At any location z within a medium specified by (k, η), the fields can be expressed as

$$\begin{aligned} V &= a\mathrm{e}^{\mathrm{i}kz} + b\mathrm{e}^{-\mathrm{i}kz}, \\ I &= \frac{1}{\eta}(a\mathrm{e}^{\mathrm{i}kz} - b\mathrm{e}^{-\mathrm{i}kz}). \end{aligned} \tag{A.11}$$

At $z = 0$, this change of basis gives

$$\begin{pmatrix} V \\ I \end{pmatrix} = \begin{pmatrix} 1 & 1 \\ \frac{1}{\eta} & -\frac{1}{\eta} \end{pmatrix}\begin{pmatrix} a \\ b \end{pmatrix}. \tag{A.12}$$

For the transfer matrix method, to relate the a and b coefficients on the two sides of a series of stratified layers along the z-direction, it is sufficient to combine the transfer matrices of three basic building blocks, as schematically illustrated in figure A.1.

Propagation through a homogeneous section of length d (figure A.1(a)) gives

$$\begin{pmatrix} a_2 \\ b_2 \end{pmatrix} = \begin{pmatrix} \mathrm{e}^{\mathrm{i}kd} & 0 \\ 0 & \mathrm{e}^{-\mathrm{i}kd} \end{pmatrix}\begin{pmatrix} a_1 \\ b_1 \end{pmatrix}. \tag{A.13}$$

At an interface between two media (k, η) and (k', η'), the boundary conditions of continuous V and I (figure A.1(b)) give

$$\begin{pmatrix} a_2 \\ b_2 \end{pmatrix} = \frac{1}{2}\begin{pmatrix} 1 + \eta'/\eta & 1 - \eta'/\eta \\ 1 - \eta'/\eta & 1 + \eta'/\eta \end{pmatrix}\begin{pmatrix} a_1 \\ b_1 \end{pmatrix}. \tag{A.14}$$

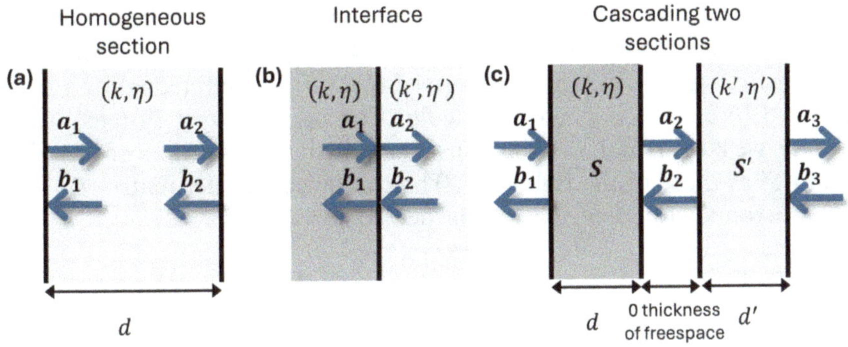

Figure A.1. Different scenarios for transfer and scattering matrices: (a) propagation through a homogeneous section, (b) an interface between two media, and (c) cascading of two slabs. Forward (a) and backward (b) amplitudes are defined at each interface.

As a simple application, for a slab (k, η, d) surrounded by free space (k_0, η_0), by cascading two interfaces and one homogeneous section (left part of figure A.1(c)), the overall transfer matrix T satisfies

$$\begin{pmatrix} a_2 \\ b_2 \end{pmatrix} = T\begin{pmatrix} a_1 \\ b_1 \end{pmatrix} = \begin{pmatrix} T_{11} & T_{12} \\ T_{21} & T_{22} \end{pmatrix}\begin{pmatrix} a_1 \\ b_1 \end{pmatrix}, \tag{A.15}$$

where

$$T = \cos(kd)\begin{pmatrix} 1 & 0 \\ 0 & 1 \end{pmatrix} + \frac{\mathrm{i}}{2}\sin(kd)\begin{pmatrix} Z + 1/Z & -Z + 1/Z \\ Z - 1/Z & -Z - 1/Z \end{pmatrix}, \tag{A.16}$$

with $Z = \eta/\eta_0$ being the impedance normalized by freespace.

Although the method is called the 'transfer matrix method,' we often prefer to compute the scattering matrix S defined by

$$\begin{pmatrix} a_2 \\ b_1 \end{pmatrix} = S\begin{pmatrix} a_1 \\ b_2 \end{pmatrix} = \begin{pmatrix} S_{11} & S_{12} \\ S_{21} & S_{22} \end{pmatrix}\begin{pmatrix} a_1 \\ b_2 \end{pmatrix}. \tag{A.17}$$

The relation between T and S (by combining equation (A.15) and (A.17)) is

$$S = \begin{pmatrix} T_{11} - T_{12}T_{22}^{-1}T_{21} & T_{12}T_{22}^{-1} \\ -T_{22}^{-1}T_{21} & T_{22}^{-1} \end{pmatrix}. \tag{A.18}$$

For the homogeneous slab, substituting equation (A.16) into equation (A.18), we have the transmission and reflection coefficients (matrix elements of S) as

$$\begin{aligned} t = S_{11} = S_{22} &= \frac{2}{2\cos(kd) - \mathrm{i}(Z + 1/Z)\sin(kd)}, \\ r = S_{12} = S_{21} &= \frac{-\mathrm{i}(Z - 1/Z)\sin(kd)}{2\cos(kd) - \mathrm{i}(Z + 1/Z)\sin(kd)}. \end{aligned} \tag{A.19}$$

Finally, when cascading two different slabs, we can reuse the previous results by purposely inserting a virtual layer of free space (k_0, η_0) between them. In this way, the T-matrix or S-matrix expressions derived for a single slab (equation (A.16) or equation (A.19)) can be applied repeatedly. For the two-layer system shown in figure A.1(c), we label T and S as the transfer and scattering matrices of the left slab, and T' and S' as those of the right slab. The combined transfer matrix T_{tot} for the two slabs, surrounded by free space, can then be written as

$$T_{\text{tot}} = T'T, \tag{A.20}$$

or join S-matrices directly:

$$S_{\text{tot}} = \begin{pmatrix} S'_{11}(I - S_{12}S'_{21})^{-1}S_{11} & S'_{12} + S'_{11}S_{12}(I - S'_{21}S_{12})^{-1}S'_{22} \\ S_{21} + S_{22}(I - S'_{21}S_{12})^{-1}S'_{21}S_{11} & S_{22}(I - S'_{21}S_{12})^{-1}S'_{22} \end{pmatrix}. \tag{A.21}$$

By repeating this process, one can construct multilayer systems. For continuously varying $\epsilon(z)$, $\mu(z)$, the system is chopped into thin layers for numerical calculation. In equation (A.21), I denotes the scalar value 1, since in the present case we are considering scalar wave propagation only.

Because evanescent waves can cause exponentially large or small matrix elements in T, it is often numerically more stable to use S-matrices, which in passive systems only contain exponentially small (but not large) terms.

Generalization: bianisotropy

The formulation can be extended to include the full 2×2 constitutive matrix with bianisotropy:

$$\mathcal{C}^{-1} = \begin{pmatrix} \epsilon & \mathrm{i}\kappa \\ \mathrm{i}\kappa' & \mu \end{pmatrix}. \tag{A.22}$$

Equivalently, the wave equation becomes

$$\Pi\,\partial_z\begin{pmatrix} V \\ \eta_0 I \end{pmatrix} = \mathrm{i}k_0\begin{pmatrix} \epsilon & \mathrm{i}\kappa \\ \mathrm{i}\kappa' & \mu \end{pmatrix}\begin{pmatrix} V \\ \eta_0 I \end{pmatrix}. \tag{A.23}$$

If $\mathcal{C}^{-1}$ is Hermitian (i.e. $\epsilon = \epsilon^*$, $\mu = \mu^*$, $\kappa = -\kappa'^*$), then the power P remains a conserved quantity along z. In this case, in equation (A.11), the forward and backward modes still have propagation constants $\pm k$ with $k^2 = k_0^2(\epsilon\mu - \kappa^2)$, but the impedance for the two modes do not have the same magnitude. This will affect the transfer and scattering matrices at both interfaces and homogeneous sections. The generalization is straightforward and will not be repeated here.

Generalization: off-oblique incidence

Let's say we are considering electromagnetic waves at oblique incidence on the z–x plane, with transverse wavevector component $k_x = k\sin\theta$ being conserved across the layers. This is just Snell's law for plane waves. θ is the angle the plane waves make with the z-axis, varying in different layers. The longitudinal wavevector component in each layer is

$$k_z = \sqrt{nk_0 - k_x}\sqrt{nk_0 + k_x},$$

where $n = \sqrt{\epsilon}\sqrt{\mu}$ is the refractive index of each layer. This comes from the isotropic dispersion relationship $k_x^2 + k_z^2 = n^2 k_0^2$. The branch of square root is chosen such that $\text{Im}(k_z) \geqslant 0$ for passive media. On the other hand, the wave impedance in each layer depends on the polarization. For TM polarization (H-field out-of-plane along y and E-field on z–x plane) as an example, there is an additional projection factor $\cos\theta = k_z/k$ for the E-field along x as E_x and H_y are the continuous fields on an interface. The impedance becomes

$$\frac{E_x}{H_y} = \eta\cos(\theta) = \frac{\eta_0 k_z}{k_0 \epsilon},$$

In equation (A.9), k is now replaced by k_z and η by $\eta_0 k_z/(k_0\epsilon)$, as in the equations used to establish the transfer and scattering matrices.

For the transmission and reflection coefficients of a homogeneous slab surrounded by freespace, equation (A.19) remains valid, but k and Z (impedance normalized by freespace) are replaced by k_z and $\frac{k_z}{\epsilon\sqrt{k_0 - k_x}\sqrt{k_0 + k_x}}$, respectively.

Generalization: beyond scalar waves

For electromagnetic waves, the two polarizations are decoupled under normal incidence on isotropic stratified media (equations (A.5) and (A.6)). However, if the medium is anisotropic or bianisotropic, the two polarizations become coupled. In this case, one must work with 4×4 transfer and scattering matrices. All eigenmodes, as well as the slab transfer and scattering matrices, need to be reformulated, and the basis transformation (equation (A.12)) must also be modified accordingly.

When combining transfer or scattering matrices, equations (A.15) and (A.17) remain valid, but a and b become two component vectors, and the scalar elements T_{ij} and S_{ij} are replaced by 2×2 sub-block matrices. The T-to-S relation (equation (A.18)) and the cascading formulas (equations (A.20) and (A.21)) also remain unchanged in matrix form. The identity I in equation (A.21) becomes a 2×2 identity matrix.

A.2 Dipole radiation and local field

A.2.1 Point dipole case

The Green's dyadic for the vacuum Maxwell's equation (with wavenumber k) is defined by

$$\overline{\overline{G}} = (\overline{\overline{I}} + \frac{1}{k^2}\nabla\nabla)\frac{e^{ikr}}{4\pi r},$$

with

$$\nabla \times \nabla \times \overline{\overline{G}} - k^2\overline{\overline{G}} = \overline{\overline{I}}\delta(\mathbf{r})$$

and the 3D position vector $\mathbf{r} = x\hat{x} + y\hat{y} + z\hat{z}$.

Then, the radiation from a point dipole (electric dipole $\mathbf{p}$ and magnetic dipole $\mathbf{m}$ at the origin) can be written as

$$\mathbf{E}_{\text{rad}}(\mathbf{r}) = -\frac{1}{3}\mathbf{p}\delta(\mathbf{r}) + k^2\overline{\overline{G}} \cdot \mathbf{p} + \mathrm{i}k(\nabla \times \overline{\overline{G}}) \cdot \mathbf{m},$$
$$\mathbf{H}_{\text{rad}}(\mathbf{r}) = -\frac{1}{3}\mathbf{m}\delta(\mathbf{r}) + k^2\overline{\overline{G}} \cdot \mathbf{m} - \mathrm{i}k(\nabla \times \overline{\overline{G}}) \cdot \mathbf{p}, \tag{A.24}$$

with

$$\nabla \times \mathbf{E}_{\text{rad}}(\mathbf{r}) = \mathrm{i}k(\mathbf{H}_{\text{rad}}(\mathbf{r}) + \mathbf{m}\delta(\mathbf{r})),$$
$$\nabla \times \mathbf{H}_{\text{rad}}(\mathbf{r}) = -\mathrm{i}k(\mathbf{E}_{\text{rad}}(\mathbf{r}) + \mathbf{p}\delta(\mathbf{r})),$$

which is also commonly written (away from origin) as

$$\begin{aligned}\mathbf{E}_{\text{rad}}(\mathbf{r}) &= \frac{k^3 e^{ikr}}{4\pi}\left(\frac{1 - \mathrm{i}kr}{k^3r^3}(3\hat{r}\hat{r} \cdot \mathbf{p} - \mathbf{p})\right.\\ &\quad \left. - \frac{1}{kr}\hat{r} \times (\hat{r} \times \mathbf{p}) - \frac{1}{k^2r^2}i\hat{r} \times \mathbf{m}(1 - \mathrm{i}kr)\right),\\ \mathbf{H}_{\text{rad}}(\mathbf{r}) &= \frac{k^3 e^{ikr}}{4\pi}\left(\frac{1 - \mathrm{i}kr}{k^3r^3}(3\hat{r}\hat{r} \cdot \mathbf{m} - \mathbf{m})\right.\\ &\quad \left. - \frac{1}{kr}\hat{r} \times (\hat{r} \times \mathbf{m}) + \frac{1}{k^2r^2}\mathrm{i}\hat{r} \times \mathbf{p}(1 - \mathrm{i}kr)\right).\end{aligned}$$

On the other hand, the local field (total field without the radiation) can be written as

$$\mathbf{E}_{\text{loc}}(\mathbf{r}) = \frac{6\pi}{k^3}(-\mathrm{i}k^2\overline{\overline{G}}_{\text{R}} \cdot \mathbf{E}_{\text{loc}}(\mathbf{0}) + k(\nabla \times \overline{\overline{G}}_{\text{R}}) \cdot \mathbf{H}_{\text{loc}}(\mathbf{0})),$$
$$\mathbf{H}_{\text{loc}}(\mathbf{r}) = \frac{6\pi}{k^3}(-\mathrm{i}k^2\overline{\overline{G}}_{\text{R}} \cdot \mathbf{H}_{\text{loc}}(\mathbf{0}) - k(\nabla \times \overline{\overline{G}}_{\text{R}}) \cdot \mathbf{E}_{\text{loc}}(\mathbf{0})). \tag{A.25}$$

Here, $\overline{\overline{G}}_{\text{R}}$ is called the regular part of the vacuum Green's dyadic (dipole radiation) given by

$$\overline{\overline{G}}_{\text{R}} = \mathrm{i}\,\mathrm{Im}(\overline{\overline{G}}) = (\overline{\overline{I}} + \frac{1}{k^2}\nabla\nabla)\frac{i\sin(kr)}{4\pi r}.$$

At the quasistatic limit, we can further approximate $\overline{\overline{G}}_{\text{R}} \approx \mathrm{i}k\overline{\overline{I}}/(6\pi)$ and $\nabla \times \overline{\overline{G}}_{\text{R}} \approx -\mathrm{i}k^3\mathbf{r} \times \overline{\overline{I}}/(12\pi)$ only up to the leading order of k. Equation (A.25) then reduces to

$$\mathbf{E}_{\text{loc}}(\mathbf{r}) \approx \mathbf{E}_{\text{loc}}(\mathbf{0}) - \frac{ik}{2}\mathbf{r} \times \mathbf{H}_{\text{loc}}(\mathbf{0}),$$
$$\mathbf{H}_{\text{loc}}(\mathbf{r}) \approx \mathbf{H}_{\text{loc}}(\mathbf{0}) + \frac{ik}{2}\mathbf{r} \times \mathbf{E}_{\text{loc}}(\mathbf{0}). \tag{A.26}$$

A.2.2 Line dipole case

The Green's dyadic for the vacuum Maxwell's equation (with z-independence) is defined by

$$\overline{\overline{G}} = (\overline{\overline{I}} + \frac{1}{k^2}\nabla\nabla)\frac{iH_0(k\rho)}{4},$$

with

$$\nabla \times \nabla \times \overline{\overline{G}} - k^2\overline{\overline{G}} = \overline{\overline{I}}\delta(\boldsymbol{\rho}),$$

and the 2D position vector $\boldsymbol{\rho} = x\hat{x} + y\hat{y}$. H_0 is the order-zero Hankel function of the first kind.

Then, the radiation (away from origin) from a line dipole (electric/magnetic dipole per unit length **p**/**m** at $\rho = 0$) can be written as

$$\begin{aligned} \mathbf{E}_{\text{rad}}(\boldsymbol{\rho}) &= k^2\overline{\overline{G}} \cdot \mathbf{p} + \mathrm{i}k(\nabla \times \overline{\overline{G}}) \cdot \mathbf{m}, \\ \mathbf{H}_{\text{rad}}(\boldsymbol{\rho}) &= k^2\overline{\overline{G}} \cdot \mathbf{m} - \mathrm{i}k(\nabla \times \overline{\overline{G}}) \cdot \mathbf{p}, \end{aligned} \tag{A.27}$$

with

$$\begin{aligned} \nabla \times \mathbf{E}_{\text{rad}}(\boldsymbol{\rho}) &= \mathrm{i}k(\mathbf{H}_{\text{rad}}(\boldsymbol{\rho}) + \mathbf{m}\delta(\boldsymbol{\rho})), \\ \nabla \times \mathbf{H}_{\text{rad}}(\boldsymbol{\rho}) &= -\mathrm{i}k(\mathbf{E}_{\text{rad}}(\boldsymbol{\rho}) + \mathbf{p}\delta(\boldsymbol{\rho})). \end{aligned}$$

On the other hand, the local field (total field without the radiation) can be written as

$$\begin{aligned} \mathbf{E}_{\text{loc}}(\boldsymbol{\rho}) = &\frac{8}{k^2}(-\mathrm{i}k^2\overline{\overline{G}}_{\text{R}} \cdot (\overline{\overline{I}} - \hat{z}\hat{z}/2) \cdot \mathbf{E}_{\text{loc}}(\mathbf{0}) \\ &+ k(\nabla \times \overline{\overline{G}}_{\text{R}}) \cdot (\overline{\overline{I}} - \hat{z}\hat{z}/2) \cdot \mathbf{H}_{\text{loc}}(\mathbf{0})), \\ \mathbf{H}_{\text{loc}}(\boldsymbol{\rho}) = &\frac{8}{k^2}(-\mathrm{i}k^2\overline{\overline{G}}_{\text{R}} \cdot (\overline{\overline{I}} - \hat{z}\hat{z}/2) \cdot \mathbf{H}_{\text{loc}}(\mathbf{0}) \\ &- k(\nabla \times \overline{\overline{G}}_{\text{R}}) \cdot (\overline{\overline{I}} - \hat{z}\hat{z}/2) \cdot \mathbf{E}_{\text{loc}}(\mathbf{0})). \end{aligned} \tag{A.28}$$

Here, $\overline{\overline{G}}_{\text{R}}$ is called the regular part of the vacuum Green's dyadic (dipole radiation) given by

$$\overline{\overline{G}}_{\text{R}} = \mathrm{i}\,\mathrm{Im}(\overline{\overline{G}}) = (\overline{\overline{I}} + \frac{1}{k^2}\nabla\nabla)\frac{\mathrm{i}J_0(k\rho)}{4}.$$

J_0 denotes the Bessel function of the first kind with order 0. At the quasistatic limit, we can further approximate $\overline{\overline{G}}_{\text{R}} \approx \mathrm{i}(\overline{\overline{I}} + \hat{z}\hat{z})/8$ and $\nabla \times \overline{\overline{G}}_{\text{R}} \approx -\mathrm{i}k^2\boldsymbol{\rho} \times \overline{\overline{I}}/8$ only up to the leading order of k. Equation (A.28) then reduces to

$$\begin{aligned} \mathbf{E}_{\text{loc}}(\mathbf{r}) &\approx \mathbf{E}_{\text{loc}}(\mathbf{0}) + \mathrm{i}k\mathbf{H}_{\text{loc}}(\mathbf{0}) \cdot (\overline{\overline{I}} - \frac{\hat{z}\hat{z}}{2}) \times \boldsymbol{\rho}, \\ \mathbf{H}_{\text{loc}}(\mathbf{r}) &\approx \mathbf{H}_{\text{loc}}(\mathbf{0}) - \mathrm{i}k\mathbf{E}_{\text{loc}}(\mathbf{0}) \cdot (\overline{\overline{I}} - \frac{\hat{z}\hat{z}}{2}) \times \boldsymbol{\rho}. \end{aligned} \tag{A.29}$$

A.3 Internal symmetry of Maxwell's equations

The Maxwell's equations in vacuum (Lorentz–Heaviside units) are written as

$$\mathrm{i}\begin{bmatrix} 0 & \nabla\times \\ \nabla\times & 0 \end{bmatrix}\begin{bmatrix} E \\ \mathrm{i}H \end{bmatrix} = (\Pi_x\partial_x + \Pi_y\partial_y + \Pi_z\partial_z)\begin{bmatrix} E \\ \mathrm{i}H \end{bmatrix} = \mathrm{i}k\begin{bmatrix} E \\ \mathrm{i}H \end{bmatrix}, \tag{A.30}$$

where

$$\Pi_x = \begin{bmatrix} 0 & \varpi_x \\ \varpi_x & 0 \end{bmatrix}, \quad \Pi_y = \begin{bmatrix} 0 & \varpi_y \\ \varpi_y & 0 \end{bmatrix}, \quad \Pi_z = \begin{bmatrix} 0 & \varpi_z \\ \varpi_z & 0 \end{bmatrix},$$

and

$$\varpi_x = \begin{bmatrix} 0 & 0 & 0 \\ 0 & 0 & -\mathrm{i} \\ 0 & \mathrm{i} & 0 \end{bmatrix}, \quad \varpi_y = \begin{bmatrix} 0 & 0 & \mathrm{i} \\ 0 & 0 & 0 \\ -\mathrm{i} & 0 & 0 \end{bmatrix}, \quad \varpi_z = \begin{bmatrix} 0 & -\mathrm{i} & 0 \\ \mathrm{i} & 0 & 0 \\ 0 & 0 & 0 \end{bmatrix}.$$

The non-bold symbols here denote matrices by default while bold symbols denote vectors. E is a column matrix holding E_x, E_y, E_z. H is a column matrix holding H_x, H_y, H_z. The vector operator $\hat{x}\Pi_x + \hat{y}\Pi_y + \hat{z}\Pi_z$ is the Poynting operator satisfying

$$\mathrm{Re}\left(\frac{1}{2}\mathbf{E}^*\times\mathbf{H}\right) = \frac{1}{4}\begin{bmatrix} E \\ \mathrm{i}H \end{bmatrix}^\dagger (\hat{x}\Pi_x + \hat{y}\Pi_y + \hat{z}\Pi_z)\begin{bmatrix} E \\ \mathrm{i}H \end{bmatrix}.$$

Now, we would like to find a constant 6×6 unitary matrix U (a global symmetry transformation) so that substituting

$$\begin{bmatrix} E \\ \mathrm{i}H \end{bmatrix} = U\begin{bmatrix} E' \\ \mathrm{i}H' \end{bmatrix}$$

still solves the same Maxwell's equation in vacuum with solution E' and H'. The form of U commuting with Π_z, after some algebra, can be written as

$$U = \begin{bmatrix} U_{11} & U_{12} & 0 & U_{14} & U_{15} & 0 \\ U_{21} & U_{22} & 0 & U_{24} & U_{25} & 0 \\ 0 & 0 & U_{33} & 0 & 0 & U_{36} \\ U_{25} & -U_{24} & 0 & U_{22} & -U_{21} & 0 \\ -U_{15} & U_{14} & 0 & -U_{12} & U_{11} & 0 \\ 0 & 0 & U_{63} & 0 & 0 & U_{66} \end{bmatrix}, \quad [U, \Pi_z] = 0. \tag{A.31}$$

The form of U for $[U, \Pi_x] = 0$ or $[U, \Pi_y] = 0$ can be easily obtained by permutation of components.

Waves in 3D

If the wave travels in all three dimensions, a global symmetry transformation U commutes with all Π_x, Π_y and Π_z. Then, we can obtain the global symmetry transformation is the duality transformation:

$$\begin{bmatrix} E \\ \mathrm{i}H \end{bmatrix} = \begin{bmatrix} \cos(\theta) & -\mathrm{i}\sin(\theta) \\ -\mathrm{i}\sin(\theta) & \cos(\theta) \end{bmatrix}\begin{bmatrix} E' \\ \mathrm{i}H' \end{bmatrix}. \tag{A.32}$$

It is called $U(1)$ gauge symmetry. It is well known that duality is a symmetry compatible with Maxwell's equations in vacuum.

Waves in 2D

If the waves travel on the x–y plane with z-invariance, a global symmetry transformation U commutes with both Π_x and Π_y. As $\partial_z \to 0$, U does not need to commute with Π_z. Then, we can obtain the global symmetry transformation as

$$\begin{bmatrix} E_z \\ \mathrm{i}H_z \end{bmatrix} = u \begin{bmatrix} E_z' \\ H_z' \end{bmatrix}, \quad \begin{bmatrix} \mathrm{i}H_\mathrm{T} \\ E_\mathrm{T} \end{bmatrix} = u \begin{bmatrix} \mathrm{i}H_\mathrm{T}' \\ E_\mathrm{T}' \end{bmatrix} \tag{A.33}$$

where u is a 2×2 unitary matrix. It is $SU(2)$ gauge symmetry. E_T is a column matrix holding E_x, E_y and H_T is a column matrix holding H_x, H_y.

A.4 Action of optical elements on Poincaré sphere

A normally incident polarization state can be written as a superposition of the two circular polarization basis states $|L\rangle = (\hat{x} + \mathrm{i}\hat{y})/\sqrt{2}$ and $|R\rangle = (\hat{x} - \mathrm{i}\hat{y})/\sqrt{2}$:

$$a_L|L\rangle + a_R|R\rangle = a\mathrm{e}^{\mathrm{i}\angle a_L}\left(\cos\frac{\Theta}{2}|L\rangle + \sin\frac{\Theta}{2}\mathrm{e}^{\mathrm{i}\Phi}|R\rangle\right). \tag{A.34}$$

A point on the Poincaré sphere (figure A.2) is specified by the polar angle Θ and azimuthal angle Φ to represent such a polarization state. An optical element described by equation (5.3), with additional rotation angle α, transforms this input into the transmitted state

$$b_L|L\rangle + b_R|R\rangle = \begin{pmatrix} |L\rangle & |R\rangle \end{pmatrix} \begin{pmatrix} t_{LL}^{(0)} & \mathrm{e}^{-\mathrm{i}2\alpha} t_{LR}^{(0)} \\ \mathrm{e}^{\mathrm{i}2\alpha} t_{RL}^{(0)} & t_{RR}^{(0)} \end{pmatrix} \begin{pmatrix} a_L \\ a_R \end{pmatrix}. \tag{A.35}$$

For many practical cases, the transmission coefficients satisfy

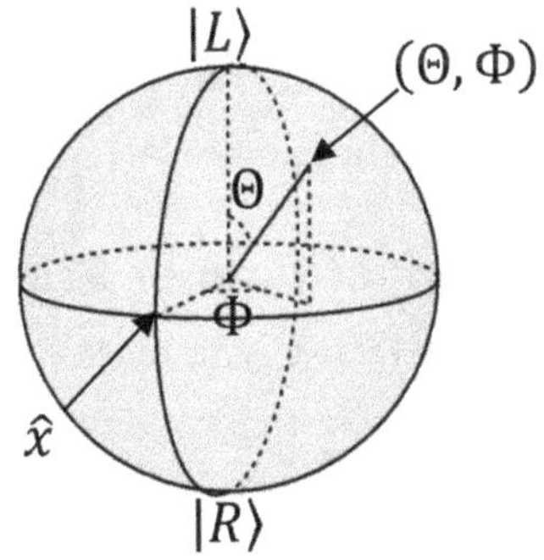

Figure A.2. Poincaré sphere representation of an incident polarization state.

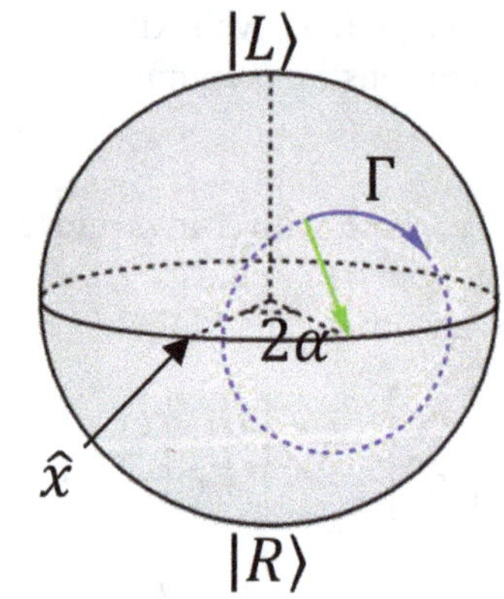

Figure A.3. Action of an optical element on the Poincaré sphere. The blue arrow shows the action of a wave plate with phase shift Γ, while the green arrow shows a linear polarizer at angle α.

$$\frac{t_{LR}^{(0)}}{t_{LL}^{(0)}} = \tan\frac{\Gamma}{2}\,\mathrm{e}^{\mathrm{i}\delta}, \quad t_{LL}^{(0)} = t_{RR}^{(0)}, \quad t_{RL}^{(0)} = t_{LR}^{(0)}, \tag{A.36}$$

with real Γ and δ. For example, a wave plate with phase delay Γ has $t_{xx}^{(0)} = t\mathrm{e}^{\mathrm{i}\Gamma/2}$, $t_{yy}^{(0)} = t\mathrm{e}^{-\mathrm{i}\Gamma/2}$, leading to $t_{LL}^{(0)} = t_{RR}^{(0)} = t\cos(\Gamma/2)$ and $t_{LR}^{(0)} = t_{RL}^{(0)} = \mathrm{i}t\sin(\Gamma/2)$, i.e., $\delta = \pi/2$. A linear x-polarizer with phase Γ gives $t_{xx}^{(0)} = 2t\mathrm{e}^{\mathrm{i}\Gamma/2}$, $t_{yy}^{(0)} = 0$, so $t_{LL}^{(0)} = t_{RR}^{(0)} = t_{LR}^{(0)} = t_{RL}^{(0)} = t\mathrm{e}^{\mathrm{i}\Gamma/2}$, i.e., $\Gamma = \pi/2$, $\delta = 0$.

By substituting equations (A.34) and (A.36) into equation (A.35), the transmitted state becomes

$$\begin{aligned} T|0\rangle = {} & \sqrt{|t_{LL}^{(0)}|^2 + |t_{LR}^{(0)}|^2}\,\mathrm{e}^{\mathrm{i}\measuredangle t_{LL}^{(0)}}\,a\,\mathrm{e}^{\mathrm{i}\measuredangle a_L}\Bigg[\left(\cos\frac{\Gamma}{2}\cos\frac{\Theta}{2} + \sin\frac{\Gamma}{2}\sin\frac{\Theta}{2}\,\mathrm{e}^{\mathrm{i}(\delta+\Phi-2\alpha)}\right)|L\rangle \\ & + \left(\cos\frac{\Gamma}{2}\sin\frac{\Theta}{2} + \sin\frac{\Gamma}{2}\cos\frac{\Theta}{2}\,\mathrm{e}^{\mathrm{i}(\delta-\Phi+2\alpha)}\right)\mathrm{e}^{\mathrm{i}\Phi}|R\rangle\Bigg]. \end{aligned} \tag{A.37}$$

For a wave plate ($\delta = \pi/2$), the effect is a rotation of the polarization state on the Poincaré sphere about the axis $(\cos 2\alpha, \sin 2\alpha, 0)$ by angle Γ (blue arrow in figure A.3). For a linear polarizer ($\Gamma = \pi/2$, $\delta = 0$), the effect is a projection of the state onto the equator at azimuthal angle 2α (green arrow in figure A.3).

A.5 Four-dimensional covariant form of Maxwell's equations

We define the Hodge dual and the Levi-Civita tensor with respect to a 4D metric g as

$$(\star\mathcal{F})^{\alpha\beta} = \frac{1}{2}\,\varepsilon^{\alpha\beta\gamma\delta}\mathcal{F}_{\gamma\delta}, \qquad (\star\mathcal{F})_{\alpha\beta} = \frac{1}{2}\,\varepsilon_{\alpha\beta\gamma\delta}\mathcal{F}^{\gamma\delta},$$

$$\varepsilon^{\alpha\beta\gamma\delta} = -\Omega^{-1}[\alpha\beta\gamma\delta], \qquad \varepsilon_{\alpha\beta\gamma\delta} = \Omega[\alpha\beta\gamma\delta],$$

where $\Omega = \sqrt{-\det g}$ and $\det g = \det([g_{\alpha\beta}])$. Here $[\alpha\beta\gamma\delta]$ denotes the Levi-Civita symbol, which evaluates to the sign of a valid permutation and 0 otherwise. The coordinates are (ct, x, y, z) or (x^0, x^1, x^2, x^3). For the Minkowski spacetime as an example, we have $[g_{\alpha\beta}] = \mathrm{diag}(1, -1, -1, -1)$, so $\det g = -1$ and $\Omega = 1$.

The Maxwell equations can be rewritten as

$$\nabla_\alpha \mathcal{D}^{\alpha\beta} = \mathcal{J}^\beta, \tag{A.38}$$

$$\nabla_\alpha (\star\mathcal{F})^{\alpha\beta} = 0, \tag{A.39}$$

where the displacement field tensor $\mathcal{D}$ stores **D** and **H** fields by

$$\mathcal{D}^{0i} = -\mathcal{D}^{i0} = -D^i, \qquad \mathcal{D}^{ij} = \varepsilon^{0ijk} H_k \;\Rightarrow\; H_i = -\frac{1}{2}\varepsilon_{0ijk}\mathcal{D}^{jk},$$

and the electromagnetic tensor $\mathcal{F}$ stores **E** and **B** fields by

$$\mathcal{F}_{0i} = -\mathcal{F}_{i0} = E_i, \qquad \mathcal{F}_{ij} = -\varepsilon_{0ijk} B^k \;\Rightarrow\; B^i = \frac{1}{2}\varepsilon^{0ijk}\mathcal{F}_{jk}.$$

The free charge density ρ and current density **J** are stored in the tensor $\mathcal{J}$:

$$\mathcal{J}^0 = \rho, \qquad \mathcal{J}^i = \frac{1}{c} J^i.$$

We adopt Heaviside–Lorentz units and use Greek indices for 4D quantities and Latin indices for 3D quantities. The symbol ∇_α represents the covariant derivative defined with respect to the metric g, while ∂_α abbreviates $\partial/\partial x^\alpha$. The field tensors in matrix form are

$$\Omega[\mathcal{D}^{\alpha\beta}] = \begin{pmatrix} 0 & -\Omega D^1 & -\Omega D^2 & -\Omega D^3 \\ \Omega D^1 & 0 & -H_3 & H_2 \\ \Omega D^2 & H_3 & 0 & -H_1 \\ \Omega D^3 & -H_2 & H_1 & 0 \end{pmatrix},$$
$$[\mathcal{F}_{\alpha\beta}] = \begin{pmatrix} 0 & E_1 & E_2 & E_3 \\ -E_1 & 0 & -\Omega B^3 & \Omega B^2 \\ -E_2 & \Omega B^3 & 0 & -\Omega B^1 \\ -E_3 & -\Omega B^2 & \Omega B^1 & 0 \end{pmatrix}. \tag{A.40}$$

3+1 decomposition

Now, we substitute the matrix elements of the field tensors (equation (A.40)) to obtain the 3+1D form. By using the divergence formula for an antisymmetric tensor:

$$\nabla_\alpha \mathcal{D}^{\alpha\beta} = \frac{1}{\Omega}\,\partial_\alpha\left(\Omega\mathcal{D}^{\alpha\beta}\right),$$
$$\nabla_\alpha (\star\mathcal{F})^{\alpha\beta} = \frac{1}{\Omega}\,\partial_\alpha\left(\Omega(\star\mathcal{F})^{\alpha\beta}\right) = -\frac{1}{2\Omega}\,\partial_\alpha\left([\alpha\beta\gamma\delta]\mathcal{F}_{\gamma\delta}\right),$$

we obtain

$$\begin{aligned} \partial_\alpha\left(\Omega\mathcal{D}^{\alpha\beta}\right) &= \Omega\mathcal{J}^\beta, \\ \partial_\alpha\left([\alpha\beta\gamma\delta]\mathcal{F}_{\gamma\delta}\right) &= 0. \end{aligned} \tag{A.41}$$

Then, the first equation becomes

$$\begin{aligned} &\partial_i(\Omega D^i) = \Omega\rho, \\ &[ijk]\,\partial_j H_k - \frac{1}{c}\,\partial_t(\Omega D^i) = \frac{1}{c}\,\Omega J^i. \end{aligned} \tag{A.42}$$

These summarize Gauss's law and Ampère's law. The second equation becomes

$$\begin{aligned} &\partial_i(\Omega B^i) = 0, \\ &[ijk]\,\partial_j E_k + \frac{1}{c}\,\partial_t(\Omega B^i) = 0. \end{aligned} \tag{A.43}$$

These summarize Gauss's law for magnetism and Faraday's law. equation (A.42) and (A.43) are compatible to the Maxwell's equations in 3+1D form (equation (3.19)). For a flat Minkowski space with $\Omega = 1$, these returns to the familiar Maxwell's equations in free space with Cartesian coordinates (equation (3.20)).

Constitutive relationship

The Maxwell equations work with the constitutive relationship

$$\mathcal{D}^{\alpha\beta} = \mathcal{K}^{\alpha\beta\gamma\delta}\mathcal{F}_{\gamma\delta}. \tag{A.44}$$

Here, the rank-4 medium tensor $\mathcal{K}$ is one way to specify the constitutive relationship. It satisfies

$$\mathcal{K}^{\alpha\beta\gamma\delta} = -\,\mathcal{K}^{\beta\alpha\gamma\delta} = -\,\mathcal{K}^{\alpha\beta\delta\gamma},$$

since $\mathcal{D}$ and $\mathcal{F}$ are antisymmetric tensors, so there are many non-zero elements. By flattening $\mathcal{D}^{\alpha\beta}$ into a column matrix of six values with $[D]$ storing D^i and $[H]$ storing H_i, and flattening $\mathcal{F}_{\alpha\beta}$ into a column matrix of six values with $[E]$ storing E_i and $[B]$

storing B^i, we obtain a more compact way to write the constitutive relationship using a 6×6 matrix:

$$\begin{aligned}\begin{pmatrix}-[\Omega D]\\ [H]\end{pmatrix} &= [K]\begin{pmatrix}[E]\\ \Omega[B]\end{pmatrix} = \begin{pmatrix}[K_A] & [K_B]\\ [K_C] & [K_D]\end{pmatrix}\begin{pmatrix}[E]\\ [\Omega B]\end{pmatrix}\\ &= \begin{pmatrix}-[\Omega\varepsilon]+[\Omega\xi]\,[\Omega\mu]^{-1}[\Omega\zeta] & -[\Omega\xi]\,[\Omega\mu]^{-1}\\ -[\Omega\mu]^{-1}[\Omega\zeta] & [\Omega\mu]^{-1}\end{pmatrix}\begin{pmatrix}[E]\\ [\Omega B]\end{pmatrix},\end{aligned} \tag{A.45}$$

where $[K_A]$, $[K_B]$, $[K_C]$, $[K_D]$ are 3×3 matrices, or equivalently,

$$\begin{aligned}\begin{pmatrix}[\Omega D]\\ [\Omega B]\end{pmatrix} &= \begin{pmatrix}[\Omega\varepsilon] & [\Omega\xi]\\ [\Omega\zeta] & [\Omega\mu]\end{pmatrix}\begin{pmatrix}[E]\\ [H]\end{pmatrix}\\ &= \begin{pmatrix}-[K_A]+[K_B][K_D]^{-1}[K_C] & -[K_B][K_D]^{-1}\\ -[K_D]^{-1}[K_C] & [K_D]^{-1}\end{pmatrix}\begin{pmatrix}[E]\\ [H]\end{pmatrix},\end{aligned} \tag{A.46}$$

where the 3×3 matrix $[\varepsilon]$ stores the contravariant components ε^{ij} of the permittivity tensor, as do the permeability matrix $[\mu]$ and the two bianisotropy matrices $[\xi]$ and $[\zeta]$.

Constitutive matrices connected to curved free space-time

For a curved free spacetime without medium, we have $\mathcal{K}^{\alpha\beta\gamma\delta} = g^{\alpha\gamma}g^{\beta\delta}$, i.e. $\mathcal{D}$ and $\mathcal{F}$ are the same tensor and $\mathcal{K}$ simply raises the indices from the covariant form of $\mathcal{F}$ to the contravariant form of $\mathcal{D}$. In this case, we have the constitutive matrices

$$\begin{aligned}[\Omega\varepsilon^{ij}] &= [\Omega\mu^{ij}] = \Omega\left(-g_{00}\,[g_{ij}] + [g_{i0}]\,[g_{0j}]^{\mathrm{T}}\right)^{-1},\\ [\Omega\xi^{ij}] &= -\,[\Omega\zeta^{ij}] = \frac{1}{g_{00}}\,[g_{0i}\times] = \frac{1}{g_{00}}\begin{pmatrix}0 & -g_{03} & g_{02}\\ g_{03} & 0 & -g_{01}\\ -g_{02} & g_{01} & 0\end{pmatrix},\end{aligned} \tag{A.47}$$

where $\Omega = \sqrt{-\det[g_{\alpha\beta}]}$. For example, plugging in the metric $[g_{ij}] = \mathrm{diag}\{1, -n^2, -n^2, -n^2\}$ gives $[\Omega\varepsilon^{ij}] = [\Omega\mu^{ij}] = n\,[I]$, where $[I]$ is the 3×3 identity matrix.

We note that scaling the entire metric $[g_{\alpha\beta}]$ by a constant leaves $\Omega K^{\alpha\beta\gamma\delta}$ (or $[K]$) invariant, and we have $\det[K] = -1$. Therefore, for a symmetric $[g_{\alpha\beta}]$ with one unimportant scaling factor, we have nine degrees of freedom (DOFs). These can be decomposed into three DOFs for the antisymmetric $[\Omega\xi^{ij}] = -[\Omega\zeta^{ij}]$ and six DOFs for the symmetric $[\Omega\varepsilon^{ij}] = [\Omega\mu^{ij}]$, giving a total of nine DOFs.

On the other hand, suppose the constitutive matrices are realized in Minkowski spacetime ($\Omega = 1$) in the form

$$[\varepsilon^{ij}] = \frac{[n^{ij}]}{\eta}, \qquad [\mu^{ij}] = [n^{ij}]\,\eta,$$
$$[\xi^{ij}] = -\,[\zeta^{ij}] = [\xi_k \times] = \begin{pmatrix} 0 & -\xi_3 & \xi_2 \\ \xi_3 & 0 & -\xi_1 \\ -\xi_2 & \xi_1 & 0 \end{pmatrix}, \tag{A.48}$$

where $[n^{ij}]$ is symmetric and $[\xi_k]$ is a column vector. We can define an effective "metric" $\mathcal{G}$ as

$$[\mathcal{G}_{i0}] = \mathcal{G}_{00}\,[\xi_i], \qquad [\mathcal{G}_{0j}] = \mathcal{G}_{00}\,[\xi_j],$$
$$[\mathcal{G}_{ij}] = -\,\mathcal{G}_{00}\Big(\det[n^{ij}]\,[n^{ij}]^{-1} + [\zeta_i]\,[\xi_j]^{\mathrm{T}}\Big), \tag{A.49}$$

and denote its inverse by

$$[\mathcal{G}^{\alpha\beta}] = [\mathcal{G}_{\alpha\beta}]^{-1}.$$

Then we have

$$\mathcal{K}^{\alpha\beta\gamma\delta} = \frac{1}{\eta}\sqrt{-\det[\mathcal{G}^{\alpha\beta}]}\;\mathcal{G}^{\alpha\gamma}\mathcal{G}^{\beta\delta}, \tag{A.50}$$

with

$$\det[K] = -\,\frac{1}{\eta^6}.$$

When the impedance η approaches unity, the constitutive matrices exactly simulate a free spacetime with metric $\mathcal{G}$. We retain η in the expression to allow implementation flexibility.

An immediate application is an impedance-matched medium in flat free space, where

$$[\varepsilon^{ij}] = [\mu^{ij}], \qquad [\xi^{ij}] = -\,[\zeta^{ij}] = 0, \tag{A.51}$$

which usually results from conventional transformation optics using space-only coordinate transformations. Such a medium can be regarded as simulating a curved space with metric

$$\mathcal{G}_{00} = 1, \qquad \mathcal{G}_{0i} = \mathcal{G}_{i0} = 0, \qquad [\mathcal{G}_{ij}] = -\,\det[\varepsilon^{ij}]\,[\varepsilon^{ij}]^{-1}. \tag{A.52}$$

The additional minus sign is due to the metric signature.

A.6 Useful formulas

Differential identities in 3D

$$
\begin{aligned}
&\nabla\times\nabla\,\psi=0,\quad \nabla\cdot\nabla\times\mathbf{a}=0,\quad \nabla\times(\nabla\times\mathbf{a})=\nabla\,\nabla\cdot\mathbf{a}-\nabla^2\,\mathbf{a}\\
&\nabla(\psi\phi)=\psi\,\nabla\,\phi+(\nabla\psi)\phi,\quad \nabla\,(\psi\mathbf{a})=\psi\,\nabla\,\mathbf{a}+(\nabla\psi)\mathbf{a},\quad \nabla\,(\psi\overline{\overline{\mu}})=\psi\,\nabla\,\overline{\overline{\mu}}+(\nabla\psi)\overline{\overline{\mu}}\\
&\nabla\cdot(\psi\mathbf{a})=\psi\,\nabla\cdot\mathbf{a}+(\nabla\psi)\cdot\mathbf{a},\quad \nabla\cdot(\psi\overline{\overline{\mu}})=\psi\,\nabla\cdot\overline{\overline{\mu}}+(\nabla\psi)\cdot\overline{\overline{\mu}}\\
&\nabla\times(\psi\mathbf{a})=\psi\,\nabla\times\mathbf{a}+(\nabla\psi)\times\mathbf{a},\quad \nabla\times(\psi\overline{\overline{\mu}})=\psi\,\nabla\times\overline{\overline{\mu}}+(\nabla\psi)\times\overline{\overline{\mu}}\\
&\nabla(\mathbf{a}\cdot\mathbf{b})=(\nabla\mathbf{a})\cdot\mathbf{b}+\mathbf{a}\cdot(\nabla\mathbf{b})^T=\mathbf{a}\cdot\nabla\,\mathbf{b}+\mathbf{b}\cdot\nabla\,\mathbf{a}+\mathbf{a}\times\nabla\times\mathbf{b}+\mathbf{b}\times\nabla\times\mathbf{a}\\
&\nabla(\mathbf{a}\cdot\overline{\overline{\mu}})=(\nabla\mathbf{a})\cdot\overline{\overline{\mu}}+\mathbf{a}\cdot(\nabla\overline{\overline{\mu}})^T\\
&\nabla(\overline{\overline{\epsilon}}\cdot\mathbf{b})=(\nabla\overline{\overline{\epsilon}})\cdot\mathbf{b}+(\nabla\mathbf{b})\cdot\overline{\overline{\epsilon}}^T\\
&\nabla(\overline{\overline{\epsilon}}\cdot\overline{\overline{\mu}})=(\nabla\overline{\overline{\epsilon}})\cdot\overline{\overline{\mu}}+\left(\overline{\overline{\epsilon}}\cdot(\nabla\overline{\overline{\mu}})^T\right)^T\\
&\mathbf{a}\times(\nabla\times\mathbf{b})=(\nabla\mathbf{b})\cdot\mathbf{a}-\mathbf{a}\cdot\nabla\,\mathbf{b}\\
&\nabla\cdot(\mathbf{a}\times\mathbf{b})=(\nabla\times\mathbf{a})\cdot\mathbf{b}-\mathbf{a}\cdot\nabla\times\mathbf{b}\\
&\nabla\times(\mathbf{a}\times\mathbf{b})=\nabla\cdot(\mathbf{ba}-\mathbf{ab})=\mathbf{a}\,\nabla\cdot\mathbf{b}-\mathbf{b}\,\nabla\cdot\mathbf{a}+\mathbf{b}\cdot\nabla\,\mathbf{a}-\mathbf{a}\cdot\nabla\,\mathbf{b}\\
&\nabla\cdot(\mathbf{ab})=\mathbf{a}\cdot\nabla\,\mathbf{b}+(\nabla\cdot\mathbf{a})\mathbf{b},\ \nabla\cdot(\mathbf{a}\overline{\overline{\mu}})=\mathbf{a}\cdot\nabla\,\overline{\overline{\mu}}+(\nabla\cdot\mathbf{a})\overline{\overline{\mu}}\\
&\nabla\cdot(\overline{\overline{\epsilon}}\mathbf{b})=\overline{\overline{\epsilon}}^T\cdot\nabla\,\mathbf{b}+(\nabla\cdot\overline{\overline{\epsilon}})\mathbf{b},\\
&\nabla\cdot(\overline{\overline{\epsilon}}\overline{\overline{\mu}})=\overline{\overline{\epsilon}}^T\cdot\nabla\,\overline{\overline{\mu}}+(\nabla\cdot\overline{\overline{\epsilon}})\overline{\overline{\mu}}\\
&\nabla^2(\psi\mathbf{a})=\psi\,\nabla^2\,\mathbf{a}+2(\nabla\psi)\cdot\nabla\,\mathbf{a}+(\nabla^2\psi)\mathbf{a}\\
&\nabla\,\nabla\cdot(\psi\mathbf{a})=\psi\,\nabla\,\nabla\cdot\mathbf{a}+(\nabla\psi)\,\nabla\cdot\mathbf{a}+(\nabla\psi)\cdot\nabla\,\mathbf{a}+\mathbf{a}\cdot\nabla\,\nabla\,\psi+(\nabla\psi)\times\nabla\times\mathbf{a}\\
&\nabla\times(\nabla\times(\psi\mathbf{a}))\\
&=\psi\,\nabla\times(\nabla\times\mathbf{a})+(\nabla\psi)\,\nabla\cdot\mathbf{a}-(\nabla\psi)\cdot\nabla\,\mathbf{a}+\mathbf{a}\cdot\nabla\,\nabla\,\psi+(\nabla\psi)\times\nabla\times\mathbf{a}-(\nabla^2\psi)\mathbf{a}
\end{aligned}
$$

A derivative operator acts on the left. e.g., $\nabla\left(\mathbf{x}^j a_j\right)=\mathbf{x}^i\mathbf{x}^j\ \nabla_i\ a_j$. In this appendix we adopt a hybrid convention. Gradient and divergence are defined using the covariant derivative ∇ with respect to the metric g and divergence ($\nabla\cdot$) has an additional contraction of the first two indices, while curl and the Laplacian are defined using the differential geometry operators: Hodge star $\star$, exterior derivative d, and codifferential δ adopted on tensors of rank m. In 3D,

$$
\begin{aligned}
(\star T)^{k_1\cdots k_{3-m}}&=\frac{1}{m!}\ T_{i_1\cdots i_m}\ \varepsilon^{i_1\cdots i_m k_1\cdots k_{3-m}},\\
(dT)_{i_0 i_1\cdots i_m}&=(m+1)\ \partial_{[i_0}T_{i_1\cdots i_m]},\\
\delta T&=(-1)^m\star d\star T,
\end{aligned}
$$

where ε^{ijk} is the Levi–Civita tensor defined in section 3.2, $[\ \cdots\ ]$ denotes antisymmetrization of indices. The wedge product between tensors T of rank m and T' of rank m' is

$$
T\wedge T' \ =\ \frac{(m+m')!}{m!\ m'!}\ \mathrm{alt}(T\otimes T'),
$$

where alt($\cdot$) denotes also antisymmetrization. Then, the curl operator is defined by $\nabla\times=\star d$. The Laplacian is defined as the Hodge Laplacian $\nabla^2=-\ \delta d-d\delta$, which, for a scalar, reduces to the familiar form of divergence of gradient. The component forms for the gradient of scalar ψ, the curl, and the divergence of vector $\mathbf{a}$ in curved space are

$$\nabla\psi = \mathbf{x}^{i}\,\partial_i\psi, \quad \nabla\cdot\mathbf{a} = \frac{1}{\sqrt{|\det g|}}\,\partial_i\left(\sqrt{|\det g|}\;a^i\right), \quad \nabla\times\mathbf{a} = \mathbf{x}_{i}\,\varepsilon^{ijk}\,\partial_j a_k.$$

These derive the formulas in the above listed differential identities. Moreover, the cross product is defined by $\mathbf{a}\times\mathbf{b} = \star(\mathbf{a}\wedge\mathbf{b})$, which agrees with the usual cross product in Euclidean $\mathbb{R}^3$ and remains valid on any oriented 3D Riemannian manifold. The following vector identities apply in a curved space:

$$\mathbf{a}\cdot\mathbf{b}\times\mathbf{c} = \mathbf{b}\cdot\mathbf{c}\times\mathbf{a} = \mathbf{c}\cdot\mathbf{a}\times\mathbf{b}$$
$$\mathbf{a}\times(\mathbf{b}\times\mathbf{c}) = (\mathbf{a}\cdot\mathbf{c})\mathbf{b} - (\mathbf{a}\cdot\mathbf{b})\mathbf{c}, \qquad (\mathbf{a}\times\mathbf{b})\times\mathbf{c} = \mathbf{b}(\mathbf{a}\cdot\mathbf{c}) - \mathbf{a}(\mathbf{b}\cdot\mathbf{c})$$
$$(\mathbf{a}\times\mathbf{b})\cdot(\mathbf{c}\times\mathbf{d}) = (\mathbf{a}\cdot\mathbf{c})(\mathbf{b}\cdot\mathbf{d}) - (\mathbf{a}\cdot\mathbf{d})(\mathbf{b}\cdot\mathbf{c})$$

Integration identities in 3D

$$\oint d\mathbf{l}\cdot(\nabla\phi) = \Delta\phi$$
$$\oint d\mathbf{l}\cdot\mathbf{a} = \int d\mathbf{S}\cdot(\nabla\times\mathbf{a})$$
$$\int d\mathbf{l}\cdot(\phi\mathbf{a}) = \int d\mathbf{S}\cdot(\phi\,\nabla\times\mathbf{a} - \mathbf{a}\times\nabla\,\phi)$$
$$\int dx^i\,\phi = \int (d\mathbf{S}\times\nabla\,\phi)^i$$
$$\int d\mathbf{S}\cdot\mathbf{a} = \int dV\,(\nabla\cdot\mathbf{a})$$
$$\int d\mathbf{S}\cdot(\phi\mathbf{a}) = \int dV\,(\phi\,\nabla\cdot\mathbf{a} + \mathbf{a}\cdot\nabla\,\phi)$$
$$\int (d\mathbf{S})_i\,\frac{\phi}{\sqrt{|\det g|}} = \int dV\,\frac{\partial_i\phi}{\sqrt{|\det g|}}$$
$$\int d\mathbf{S}\cdot(\mathbf{a}\times\mathbf{b}) = \int dV\,(\mathbf{b}\cdot\nabla\times\mathbf{a} - \mathbf{a}\cdot\nabla\times\mathbf{b})$$
$$\int (d\mathbf{S}\times\mathbf{a})^i = \int (\nabla\times\mathbf{a})^i\,dV$$
$$\int d\mathbf{S}\cdot(\phi\,\nabla\,\psi) = \int dV\,(\phi\,\nabla\cdot\nabla\,\psi + \nabla\,\psi\cdot\nabla\,\phi)$$
$$\int d\mathbf{S}\cdot(\phi\,\nabla\,\psi - \psi\,\nabla\,\phi) = \int dV\,(\phi\,\nabla\cdot\nabla\,\psi - \psi\,\nabla\cdot\nabla\,\phi)$$
$$\int d\mathbf{S}\cdot(\mathbf{a}\times\nabla\times\mathbf{b}) = \int dV\,((\nabla\times\mathbf{b})\cdot(\nabla\times\mathbf{a}) - \mathbf{a}\cdot\nabla\times\nabla\times\mathbf{b})$$
$$\int d\mathbf{S}\cdot(\mathbf{a}\times\nabla\times\mathbf{b} - \mathbf{b}\times\nabla\times\mathbf{a}) = \int dV\,(\mathbf{b}\cdot\nabla\times\nabla\times\mathbf{a} - \mathbf{a}\cdot\nabla\times\nabla\times\mathbf{b})$$

Line, surface, and volume elements with basis vectors $\mathbf{x}_i$ and dual basis vectors $\mathbf{x}^i$:

$$\begin{aligned} d\mathbf{l} &= dx^i\mathbf{x}_i, \\ d\mathbf{S} &= \sqrt{|\det g|}\left(\mathbf{x}^1\,dx^2\wedge dx^3 + \mathbf{x}^2\,dx^3\wedge dx^1 + \mathbf{x}^3\,dx^1\wedge dx^2\right), \\ dV &= \sqrt{|\det g|}\;dx^1\wedge dx^2\wedge dx^3 \end{aligned}$$

www.ingramcontent.com/pod-product-compliance
Lightning Source LLC
LaVergne TN
LVHW062238200926
848484LV00006B/1565